The pioneering work of Edwin T. Jaynes in the fields of statistical physics, quantum optics, and probability theory has had a significant and lasting effect on the study of many physical problems, ranging from fundamental theoretical questions through to practical applications such as optical image restoration. *Physics and Probability* is a collection of papers in these areas by some of his many colleagues and former students, based largely on lectures given at a symposium celebrating Jaynes' contributions, on the occasion of his seventieth birthday and retirement as Wayman Crow Professor of Physics at Washington University.

The collection contains several authoritative overviews of current research on maximum entropy and quantum optics, where Jaynes' work has been particularly influential, as well as reports on a number of related topics. In the concluding paper, Jaynes looks back over his career, and gives encouragement and sound advice to young scientists.

All those engaged in research on any of the topics discussed in these papers will find this a useful and fascinating collection, and a fitting tribute to an outstanding and innovative scientist.

Physics and Probability

Essays in honor of Edwin T. Jaynes

Physics and Probability

Essays in honor of Edwin T. Jaynes

Edited by

W. T. GRANDY, JR
Department of Physics and Astronomy
University of Wyoming

and

P. W. MILONNI
Theoretical Division
Los Alamos National Laboratory

CAMBRIDGE
UNIVERSITY PRESS

PUBLISHED BY THE PRESS SYNDICATE OF THE UNIVERSITY OF CAMBRIDGE
The Pitt Building, Trumpington Street, Cambridge, United Kingdom

CAMBRIDGE UNIVERSITY PRESS
The Edinburgh Building, Cambridge CB2 2RU, UK
40 West 20th Street, New York NY 10011–4211, USA
477 Williamstown Road, Port Melbourne, VIC 3207, Australia
Ruiz de Alarcón 13, 28014 Madrid, Spain
Dock House, The Waterfront, Cape Town 8001, South Africa

http://www.cambridge.org

First published 1993
First paperback edition 2004

A catalogue record for this book is available from the British Library

ISBN 0 521 43471 8 hardback
ISBN 0 521 61710 3 paperback

CONTENTS

PREFACE

Most physicists, most of the time, are engaged in what Thomas Kuhn calls "normal science," the process of solving problems under a prevailing paradigm. Part of Edwin Jaynes' career has been profitably devoted to this enterprise, but his best known work may aptly be called "abnormal" in this sense. We organized the Symposium on Physics and Probability not only to recognize Jaynes' accomplishments, but also to celebrate the scientific style and integrity that have been an inspiration to so many of his students and colleagues.

The articles in this volume are based, with a few exceptions, on lectures given at the Symposium, which was held 15-16 May 1992 at the University of Wyoming. The occasion for the Symposium was Jaynes' official retirement from his position as Wayman Crow Professor of Physics at Washington University — as well as to commemorate his seventieth birthday. The authors are former graduate students of Professor Jaynes, or colleagues, or people whose work has been so directly influenced by Jaynes that it was deemed appropriate to invite them to speak at the Symposium.

Jaynes is best known for his work in statistical physics and quantum optics. His seminal papers on information theory and statistical mechanics formulated the latter (in the spirit of Gibbs) without an ergodic hypothesis, relying instead on the maximum entropy principle to assign probabilities in a manner least biased with respect to the available information. The "Maxent" principle, along with Jaynes' broader work in Bayesian statistics, has more recently been gainfully applied to a growing number of practical problems including optical image restoration, radar target identification, model selection in economics, and others described in this volume. The evolution of these applications from Jaynes' original, purely academic motivations nicely illustrates the truth in Richard Feynman's remark that "no matter how small a thing is, if it has physical interest and is thought about carefully enough, you're bound to think of something that's good for something else."

Jaynes' work in quantum optics, on the other hand, seems to have evolved from a practical concern with maser problems to fundamental questions about field quantization. The Jaynes-Cummings paper showed that semiclassical radiation theory could be an excellent approximation even under conditions where the average number of photons is small. This result led to Jaynes' neoclassical theory, where even spontaneous emission and the Lamb shift were approached without quantization of the field. The aim of this theory was to see just how far one could take semiclassical theory without running into conflict with experiment; although it has not yet proved a viable substitute for quantum electrodynamics, it certainly resulted in a much better understanding of semiclassical theory and field quantization in quantum optics. After three decades the Jaynes-Cummings model, to borrow a phrase from Pierre Meystre, remains alive and well, as the papers in this volume amply demonstrate.

Over the years Jaynes has remained true to his conviction that "progress in science goes forward on the shoulders of doubters, not believers." In the final article of this volume we are reminded again just why it is he feels so strongly about this and other matters of concern to the academic community. His unhappiness with the conventional interpretation of quantum theory — a way of reasoning that would be considered a "psychiatric disorder" in any other field — is here reiterated succinctly, illustrating once more the generating force behind much of the work discussed in this volume.

We wish to thank all the contributors to this volume for their enthusiastic response that made the Symposium a very pleasurable and memorable occasion. Thanks also go to other people, including Freeman Dyson, Erwin Hahn, Peter Knight, and John Toll, who could not attend owing to prior commitments but nevertheless took the time to express encouragement and their admiration for Edwin Jaynes. Finally we are grateful to the Department of Physics and Astronomy of the University of Wyoming for the excellent facilities and services provided during the Symposium, to Ms. Kim Bella for her efficient polishing of the various manuscripts, and to Cambridge University Press for their enthusiastic willingness to publish this collection.

W.T. Grandy, Jr.
Laramie, Wyoming

P.W. Milonni
Los Alamos, New Mexico

RECOLLECTIONS OF AN INDEPENDENT THINKER

Joel A. Snow
Argonne National Laboratory
Argonne, Illinois 60439

Let me first point out that my title is consciously ambiguous. It reflects the marvelous imprecision that is possible with language. Am I the independent thinker or is someone else? Whose recollections are these anyway? And is the "of" equivalent to "about" or to "by"? And who here is "independent," of what, or of whom?

In apposition to this observation, let me quote one for whom ambiguity is anathema:

"*This may seem like an inflexible, cavalier attitude; I am convinced that nothing short of it can ever remove the ambiguity of 'what is the problem?' that has plagued probability theory for two centuries.*"
                                                                                    E.T. Jaynes

I will touch upon the work of several independent thinkers tonight, but what I have to say is mostly, of course, about E. T. Jaynes. Those in this roomful of independent thinkers surely recognize both his independence and his originality. He is a man who has marched to a different drummer upon a road less traveled by. Those of us gathered here tonight, and many others in the world of science and engineering, now find themselves following in his footsteps.

Where many of us have had one career in one field, Ed Jaynes has had several. The broad collection of expertise from a wide variety of different disciplines in this gathering reflects this diversity of his ideas and their applications. Part of what I have to say is verifiable fact, part is subjective speculation, and part is derived from the recollection of others – however suitably filtered. I trust that Ed will correct the more egregious of my mistakes.

I came to Washington University in the fall of 1961 as a green graduate student, after three years in the Navy's Nuclear Submarine Program in New London, Connecticut. I was strongly motivated toward theoretical physics. I had dabbled in theater, English literature, philosophy and journalism before ending up in physics. My mentor, Eugen Merzbacher, had impressed me with his mastery of many fields and with the power and beauty of the interplay of theory and experiment in physics. The resident theorists at Washington University were Edward Condon, Eugene Feenberg, and Ed Jaynes – all three notably independent thinkers with unique and original things to say. All three, by chance, had a personal linkage to Princeton and to Eugene Wigner.

Ed had come to Washington University in 1960 after ten years at Stanford, after graduate study first at Berkeley and later at Princeton. This brief summary masks, however, the story of his intellectual development and his contact with some of the leading eminences of contemporary physics. He spent the war years first at the Sperry Gyroscope Company and later at the Naval Research Laboratory working on the development of microwave radar equipment. He went to Berkeley to work with J. Robert Oppenheimer, drawn as were many other bright, young men just after the War by the legendary reputation of the leading contemporary figure in science and public affairs. When Oppenheimer moved to the Institute for Advanced Study in Princeton, only three graduate students were taken with him. Oppenheimer wrote personal "Dear Harry" letters to Harry Smythe, the chairman of the Princeton Physics Department and author of the famous *Smythe Report* about the development of nuclear weapons, asking that these students be admitted to the Princeton Physics Department. Others who wished to follow had to make it as best they could. Ed Jaynes was one of the three selected. According to his contemporaries, he already exhibited characteristics that we now widely recognize. He was quiet, private, conservative, and very independent in his thinking. He harbored some skepticism about the quantum theory and its conceptual and philosophical foundations.

Apparently none of the students who followed Oppenheimer to Princeton ended up finishing their dissertations with him as their advisor. Some gravitated to David Bohm; others to Eugene Wigner, both of whom, themselves, were highly independent and original thinkers and respected those qualities in others. Both Bohm and Wigner had by that time made important and seminal contributions to the development of quantum theory. One contemporary reports that "Ed Jaynes was always interested in things from first principles and skeptical of many of the things repeated in the textbooks of the times, but was too polite to be out-and-out disagreeable about it."

As was the case with John Bardeen ten years earlier, the problem Jaynes worked on with Wigner was highly contemporary and also involved a quite demanding calculation. The title of his first publication, which was on the displacement of the oxygen in barium titanate, hardly reveals that this was the basis for explaining the fundamental properties of the whole class of materials called ferroelectrics. These materials are crystalline substances that have a permanent, spontaneous electric polarization; that is, an electric dipole moment per unit volume, that can be reversed by an electric field. This is a pretty odd property, since this internal electric polarization implies a charge separation that in most materials would be wiped out due to the enormous mobility of electrons. Ferroelectric materials have not become as important to technology as the analogous ferromagnetic materials, but Jaynes' calculations on barium titanate were evidently a *tour de force*. The description of the occurrence of this phenomenon, which is due to minute changes in the location of ions within the crystal, proved highly convincing and in good agreement with a variety of electrical and thermodynamic data. This work led directly to Jaynes' first book, *Ferroelectricity*, published by Princeton University Press in 1953 and was Volume I in a prestigious new series called *Investigations in Physics*. Ten years later – when I took solid state physics from him – this book was still the principal work in the field.

A review in *Physics Today* stated, "... an excellent survey of the diverse theories of ferroelectricity of $BaTiO_3$, with a particular emphasis on the electronic theory first proposed by Wigner, and later more fully developed by Jaynes. Of particular merit is the objectivity and conscientiousness with which the author evaluates the pros and cons of various theories and their experimental justification." That's pretty fine praise for a first effort.

His book was also notable by the company it kept. Volume II in this series was *The Mathematical Foundations of Quantum Mechanics* by John von Neumann; Volume III,

The Shell Theory of the Nucleus by Eugene Feenberg; Volume IV, *Angular Momentum* by A.R. Edmonds; and Volume V, *Nuclear Structure* by Leonard Eisenbud and E.P. Wigner. This was an assortment of leading monographs by major figures at the time in their fields of physics. As one of his contemporaries has commented, "Ed was the only one of that group of Princeton graduate students who got a good job at a good institution."

Jaynes' thesis problem has a remarkable contemporary flavor. Barium titanate and its close relatives occur in what is called the perovskite structure. The ferroelectricity corresponds to a displacement of oxygen ions by only about 1% of the dimension of the crystalline unit cell. The resulting phase transition results from an instability – a softening – of one of the normal lattice vibration modes, a property that is associated with pairing in strong-coupling superconductors. Somewhat related electronic phenomena may also occur in the much more recently discovered high-temperature superconductors that also have the perovskite structure. Jaynes' advisor Eugene Wigner, himself one of the truly creative and independent thinkers of our time, is said to have later characterized Jaynes as "one of the two most under-appreciated people in physics." My source for that comment also said that "people who have their own way of doing things always got along best with Wigner."

One of the attractions of Stanford for Jaynes and Jaynes for Stanford, may well have been due to his war-time experience developing microwave devices. He had spent a summer working in the microwave laboratory of the W.W. Hanson Laboratories of Physics at Stanford before going to Princeton. Jaynes became an assistant professor in applied physics, and several publications from this period deal with nonlinear dielectric materials, electromagnetic propagation, and waveguides. He was extensively involved as a consultant, particularly with the Varian brothers. He was the resident guru on the behavior of electrons in cavity resonators and other phenomena and also worked on magnetic resonance. His calculations were only replaced much later with the advent of powerful computers. At the same time, he must have been well along in the development of the information theoretic approach to statistical mechanics, since the two very significant and profound papers on that subject were published in *The Physical Review* in 1957. Although some of the older faculty were apparently not enamored of Ed Jaynes' formulation of statistical mechanics, a lot of younger faculty and graduate students found it much more appealing than more standard approaches.

Electromagnetic theory and statistical mechanics might seem like very different fields. They are linked, however, through the fundamental concept of radar – to launch electromagnetic radiation which propagates through various media, is scattered, and the scattered radiation is detected. The resulting inverse scattering problem is a natural arena for Bayesian methods.

By about this time, Jaynes' work with Varian resulted in his buying a quite large house. When he moved to St. Louis, he purchased an even larger house, quite close to the university campus, which he generously shared with visiting friends, and on occasion with graduate students. When George Pake returned to Washington University from Stanford as Provost in 1962, Ed invited the entire Pake family to come and stay with him until they found a place to live. The Pake children explored the house from top to bottom and, in the process, it is claimed that they found a bathroom in the house that Ed did not even know that he had. By the time I began working with Ed in 1964, one graduate student's family was in residence helping around the property, and Ed was inviting other students and their families to join him on Friday afternoons around the pool. It was a warm and much appreciated gesture.

Ed was an extraordinarily effective teacher. He was quite the antithesis of the professor whose well-worn lecture notes finally are transformed into a textbook summation of the standard dogmas of the day. I can recall taking quantum mechanics, solid state physics, statistical mechanics – twice–probability theory, and quantum electronics during my six years as a graduate student and auditing many of his individual lectures as well. Even when covering seemingly well-known standard material, Ed's lectures always had a distinct flavor of having been freshly rethought. Though mathematical techniques were there as needed, discussions of the physics were always firmly rooted in experiment or in logical developments from first principles. This was particularly the case in his approach to statistical mechanics, where a whole variety of experiments of a purely macroscopic character by 1900 had exposed the inadequacy of the classical description of matter. Gibbs' formulation of statistical mechanics provided the natural linkage between the various macroscopic experiments and the behavior of particles at the microscopic level. Indeed one commentator on the work of Gibbs some 30 years later observed, "Gibbs' definition of the phase of the statistical system (in counter distinction to that given by eminent contemporary Boltzmann) is the one which clearly leads to the modern quantum statistics as created on the one hand by Einstein and Bose and on the other by Fermi and Dirac."

Ed's lectures invariably began with an historical perspective developing the fundamental ideas in the context of their time of origin, with the line of argument often seeming transparently simple. In retrospect I can see in Jaynes' lectures of the early 1960's the well-developed roots of his major lines of investigation that have been so original and so fruitful. His early skepticism about some aspects of the quantum theory of electromagnetic processes led to a whole series of masterful analyses of quantum electronic systems on semiclassical grounds. His highly original formulation of statistical mechanics, using information theoretic arguments, directly extended the fundamental ideas of Gibbs and provided the basis of a whole series of fundamental investigations of statistical mechanics. Along the way he clarified a number of puzzling, and sometimes out-and-out incorrect, conclusions that have propagated in the literature. His investigations of issues connected with statistical inference in science, rooted in many respects in the work of Sir Harold Jeffreys, not only plowed a lot of new ground with regard to principles in physics, but also developed powerful techniques usable in many other areas of science. Much of his work on the foundations of mathematical statistics, particularly on the generalized inverse Bayesian technique, touches on many of the most interesting issues involving the use of incompletely characterized or simply incomplete data for the purpose of drawing practically useful conclusions. We thus see this Princeton-trained physicist lecturing at companies and in business schools to economists and mathematical statisticians on the foundations of their own subjects. It is a tribute to the clarity of Jaynes' thinking and his recognition of the universality of the principles with which he has dealt that he has been able to translate so much of his work into other domains, totally different from those that originally motivated his work.

Jaynes' papers are invariably lucid and often seem more in the manner of a friendly blackboard chat than the somewhat terse, forbidding, and stylized presentations affected by many physicists. He does not simply write up every new idea, but thinks things through, often seeking the more general, more elegant, or clearer way. Some of his speculative pastries never get fully baked. I recall a draft about formulating theories in macroeconomics and another about the muscle as an engine that were privately circulated and never seem to have made it through his filter.[†] His extension of the information theory formalism to irreversible processes ("Nonlocal Transport Theory" with Douglas Scalapino) never seems to have been published, although it was included in his lectures. Some simple calculations applying this technique to sound attenuation in ideal and weakly interacting quantum gases were included in my thesis and also not published. Analyses that took several weeks of work in 1964 with an IBM-704 can nowadays be done as student problems with a hand-held HP in an afternoon, so posterity is no less rich. Thoughts of Jaynes' that I recall on the approach to equilibrium and on the degree to which reversible thermodynamic cycles are unattainable, I saw later much more fully developed by the very independent thinking chemist Steve Berry.

Jaynes is a member of that illustrious company of independent thinkers who contradict the notion that a theorist should properly be concerned only with fundamental principles and their mathematical elaboration and thereby eschew practical problems, realistic situations, and experimental realities. Indeed, many of the truly great theorists have been involved extensively with the design and analysis of experiments. Einstein, after all, formulated much of his statistical mechanics to analyze a series of experiments which allowed him, among other things, to determine Avogadro's number in several different ways, establish concretely the atomistic nature of matter, and build a goodly part of the foundations for the development of quantum mechanics.

Perhaps more surprisingly, Einstein and Leo Szilard in 1928 applied for a patent for a refrigerator using a pumped liquid metal as a working fluid. Such electromagnetic pumps are now incorporated in the cooling circuit in liquid-metal-cooled reactors. Szilard apparently originated the notion of a liquid-metal-cooled pile, which he pushed as a candidate for the first plutonium production reactors. The official AEC historian has noted drily: "For Szilard, the new and unusual held no cause for hesitation." (Szilard, very much an independent thinker, also explored the relationship between information and statistical mechanics, but did not fully recognize the advantage of generalizing the approach of Gibbs that was later discovered by Jaynes.) Eugene Wigner, although renowned for his work on group theory and quantum mechanics, on nuclear structure, on R-matrix theory, other fundamental developments in nuclear and particle physics, and helping to found modern solid state theory, also was the principal designer of the plutonium production reactors that were built at Hanford during World War II. These were an enormous scale-up from very small scale experiments to what nowadays would be multibillion dollar investments. He also designed the "Clinton pile" at Oak Ridge. Wigner's original training was in chemical engineering, although his Nobel Prize was for work in theoretical physics. John Bardeen, also a Wigner student, was almost as deeply involved in electrical engineering as he was in theoretical physics, as attested to by articles in the April 1992 issue of *Physics Today*. Bardeen, as did Wigner, worked and consulted with a variety of industrial companies while still advancing the frontiers of theory.

Jaynes, as noted before, had practical experience with radar and other microwave devices early in his career and has maintained a continuing interest in quantum electronics. Concurrently, he has maintained an interest for many years in the physics of music.

Even Jaynes' intellectual forebear, J. Willard Gibbs, had a practical side. The scientific writings of J. Willard Gibbs were noted for their conciseness and difficulty and were

---

[†] *Editors' note:* The work on the muscle as an engine finally appeared in 1989, included in reference [64] in the Jaynes bibliography at the end of this volume.

often published in little-known local journals. Some 33 years after Gibbs' death in 1903, Yale University Press published an extended commentary on his scientific writings. It was considerably longer than the two bound volumes of the scientific works themselves. Typical statements by the distinguished authors include, "In this immortal work, . . . the originality, power, and beauty of Gibb's work in the domain of thermodynamics have never been surpassed" and "the greatest mathematical physicist of the American nation."

But the earliest work of J. Willard Gibbs, which includes his Ph.D. thesis, is of a rather different character. His dissertation bears the title "Of the Form of the Teeth of Wheels in Spur Gearing." It is primarily an exercise in applied geometry. His 1863 Ph.D. was the fifth such degree granted in the United States, the first in engineering, and arguably the second in science. Immediately after receiving his degree, Gibbs served as a tutor at Yale, initially in Latin and only later in natural philosophy. The only original work that he seems to have completed during this period was an improved railway car brake, which was patented in 1866.

Shortly after receipt of this patent, Gibbs left for a three-year sojourn in Europe, which brought him into contact with the most advanced thinking in the most advanced scientific capitals in the world. Nonetheless, upon returning to Yale, Gibbs focused his attention on improving the conical pendulum governor for steam engines, originally invented by James Watt. A model embodying this improvement was apparently made in about 1872 and still exists in New Haven. (It is noteworthy in these times that Gibbs was initially appointed professor of mathematical physics without salary in 1871 and only received a formal salary from Yale after receiving an offer from Bowdoin College of $1800 per year for a "chair of mathematics, or of physics.")

It is perhaps inevitable that the work of a truly independent thinker may take some time to be fully recognized. As the diversity of participants at this meeting attests, Edwin T. Jaynes' work has become increasingly recognized as well as respected. In their introduction to the book *Maximum Entropy and Bayesian Methods in Inverse Problems*, Smith and Grandy point out, reflecting work up to about ten years ago, "The sheer number of citations of his works indicates the considerable influence that he has had on this subject. His influence through personal interaction is even greater." More recent comments reinforce this view. Half a dozen years earlier in an essay in tribute to another truly and original independent thinker, George Gamow, Wolfgang Yourgrau and Alwyn Van der Merwe state, "Finally, a decade after the birth of information theory, Edwin Jaynes drew from it the one important lesson for physics, which has since given rise to highly significant simplified treatment of the foundation of statistical mechanics." These authors began the last paragraph in their 37-page paean to the information theory approach: "But more important than the didactical advantages is the unification, simplification, and clarification which information theory has wrought of the foundation of statistical mechanics." They end with the statement, "To us with a philosophic bias in scientific matters, this in itself is a genuine achievement, although the pragmatic physical scientist will rightly discern the crowning glory of the information approach is in the fresh answers to old problems, which it is starting to provide in the study of irreversible phenomena." Such praise is much the sort that, some thirty years after his death, devolved upon J.Willard Gibbs. Ed Jaynes fully deserves to be recognized as Gibbs' modern heir.

Some of the nuances must, however, get lost in translation. In a work entitled *Energy and Entropy* by G.N. Alekseev, first published in Moscow in 1978 (in English in 1987), it is stated, ". . . difficulties stem from the application of a method, based on the superficial similarity between information and Boltzmann's formula for entropy to areas for which the latter was never intended. . . It is thus much easier for theorists to proclaim a relationship between entropy and information than to demonstrate it by practical numerical examples. Nevertheless, 22 or 27 years ago this relation was so thoroughly developed that information theory became the basis (not vice versa) for the development of a sophisticated system of universal thermodynamics which was derived from a group of original equations. (The theory belongs to M. Tribus, an American scientist.)"

Finally, it should be said that Ed Jaynes encouraged his students, both by example and sometimes by his comments, to indulge in some of his extracurricular habits. One of these is certainly music and the appreciation of the physics of music; another is an appreciation of history and of old books, wherever they may be found. In the thirty years since I first met Ed Jaynes, I have suffered from the rampant spread of the old-book disease. Scratching that old-book itch has led me to many classics, including a little volume on the early work of J. Willard Gibbs, including the aforementioned thesis, which I have brought here for Ed tonight. I would like to present it to Ed at this time, with the suggestion that his work is quite likely to have the same kind of long history and the same kind of future as that of his illustrious predecessor.

A LOOK BACK:
EARLY APPLICATIONS OF MAXIMUM ENTROPY ESTIMATION
TO QUANTUM STATISTICAL MECHANICS

D.J. Scalapino
Department of Physics
University of California
Santa Barbara, California 93106–9530

ABSTRACT. E.T. Jaynes has been the central figure for over three decades in showing how maximum entropy estimation (MEE) provides an extension of logic to cases where one cannot carry out Aristotelian deductive reasoning. Here I will review two early applications of MEE which I hope will provide some insight into how these ideas were being used in quantum and statistical mechanics around 1960.

In May of '61 I turned in my Ph.D. thesis (Scalapino, 1961) to Stanford University and, following a well-known dictum, went on to work on other problems. Now, three decades later, on this, the occasion of Ed Jaynes' seventieth birthday, I decided to look back to see what we were doing when I was first getting to know Ed and his special approach to problems.

Not surprisingly, my thesis contains several applications of the principle of Maximum Entropy Estimation (MEE). In 1957 Ed had written two seminal articles (Jaynes, 1957(a), 1957(b)) showing how one could use Shannon's (1948) Information Theory to construct density matrices for a variety of different problems in equilibrium statistical mechanics. I had been very much taken with this work and wanted to understand how it could be applied to other systems.

Ed had shown how the MEE formalism could be used to generate ensembles in statistical mechanics when there was only partial knowledge available about a system with many degrees of freedom. It seemed natural to wonder what this formalism would predict for a microscopic quantum system with a small number of degrees of freedom, particularly when the measurement provided sufficient information to specify the state uniquely. Ed may have known, but I had no idea what to expect when I started out to apply MEE for the first time.

The problem was certainly basic enough. Consider a spinless particle moving in one dimension, and imagine that experimentally one knows that

$$\text{Tr}\,\rho x = \text{Tr}\,\rho p = 0$$
$$\text{Tr}\,\rho x^2 = \Delta x^2, \qquad \text{Tr}\,\rho p^2 = \Delta p^2. \tag{1}$$

According to the MEE formalism, ρ is determined by maximizing $S = -\operatorname{Tr}\rho\ln\rho$, subject to the constraints of Eq. (1). This leads in the usual way to the density matrix

$$\rho = \frac{e^{-\lambda_1 x^2 - \lambda_2 p^2 - \lambda_3 x - \lambda_4 p}}{Z(\lambda_i)} \tag{2}$$

with

$$Z(\lambda_i) = \operatorname{Tr}\rho. \tag{3}$$

The λ_i are determined so as to satisfy the constraints of Eq. (1),

$$0 = \frac{\partial}{\partial\lambda_3}\ln Z = \frac{\partial}{\partial\lambda_4}\ln Z$$

$$\Delta x^2 = -\frac{\partial}{\partial\lambda_1}\ln Z, \qquad \Delta p^2 = -\frac{\partial}{\partial\lambda_2}\ln Z. \tag{4}$$

Now λ_3 and λ_4 vanish by inspection, and using a harmonic oscillator basis in which

$$\left(\lambda_1 x^2 + \lambda_2 p^2\right)|n\rangle = 2\left(\lambda_1\lambda_2\right)^{1/2}\hbar\left(n + \frac{1}{2}\right)|n\rangle, \tag{5}$$

one can show that

$$Z = \sum_n e^{-2(\lambda_1\lambda_2)^{1/2}\hbar(n+\frac{1}{2})} = \frac{1}{2\sinh\left(\hbar(\lambda_1\lambda_2)^{1/2}\right)}. \tag{6}$$

Then solving Eq. (4), one obtains

$$\lambda_1 = \frac{1}{\hbar}\frac{\Delta p}{\Delta x}\coth^{-1}\left(\frac{2}{\hbar}\Delta p\Delta x\right)$$

$$\lambda_2 = \frac{1}{\hbar}\frac{\Delta x}{\Delta p}\coth^{-1}\left(\frac{2}{\hbar}\Delta p\Delta x\right) \tag{7}$$

so that

$$Z = \left(\left(\frac{2}{\hbar}\right)^2\Delta x^2\Delta p^2 - 1\right)^{1/2}. \tag{8}$$

Now the probability of being in the nth harmonic oscillator state is

$$P_n = \langle n|\rho|n\rangle = \frac{1}{Z}e^{-2(\lambda_1\lambda_2)^{1/2}\hbar(n+\frac{1}{2})} = \frac{\left(\frac{2}{\hbar}\Delta p\Delta x - 1\right)^n}{\left(\frac{2}{\hbar}\Delta p\Delta x + 1\right)^{n+1}}. \tag{9}$$

In the limit of a *complete measurement* in which Δx and Δp are determined as accurately as the Heisenberg uncertainty principle allows so that $\Delta x\Delta p = \hbar/2$, Eq. (9) becomes

$$P_n = \delta_{n,0}. \tag{10}$$

Thus in this case the MEE density matrix $\rho = |0\rangle\langle 0|$ corresponds to the pure state

$$|0\rangle = \frac{1}{(2\pi)^{1/4}(\Delta x)^{1/2}}\, e^{-x^2/4\Delta x^2}. \tag{11}$$

One can note that if $\Delta x \Delta p < \hbar/2$, the probabilities associated with the odd n states are negative, clearly indicating the fact that one has moved into an unphysical regime.

Finally, evaluating the entropy one obtains

$$S = \left(\frac{\Delta x \Delta p}{\hbar} + \frac{1}{2}\right)\ln\left(\frac{\Delta x \Delta p}{\hbar} + \frac{1}{2}\right) - \left(\frac{\Delta x \Delta p}{\hbar} - \frac{1}{2}\right)\ln\left(\frac{\Delta x \Delta p}{\hbar} - \frac{1}{2}\right). \tag{12}$$

For $\Delta x \Delta p = \hbar/2$, $S = 0$ as expected, while for $\Delta x \Delta p < \hbar/2$, S becomes complex. For $\Delta x \Delta p \gg \hbar/2$,

$$S \simeq \ln\left(\frac{\Delta x \Delta p}{\hbar}\right). \tag{13}$$

This is just the usual Boltzmann results in which S is the log of the number of states in the reasonably probable part of phase space.

All of this seems obvious now, but I can assure you, for a graduate student trying to understand Ed's MEE formalism, it was very reassuring.

With this background, I turned to the main problem of interest, which was extending the MEE formalism to non-equilibrium problems. Here the goal was to find an algorithm for constructing density matrices appropriate to non-equilibrium conditions. If this could be done, then, for example, transport properties could be obtained by simply tracing the desired operators over the density matrix exactly as in equilibrium statistical mechanics. In particular, there would be no need to integrate ρ forward over an "induction time" or to carry out some coarse-graining procedure (Mori, 1958). As it was to turn out, this algorithm had already been stated by Gibbs (1960): find the distribution which, while agreeing with what is known, "gives the least value of the average index of probability of phase." Ed had clearly understood how this applied to equilibrium statistical mechanics, but what should be done in the non-equilibrium case? I began by studying a steady-state non-equilibrium problem.

Suppose one knows in addition to $\langle H \rangle$, the average value of another operator $\langle F \rangle$; then applying MEE to determine the density matrix gives

$$\rho = \frac{e^{-\beta H - \lambda F}}{Z(\beta, \lambda)} \tag{14}$$

with

$$Z = \operatorname{Tr} e^{-\beta H - \lambda F}. \tag{15}$$

Here β and λ are determined, so that one obtains the known expectation values of H and F. However, if one also knows that the system is in a steady state but that F does not commute with H, one has a new and interesting problem. Clearly in this case the expression for ρ, given by Eq. (15), is not stationary, so that the expectation values of typical operators will evolve in time. In addition, they will be found to have been evolving in the past. Thus $\langle F(t) \rangle = \operatorname{Tr} \rho \left(e^{iHt} F e^{-iHt}\right)$ will not have been stationary for $t < 0$, contrary to the information one has.

Ed and I wondered what to do in order to describe such a stationary but non-equilibrium system. One approach we tried was to maximize the entropy $S = -\text{Tr}\,\rho \ln \rho$ over the manifold of density matrices which *commuted* with H and satisfied the constraints. This led to an interesting form with

$$\rho = \frac{\exp\left[-\beta H - \lambda F_d\right]}{Z[\beta,\lambda]}. \tag{16}$$

Here the *diagonal* operator F_d is the d.c. part of the Heisenberg operator $F(t) = e^{iHt}Fe^{-iHt}$:

$$F_d = \lim_{T\to\infty} \frac{1}{T} \int_{-\infty}^{0} dt\, e^{t/T} F(t). \tag{17}$$

Note that here and in the following, T is time, not temperature which is β^{-1}.

To test this idea, I studied a many-particle system described by an average energy $\langle H \rangle$, an average particle number $\langle N \rangle$, and a steady-state density gradient. The goal was to construct a steady-state density matrix ρ from which the current density $j_\mu(x)$ could be calculated simply as

$$\langle j_\mu(x) \rangle = \text{Tr}\,\rho j_\mu(x), \tag{18}$$

and then in the long-wavelength limit extract the expression for the diffusion coefficient.

Maximizing the entropy over the manifold of density matrices which commuted with $H - \mu N$, a steady-state density matrix was obtained,

$$\rho = \left(\tfrac{1}{Z}\right) \exp\left(-\beta(H - \mu N) + \beta \int d^3x\, \Delta\mu(x) \int_{-\infty}^{0} dt \frac{e^{t/T}}{T} n(x,t)\right). \tag{19}$$

Here $\Delta\mu(x)$ was determined so as to give the steady state density gradient, and the limit $T \to \infty$ was to be taken at the end of the calculation. The current density is then given by Eq. (18). Expanding this to determine the linear departure from equilibrium,

$$\langle j_\mu(x) \rangle = \int d^3x'\, \Delta\mu(x') \int_{-\infty}^{0} dt' \frac{e^{t'/T}}{T} \langle \tilde{n}(x',t')j_\mu(x)\rangle_0 , \tag{20}$$

with

$$\langle 0 \rangle_0 = \frac{\text{Tr}\, e^{-\beta(H-\mu N)}0}{Z} \tag{21}$$

and

$$\tilde{n}(x,t) = \int_{0}^{\beta} d\tau\, n(x, t - i\tau). \tag{22}$$

Setting $\Delta\mu(x) = x_\nu \frac{\partial \mu}{\partial x_\nu}$ and integrating Eq. (20) by parts leads, after a bit of algebra, to

$$\langle j_\mu(x,0)\rangle = -\int d^3x' \frac{\partial \mu(x')}{\partial x'_\nu} \int_{-\infty}^{0} dt'\, e^{t'/T} \int_{0}^{\beta} d\tau \langle j_\nu(x', t' - i\tau)j_\mu(x,0)\rangle_0 , \tag{23}$$

or, introducing the diffusion kernel $D_{\nu\mu}$,

$$\langle j_\mu(x,0)\rangle = -\int d^3x' \frac{\partial \Delta\mu(x')}{\partial x'_\nu} D_{\nu\mu}(x - x'). \tag{24}$$

Here

$$D_{\nu\mu}(x - x') = \lim_{T \to \infty} \int_{-\infty}^{0} dt'\, e^{t'/T} \int_{0}^{\beta} d\tau\, \langle j_\nu(x', t' - i\tau) j_\mu(x, 0) \rangle_0 , \qquad (25)$$

which was equal to the well known Kubo form (Kubo, 1957).

I remember being pleased that this had worked out and feeling that the notion of making a MEE over the manifold of density matrices which commuted with H seemed a sensible way to describe steady-state phenomena. However, when I took my oral exam at Stanford, professor J.D. Bjorken listened to my presentation and asked why hadn't I simply followed the MEE procedure to its logical conclusion by demanding that $\langle F(t) \rangle$ be constant over past times. Bjorken's suggestion was to provide the key to a much more general approach. I remember coming back to Washington University* and telling Ed about Bjorken's question.

Some time passed, and in the fall of '62 when Ed came to visit me at the University of Pennsylvania where I was a postdoc working with J.R. Schrieffer, he had a manuscript, "Nonlocal Transport theory" (Jaynes and Scalapino, 1963), in which a general algorithm for calculating irreversible processes was described. It was a generalization of the Gibbs construction of ensembles extended to irreversible processes.

The key feature was to use the MEE procedure when the information on F was obtained during a time period $(-T, 0)$,

$$\langle F(t) \rangle = \text{Tr}\, \rho F(t), \qquad -T \leq t \leq 0. \qquad (26)$$

When this was combined with the knowledge of $\langle H \rangle$ and $\langle N \rangle$, and the entropy maximized subject to the constraints, one obtained a density matrix

$$\rho = \frac{e^{-\beta(H - \mu N) - \int_{-T}^{0} dt\, \lambda(t) F(t)}}{Z[\beta, \mu, \lambda(t)]} , \qquad (27)$$

with $F(t)$ the Heisenberg operator. The partition function became a functional of $\lambda(t)$,

$$Z[\beta, \mu, \lambda(t)] = \text{Tr}\, e^{-\beta(H - \mu N) - \int_{-T}^{0} dt\, \lambda(t) F(t)}, \qquad (28)$$

with $\lambda(t)$ determined by

$$\langle F(t) \rangle = -\frac{\partial}{\partial \lambda(t)} \ln(Z[\beta, \mu, \lambda(t)]), \qquad -T \leq t \leq 0. \qquad (29)$$

For the case in which the information-gathering time becomes long, $T \to \infty$, one retrieves the d.c. result in leading order. However, Eqs. (27)–(29) are quite general and represented the solution to the problem of extending the MEE formalism to non-equilibrium phenomena. To give you a flavor of this, I'll conclude by quoting the abstract of the manuscript (Jaynes and Scalapino, 1963) presented at the first meeting of the Society for Natural Philosophy held in Baltimore in March of 1963.

---

* In 1960 Jaynes had accepted a professorship at Washington University in St. Louis, and I had gone with him to complete my Ph.D. work.

A general algorithm for calculation of irreversible processes, analogous to the partition sum of equilibrium theory, is believed to be at hand. Mathematically, it is a generalization of the procedure by which Gibbs constructed his canonical and grand canonical ensembles, involving the extension of the partition function to a partition functional. Conceptually, it has a simple interpretation in terms of Information Theory. We illustrate its use by developing a theory of transport phenomena along the lines initiated by Green, Kubo, and Mori, in which the following features emerge: (1) In the new method of constructing ensembles, steady-state dissipative effects are obtained by direct quadratures over the initial ensemble, with no need for the operations of forward integration and time-smoothing characteristic of previous treatments. (2) The formalism leads automatically to a nonlocal theory, similar to Pippard's modification of the London equations for superconductivity.

ACKNOWLEDGMENTS. I want to acknowledge the special way E.T. Jaynes taught one to think about physics and to thank him for sharing his deep insight into basic problems. It seems also appropriate to acknowledge my fellow Stanford graduate student colleagues R. Blankenbecler and F. Cummings, and Washington University graduate student colleagues W. Jackson, W. Massey, C. Williams, and F. Wu, for all the times we shared in those days discussing and learning physics.

REFERENCES

Gibbs, J.W.: 1960, *Elementary Principles in Statistical Mechanics*, Dover Publications, New York.

Jaynes, E.T. and D.J. Scalapino: 1963, 'Nonlocal Transport Theory,' presented at the first meeting of the Society for Natural Philosophy, Baltimore, Md.

Jaynes, E.T.: 1957(b), 'Information Theory and Statistical Mechanics', *Phys. Rev.* **106**, 620.

Jaynes, E.T.: 1957(a), 'Information Theory and Statistical Mechanics II', *Phys. Rev.* **108**, 171.

Kubo, R., M. Yokota, and S. Nakajima,: 1957, *J. Phys. Soc. Japan* **12**, 1203.

Mori, M.: 1958, 'Statistical-Mechanical Theory of Transport in Fluids', *Phys. Rev.* **112**, 1829.

Scalapino, D.J.: 1961, 'Irreversible Statistical Mechanics and the Principle of Maximum Entropy,' Ph.D. dissertation, Stanford University.

Shannon, C.E.: 1948, *Bell Syst. Tech. J.* **27**, 379, 623.

THE JAYNES-CUMMINGS REVIVAL

B.W. Shore
Lawrence Livermore National Laboratory
Livermore, CA 94550

P.L. Knight
Blackett Laboratory,
Imperial College
London SW7 2BZ, UK

ABSTRACT. The Jaynes-Cummings Model (JCM), a soluble fully quantum mechanical model of an atom in a field, was first used (in 1963) to examine the classical aspects of spontaneous emission and to reveal the existence of Rabi oscillations in atomic excitation probability for fields with sharply defined energy (or photon number). For fields having a statistical distribution of photon numbers the oscillations collapse to an expected steady value. In 1980 it was discovered that with appropriate initial conditions (e.g. a near-classical field), the Rabi oscillations would eventually revive – only to collapse and revive repeatedly in a complicated pattern. The existence of these revivals, present in the analytic solutions of the JCM, provided direct evidence for discreteness of field excitation (photons) and hence for the truly quantum nature of radiation. Subsequent study revealed further nonclassical properties of the JCM field, such as a tendency of the photons to antibunch. Within the last two years it has been found that during the quiescent intervals of collapsed Rabi oscillations the atom and field exist in a macroscopic superposition state (a Schrödinger cat). This discovery offers the opportunity to use the JCM to elucidate the basic properties of quantum correlation (entanglement) and to explore still further the relationship between classical and quantum physics. In tribute to E. T. Jaynes, who first recognized the importance of the JCM for clarifying the differences and similarities between quantum and classical physics, we here present an overview of the theory of the JCM and some of the many remarkable discoveries about it.

1. Introduction

When one surveys contemporary optics one cannot avoid encountering what is variously termed the hydrogen atom of quantum optics, the simplest exactly soluble problem of a self consistent quantum theory of matter interacting with radiation, or simply the Jaynes-Cummings model (JCM).

As originally defined (Jaynes and Cummings, 1963), the JCM comprised a single two-state atom ("molecule") interacting with a single near-resonant quantized cavity mode

15

of the electromagnetic field. The model was first used to examine the classical aspects of spontaneous emission. It was subsequently discovered that the JCM atomic population histories presented direct evidence for the discreteness of photons. It was also found that the cavity field, once modified by the JCM interaction with the atom, had statistical properties not found in classical fields. More recently, the JCM has been used to elucidate quantum correlation and the formation of macroscopic quantum states.

In the three decades that have passed since Jaynes and Cummings presented the JCM (or four decades since the original unpublished Stanford Microwave Laboratory report (Jaynes, 1958) in which Jaynes first examined this model), there has arisen a sizable scientific industry devoted to exploiting and extending this work (for reviews see Yoo and Eberly, 1985; Haroche and Raimond, 1985; Barnett et al, 1986; Milonni and Singh, 1991; Meystre, 1992). It is appropriate, on this occasion commemorating the work of Jaynes, to recall the origin of this enterprise, to note some of the most remarkable and startling subsequent discoveries about the model.

2. The Work of Jaynes and Cummings

In reading today the original 1963 article by Jaynes and Cummings, one is struck by several things. First, the clarity of presentation stands as a model of scientific exposition. One follows quite easily the straightforward analysis that leads to simple analytic expressions for the time dependence of the Schrödinger statevector and the density matrix of the atom-field system. With hindsight, it is also interesting to see what Jaynes and Cummings thought was important in their work.

According to their abstract, they sought two things. Their secondary objective was to understand some properties of a particular device, the ammonia maser. But their primary objective was

"...to clarify the relationship between the quantum theory of radiation, where the electromagnetic field expansion coefficients satisfy commutation relations, and the semiclassical theory where the electromagnetic field is considered as a definite function of time rather than as an operator."

To appreciate the ambition of this objective, one must recall that the conventional wisdom of that time considered semiclassical theory to be the natural and inevitable limiting case of quantum theory taken to large numbers of quanta. Furthermore, it was generally believed that a quantum theory of radiation was a necessary prerequisite for a description of spontaneous emission. Keeping that view in mind, we are impressed by the results of the Jaynes-Cummings paper, wherein

" ...it is shown that the semiclassical theory, when extended to take into account both the effect of the field on the molecules and the effect of the molecules on the field, reproduces almost quantitatively the same laws of energy exchange and coherence properties as the quantized field theory, even in the limit of one or a few quanta in the field mode. In particular, the semiclassical theory is shown to lead to a prediction of spontaneous emission, with exactly the same decay rate as given by quantum electrodynamics, as described by the Einstein A coefficients."

What Jaynes and Cummings termed semiclassical radiation theory became known as neoclassical theory – a theory in which there are no photons and no field commutators, but in which the back reaction of the atom upon the field is included, just as in the radiative reaction theory of Lorentz (Mandel, 1976; Milonni, 1976). The insights obtained through the study of semiclassical theory, especially those obtained using the Heisenberg picture to examine the quantum theory of radiation reaction, have transformed our view of quantum electrodynamics. Nowadays we have accepted Jaynes' argument that much of radiation theory has a classical interpretation, but interest in neoclassical theory per se has waned. Nevertheless, interest in the Jaynes-Cummings model continues. In part, the interest has been stimulated by the discovery of a number of remarkable properties of the model, along with the possibility of finding solutions (often exact) to fundamental models of a quantum theory of interacting fields and atoms. Perhaps equally important as a motivating force has been the remarkable advance in experiments involving single atoms within single-mode cavities (the micromaser, see Meschede et al, 1985; Haroche and Raimond, 1985; Rempe et al, 1986,1987; Brune et al, 1987) which have turned the theory from an academic curiosity into a useful and testable enterprise. These motives have certainly provided incentive for generalizing what is still called the Jaynes-Cummings model. The following sections comment on a few of these generalizations, particularly a variety of initial conditions and dissipation and damping.

3. The JCM Hamiltonian

The original JCM Hamiltonian can be expressed in the form

$$H = \hbar\omega\hat{a}^\dagger\hat{a} + E_1\hat{S}_{11} + E_2\hat{S}_{22} + \frac{\hbar}{2}\Omega_1[\hat{a}^\dagger\hat{S}_{12} + \hat{S}_{21}\hat{a}] \tag{1}$$

where E_k is the energy of atomic state ψ_k, the atom-field coupling constant Ω_1 is the vacuum (or single-photon) Rabi frequency, and $\hat{S}_{jk}$ is a transition operator acting on atomic states, definable by the expressions

$$\hat{S}_{jk}|\psi_n\rangle = \delta_{kn}|\psi_j\rangle \quad \text{or} \quad \hat{S}_{jk}\hat{S}_{nm} = \delta_{kn}\hat{S}_{jm}. \tag{2}$$

The photon creation and annihilation operators $\hat{a}^\dagger$ and $\hat{a}$ obey the commutator relation

$$[\hat{a}, \hat{a}^\dagger] = 1 \tag{3}$$

and the field is described by the photon number states $|n\rangle$, eigenstates of the photon number operator $\hat{a}^\dagger\hat{a}$,

$$\hat{a}^\dagger\hat{a}|n\rangle = n|n\rangle, \quad \hat{a}^\dagger|n\rangle = \sqrt{n+1}|n+1\rangle, \quad \hat{a}|n\rangle = \sqrt{n}|n-1\rangle. \tag{4}$$

The field frequency ω, the atomic energies E_k, and the (vacuum) Rabi frequency Ω_1 are fixed by physical considerations (e.g. cavity volume V and atomic transition moment d, as in the expression $|\Omega_1|^2 = 4d^2\omega/\hbar V\varepsilon_0$). To place this theory into an experimental context one may consider the micromaser (Meschede et al, 1985): a stream of velocity-selected excited Rydberg atoms pass through a cold high-Q single-mode microwave cavity at a rate that allows only a single atom in the cavity at any time. Time $t = 0$ marks the entrance of an atom into the cavity. Missing from the JCM Hamiltonian of Eqn. (1) are such effects as cavity loss, multiple cavity modes, atomic sublevel degeneracy and atomic polarizability (leading to dynamic Stark shifts), all of which have subsequently been examined.

4. The Atomic JCM Hamiltonian

The construction of the JCM Hamiltonian is such that each photon creation accompanies an atomic deexcitation, and each photon annihilation accompanies atomic excitation. Therefore in addition to the conservation of atomic probability,

$$\langle \hat{S}_{11} \rangle + \langle \hat{S}_{22} \rangle = 1 \tag{5}$$

there occurs a conservation of excitation

$$\langle \hat{a}^\dagger \hat{a} \rangle + \langle \hat{S}_{22} \rangle = \text{constant} \tag{6}$$

Thus, as Jaynes and Cummings pointed out, the problem becomes that of an infinite set of uncoupled two-state Schrödinger equations, each pair identified by the number of photons that are present when the atom is in the lowest-energy state. We can write the atom-field statevector for specified photon number as a combination of two basis states,

$$\Psi(n,t) = \exp(-in\omega t - iE_1 t/\hbar)[C_1(n,t)\phi_1(n) + C_2(n,t)\phi_2(n)] \tag{7}$$

where $\phi_k(n)$ is the atom-field product state

$$\phi_1(n) = |n\rangle \psi_1, \qquad \phi_2(n) = |n-1\rangle \psi_2. \tag{8}$$

The 2×2 Hamiltonian matrix of one such pair has the form

$$H(n) = \frac{\hbar}{2} \begin{bmatrix} 0 & \Omega(n) \\ \Omega(n) & 2\Delta \end{bmatrix}, \quad \Psi(n,t) = \exp(-in\omega t - iE_1 t/\hbar) \begin{bmatrix} C_1(n,t) \\ C_2(n,t) \end{bmatrix} \tag{9}$$

involving the cavity-atom detuning Δ and the n-photon Rabi frequency $\Omega(n)$

$$\hbar\Delta = E_2 - E_1 - \hbar\omega, \qquad \Omega(n) = \sqrt{n}\,\Omega_1. \tag{10}$$

5. The JCM Solutions for Population

The separate paired equations of the simplified JCM Hamiltonian (9) are just those of a two-state atom in a monochromatic field (Allen and Eberly, 1975; Meystre and Sargent, 1990). As with their earlier occurrence in magnetic resonance (Rabi, 1937), they predict sinusoidal changes in populations now known as Rabi oscillations. In particular, if the atom is initially unexcited while the field has exactly n photons, then the population inversion $w(t) = P_2(t) - P_1(t)$ at time t is (cf. Shore, 1990)

$$w(t) = |C_2(n,t)|^2 - |C_1(n,t)|^2 = 1 - \sin^2(2\theta_n)[1 - \cos(\lambda_n t)]. \tag{11}$$

This expression uses the mixing angle θ_n and the flopping frequency λ_n

$$\tan(2\theta_n) = \Omega(n)/\Delta, \qquad \lambda_n = \sqrt{\Delta^2 + \Omega(n)^2}. \tag{12}$$

The assertion that the JCM is analytically soluble is equivalent to the observation that there are algebraic expressions for the (two) probability amplitudes $C_k(n,t)$ or for the (two) eigenvalues of the matrix $H(n)$ and the corresponding pair of eigenvectors, the two – component dressed states $\Phi_k(n)$,

$$H(n)\Phi_k(n) = \hbar\Lambda_k(n)\ \Phi_k(n). \tag{13}$$

When the system is resonant ($\Delta = 0$) the two dressed states are independent of photon number. The resulting superposition of atomic states is

$$\psi_\pm = \frac{1}{\sqrt{2}}[e^{i\varphi}\psi_1 \pm \psi_2] \tag{14}$$

where φ is the phase of the field. Given dressed states of $H(n)$ we construct the atom-field statevector at any time t as

$$\Psi(n,t) = \exp(-in\omega t - iE_1 t/\hbar)\sum_k c_k(n)\Phi_k(n)\exp[-i\Lambda_k(n)t] \tag{15}$$

where the numbers $c_k(n)$ are chosen to reproduce specified initial conditions $\Psi(n,0)$. Following the lead of Ackerhalt (1974) and Ackerhalt and Rzazewski (1975), who presented exact solutions for the Heisenberg picture operators of the JCM, we have here exploited the conservation of excitation number to generate exact Schrodinger picture solutions.

6. The Density Matrix

The density matrix offers a simple way of incorporating statistical distributions of initial conditions into quantum mechanics. More importantly, the Liouville equation for the density matrix $\rho(t)$

$$\hbar\frac{\partial}{\partial t}\rho(t) = -i[H, \rho(t)] \tag{16}$$

whose solution is used to evaluate expectation values as a trace

$$\langle M(t)\rangle = Tr[\hat{M}\rho(t)], \tag{17}$$

generalizes to permit incorporation of dissipative mechanism and homogeneous relaxation, as noted below.

When the system comprises two parts, the field and atom components of the JCM, then the trace consists of a sum over atom attributes and field attributes. By carrying out the separate sums we define reduced density matrices for the atom and field subsystem,

$$\rho^A(t) = Tr_F\rho(t), \qquad \rho^F(t) = Tr_A\rho(t) \tag{18}$$

with which to evaluate expectation values of atom operators $\hat{A}$ and field operators $\hat{F}$

$$\langle\hat{A}\rangle = Tr_A[\rho^A(t)\hat{A}], \qquad \langle\hat{F}\rangle = Tr_F[\rho^F(t)\hat{F}]. \tag{19}$$

When applied to the traditional JCM the atom density matrix has dimension 2, and so it is quite manageable. When the field description is by means of photon numbers the field density matrix has denumerably infinite dimension.

7. Atom Observables

The probabilities $P_k(t)$ of finding the atom in atom state ψ_k at time t is the expectation value of the atomic operator $\hat{S}_{kk}$

$$P_k(t) = \langle \hat{S}_{kk} \rangle. \qquad (20)$$

When the overall field-atom system is in a pure quantum state $\Psi(t)$ then this probability is

$$P_k(t) = \| \langle \psi_k | \Psi(t) \rangle \|^2 . \qquad (21)$$

More typically these population histories are ensemble averages over initial distributions of atom and field properties.

Perhaps the most notable aspect of a two-state atom interacting coherently with a monochromatic classical field is the periodic shift of population between the ground state and the excited state (Knight and Milonni, 1980). Jaynes and Cummings showed that these same Rabi oscillations will occur for a quantized field, if the field is initially in a photon-number state, and they showed that almost indistinguishable oscillations occur in neoclassical theory. In particular, an excited atom entering a vacuum cavity undergoes spontaneous Rabi oscillations.

The Rabi oscillations of the JCM differ dramatically from the irreversible exponential decay that occurs in free space. This is because an atom within a cavity undergoes reversible spontaneous emission, as it repeatedly emits and then reabsorbs radiation. This simplest example occurs when an excited atom enters a cold dark cavity (the photon vacuum). The JCM predicts exactly periodic interchange of energy between atom and field as the atom repeatedly emits and then reabsorbs the single quantum of energy. When the cavity and Bohr frequencies coincide, the population oscillations occur at the vacuum Rabi frequency (the product of atom dipole transition moment and the single-photon electric field of the cavity). If the atom leaves in its unexcited state, then we can infer that the cavity contains exactly one photon.

8. Field Observables; Fluctuations

The Rabi oscillations of atomic population are complemented by oscillations of field energy, as expressed by the mean photon number,

$$\langle \hat{n} \rangle = \langle \hat{a}^\dagger \hat{a} \rangle \qquad (22)$$

and so the JCM offers a soluble case of a quantum field that changes with time. The distribution of possible field energies is expressed by the probability $p_n(t)$ of finding n photons at time t.

$$p_n(t) = \| \langle n | \Psi(t) \rangle \|^2 . \qquad (23)$$

The photon number variance is

$$\langle (\Delta \hat{n})^2 \rangle = \langle \hat{a}^\dagger \hat{a} \hat{a}^\dagger \hat{a} \rangle - \langle \hat{a}^\dagger \hat{a} \rangle^2 = \langle \hat{n}^2 \rangle - \langle \hat{n} \rangle^2 \qquad (24)$$

and the Mandel Q parameter (Mandel, 1979) is

$$Q(t) = \frac{\langle \hat{a}^\dagger \hat{a}^\dagger \hat{a} \hat{a} \rangle - \langle \hat{a}^\dagger \hat{a} \rangle^2}{\langle \hat{a}^\dagger \hat{a} \rangle} = \frac{\langle (\Delta \hat{n})^2 \rangle - \langle \hat{n} \rangle}{\langle \hat{n} \rangle}. \qquad (25)$$

When $Q > 0$ the photons are bunched (super-Poisson), while for $Q < 0$ the photons are antibunched (sub-Poisson, a purely quantum regime).

The field generated by a two-state atom has nonclassical statistics: the antibunching originates in the requirement that before an atom can emit a second photon it must be reexcited. The JCM field exhibits nonclassical effects, even when the cavity begins in a near-classical state.

The electric and magnetic fields of the JCM are proportional to the quadrature operators

$$\hat{X} = \frac{1}{2}(\hat{a} + \hat{a}^\dagger), \qquad \hat{Y} = \frac{1}{2i}(\hat{a} - \hat{a}^\dagger). \tag{26}$$

These have the variances

$$\langle(\Delta X)^2\rangle = \frac{1}{2}\langle\hat{a}^\dagger\hat{a}\rangle + \frac{1}{4} + \frac{1}{4}\langle\hat{a}\hat{a}\rangle + \frac{1}{4}\langle\hat{a}^\dagger\hat{a}^\dagger\rangle - \frac{1}{4}\langle\hat{a}^\dagger + \hat{a}\rangle^2 \tag{27}$$

$$\langle(\Delta Y)^2\rangle = \frac{1}{2}\langle\hat{a}^\dagger\hat{a}\rangle + \frac{1}{4} - \frac{1}{4}\langle\hat{a}\hat{a}\rangle - \frac{1}{4}\langle\hat{a}^\dagger\hat{a}^\dagger\rangle + \frac{1}{4}\langle\hat{a}^\dagger + \hat{a}\rangle^2. \tag{28}$$

When either of these variances is less than 1/4, the field is squeezed (Loudon and Knight, 1987). The cavity field interacting with a two-state atom, initially excited, exhibits a time-varying pattern of squeezing (Meystre and Zubairy, 1982; Aravind and Hu, 1988; Hillery, 1989)

9. Initial Conditions: Incoherent Superpositions

With the introduction of a Hamiltonian one has established the time evolution of a quantum system. To complete the definition of a model one must also specify initial conditions. The mathematics is simplest when one proposes a single atom of precisely known energy (i.e. a particular excitation state) brought suddenly into a cavity with precisely known energy (i.e. a definite number of photons). The more usual condition is a cavity for which one can only specify statistical properties of the single-mode field. Over the years there have been studies of the JCM interacting with a great variety of initial single-mode fields. These studies have shown that there can be remarkable differences in the behavior of the system with different initial conditions.

When the original field has some unknown photon number, specified only by a probability $\wp_m$ for observing m photons, then the atomic excitation probabilities and photon distributions are calculated as the sums

$$P_k(t) = \sum_{m=0}^{\infty} \wp_m \parallel \langle\psi_k|\Psi(m,t)\rangle \parallel^2 = \sum_{m=0}^{\infty} \wp_m |C_k(m,t)|^2 \tag{29}$$

$$p_n(t) = \sum_{m=0}^{\infty} \wp_m \parallel \langle n|\Psi(m,t)\rangle \parallel^2 = \sum_{k} \wp_{n+k-1} |C_k(n+k-1,t)|^2. \tag{30}$$

Superpositions of periodic solutions produce damped oscillations (Cummings 1965). An important example occurs when two-state atoms encounter a cavity maintained at a finite temperature T, so that the photon number distribution is that of one mode of black-body radiation, the Bose – Einstein probability distribution

$$\wp_m(T) = \frac{1}{1 + \langle n\rangle}\left(\frac{\langle n\rangle}{1 + \langle n\rangle}\right)^m, \tag{31}$$

for which $\langle n \rangle = [\exp(\hbar\omega/kT) - 1]^{-1}$ and $\langle \Delta n^2 \rangle = \langle n \rangle^2 + \langle n \rangle$. Figure 1 illustrates this effect of incoherent superpositions by plotting the population inversion $w(t)$ for an initially excited two state atom interacting with a single-mode cavity field in which there is a thermal distribution of photon numbers. The uppermost frame shows the vacuum Rabi oscillations that occur when the atom enters an initially empty cavity. Subsequent frames show the atomic behavior as the initial mean photon number increases. The wide range of photon numbers that are present in such a thermal field give such a broad distribution of Rabi frequencies that there is almost no trace of population oscillations visible in an ensemble average (Cummings, 1965; Knight and Radmore, 1982; Haroche and Raimond, 1985). The collapse time for a thermal field with large $\langle n \rangle$ is (Barnett et al, 1986)

$$(t_c)^{-1} = \frac{1}{2}\Omega_1 \sqrt{\langle n \rangle} = \frac{1}{2}\Omega(\langle n \rangle) \quad \text{for} \quad \langle n \rangle >> 1. \tag{32}$$

10. Initial Field as Coherent Superposition

When the initial field state is expressible as a coherent superposition of photon-number states and the atom is initially unexcited, then the initial statevector is

$$\Psi(0) = \sum_n f_n |n\rangle \psi_1 = \sum_n f_n \Psi(n,0), \quad \wp_m = |f_m|^2. \tag{33}$$

Each statevector $\Psi(n,t)$ evolves independently, and so we obtain the expression

$$\Psi(t) = \sum_n f_n \Psi(n,t) = \sum_{nk} f_n C_k(n,t)|n - k + 1\rangle \psi_k. \tag{34}$$

One can obtain any required property of the atom or the field from values of the Schrödinger amplitudes $C_k(n,t)$.

Following Glauber (1963) it was known that the most classical of single-mode quantum states is a coherent state,

$$|\alpha\rangle = \exp(\alpha \hat{a}^\dagger - \alpha^* \hat{a})|n = 0\rangle \equiv D(\alpha)|0\rangle = \sum_{n=0}^{\infty} f_n(\alpha)|n\rangle, \tag{35}$$

$$f_n(\alpha) \equiv \langle n|\alpha\rangle = \frac{\alpha^m}{\sqrt{n!}} \exp\left(-\frac{1}{2}|\alpha|^2\right) \tag{36}$$

whose photon probabilities follow a Poisson distribution,

$$\wp_m(\alpha) = |f_m(\alpha)|^2 = \frac{\langle n \rangle^m}{m!} \exp[-\langle n \rangle], \tag{37}$$

for which $\langle n \rangle = |\alpha|^2$ and $\langle (\Delta n)^2 \rangle = \langle n \rangle$. Thus it was natural to investigate (e.g. Stenholm, 1973; Meystre et al 1975; von Foerster, 1975), the behavior of the two-state atom when it encountered a coherent state of a single-mode quantum field. As with a thermal cavity, the Rabi oscillations did not persist indefinitely. Instead, the oscillations collapsed to yield constant populations (see Figure 2). As the field becomes more intense, and the mean

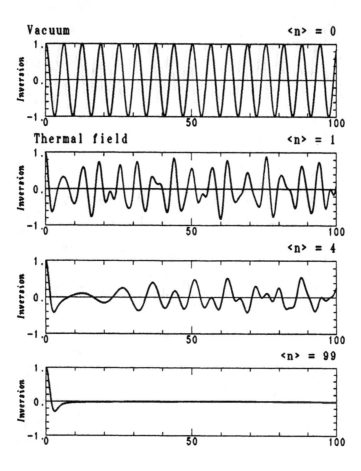

Fig. 1. The population inversion $w(t)$ for an initially excited two state atom interacting with a single-mode cavity field in which there is a thermal distribution of photon numbers. Mean photon numbers of the initial field are $\langle n \rangle = 0$, 1, 4 and 99. Times are in units of mean Rabi frequency. Note the lack of Rabi oscillations; instead there occur irregular fluctuations.

photon number larger, the Rabi oscillations persist for longer intervals, but inevitably the incommensurability of Rabi oscillations for different photon numbers washes out the periodicity of population transfers. The envelope surrounding the oscillations is a Gaussian (Cummings, 1965) and the collapse time is (Eberly et al, 1980)

$$(t_c)^{-1} = \frac{1}{2}\Omega_1 = \frac{\Omega(\langle n \rangle)}{2\sqrt{\langle n \rangle}}, \tag{38}$$

independent of photon number.

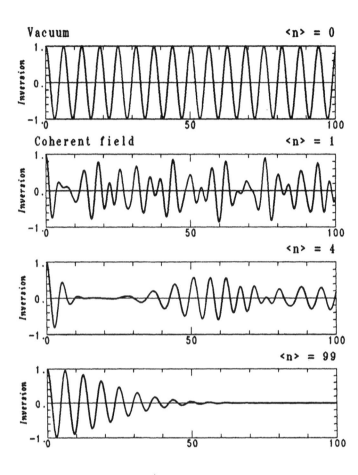

Fig. 2. Population inversion, as in Fig. 1, for a cavity field that is initially in a coherent state. Mean photon numbers of the initial field are $\langle n \rangle = 0$, 1, 4 and 99. Times are in units of mean Rabi frequency. Note the Cummings collapse of Rabi oscillations, followed by quiescence.

What was more remarkable than the collapses was that the Rabi oscillations should revive after a quiescent interval. Thanks to the work of Eberly and his coworkers (Eberly et al, 1980; Narozhny et al, 1981; Yoo et al, 1981), this phenomena is now understood. This revival time is essentially the inverse of the separation between distinct Rabi frequencies (Narozhny et al, 1981),

$$t_R = t_c 2\pi \sqrt{\langle n \rangle}. \tag{39}$$

Thus the behavior of a two-state atom interacting via one-photon transitions (in the rotating wave approximation) with a single-mode coherent state, is surprisingly irregular: Rabi oscillations that collapse, remain quiescent, revive, and collapse again repeatedly.

Figure 3 displays the collapse and revival phenomena in population inversion, as well as in statistical properties of the field: the mean photon number, the Mandel Q parameter, and the normalized electric field variance. These show the nonclassical behavior of the field, specifically sub-Poisson statistics (from Q) and squeezing (from the field variance). The squeezing (Meystre and Zubairy, 1982) is slight when the mean photon number is fairly small (only 16 in the present example). However, for sufficiently large initial photon numbers n_0, the bound on $\Delta X > 1/4\sqrt{n_0}$ can become arbitrarily small, and squeezing can become complete periodically (Kuklinski and Madajczyk, 1988; Hillery, 1989).

Plots of population or inversion give the impression that the system is dormant following the collapse. This is not correct. As Narozhny et al (1981) showed, the atomic dipole moment continues to change during this interval, reaching a broad extremum at the half revival time. The field, as well as the atom, continues to evolve even though the populations remain unchanged.

A variety of other initial fields have been studied, including various squeezed states (Milburn, 1984; Puri and Agarwal, 1987; Agarwal and Puri, 1988).

11. Phase Space

Full characterization of a field requires more than a specification of mean and variance of photon number. One requires an infinite number of correlation functions. Alternatively, one can examine phase-space properties of the single mode field. A useful tool for such studies is the Q function, defined as the expectation value of the field density operator in a coherent state basis, $Q(\alpha, t) = \langle \alpha | \rho^F | \alpha \rangle$. For a system that begins in a definite atomic state one can obtain the Q function from the (two) probability amplitudes $C_1(n, t)$ and $C_2(n, t)$ of eqn. (8),

$$Q(\alpha, t) = \sum_n | \sum_k f_{n-k+1}(\alpha) f_n C_k(n, t)|^2. \qquad (40)$$

The Q function has useful pictorial properties (cf. Kim et al 1989). For a number state, it is a circular doughnut centered at the origin with mean radius $\sqrt{\langle n \rangle}$. A coherent state appears as a Gaussian mound centered at some point in the complex alpha plane. Squeezing appears as an elongation of circular contours.

For a two-state atom interacting with a coherent state, the field quasiprobability function $Q(a, t)$ appears initially as a Gaussian distribution in the complex a plane. With the usual choice of real α, the distribution is centered on the horizontal axis, at a distance $\sqrt{\langle n \rangle}$ from the origin. As shown by Eiselt and Risken (1989, 1991) this distribution subsequently bifurcates into two peaks that move in opposite directions around a circular path whose radius is $\sqrt{\langle n \rangle}$. As the peaks meet at the opposite side of the circle from the start, they produce a revival of Rabi oscillations. Further revivals occur as the peaks repeatedly pass each other along the circular track .

Figure 4 shows the appearance of the Q function of the JCM for a coherent-state initial field having mean photon number of 9. The topmost frame is the initial Gaussian. The next depicts the Q function during the collapse. At this time the density matrix is a coherent superposition of two distinct states. The final frame shows the function during the first revival.

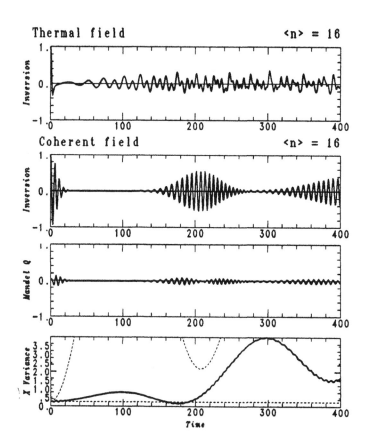

Fig. 3. Population inversion over longer time interval, for initial mean photon number of $\langle n \rangle =$ 16. Top frame shows thermal field, lower frames show a coherent field. Also shown for the coherent field are the mean photon number, the Mandel Q parameter, and the field variance ΔX (solid line; the dashed line shows ΔY). Note the Eberly revival of Rabi oscillations, followed by collapse, also the sub-Poisson statistics and squeezing.

12. Initial Atomic Coherence

One can also consider atoms that are initially in a coherent superposition of the two states (Krause et al, 1986; Zaheer and Zubairy, 1990). The initial condition, for a field in a superposition of photon number states and the atom in coherent superposition of energy states ψ_k reads

$$\Psi(0) = \Phi^F \psi^A, \quad \Phi^F = \sum_{n=0}^{\infty} f_n |n\rangle, \quad \psi^A = \sum_{k=1}^{N} g_k \psi_k. \tag{41}$$

When the atom arrives at the cavity in such a state then the subsequent behavior is sensitive

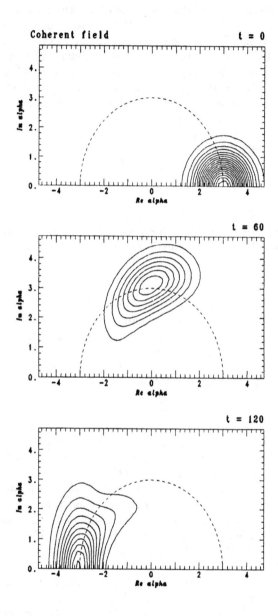

Fig. 4. Phase plane plot of $Q(\alpha)$ function at various times, for an initially excited atom interacting with a coherent field, with initial mean photon number $\langle n \rangle = 9$. The figure shows only the upper half of the complex plane; the lower half is a mirror image of this. Times are $t = 0$, $t = 60$ (during collapse) and $t = 120$ (during revival). Note the formation of two wavepackets, moving in opposite directions along the circular path with constant mean photon number (dashed semicircle).

to the relative phase between the cavity field and the atom superposition (Puri and Agarwal, 1987; Agarwal and Puri, 1988; Joshi and Puri, 1989). More remarkably, the collapse and revival phenomena are absent if the atom is initially in one of the dressed states ψ_+ or ψ_- defined by Eqn. (14). These are states that, in the semiclassical treatment of a specified field interacting with a two-state atom, remain constant in time (Bai et al, 1985; Lu et al, 1986). By starting with an atom in an excited (or unexcited) state, one deals with a coherent superposition of dressed states. These are the two amplitudes that evolve as the bifurcation of the Q function. Subsequent time evolution will entangle the field and atom states. However, at the half-revival time the atom is again nearly in a pure state (Gea-Banacloche, 1991, 1992; Phoenix and Knight, 1991). At those moments the field is a macroscopic superposition state composed of two field states that have the same amplitude but opposite phase.

13. Entropy and Entanglement

Entropy offers a quantitative measure of the disorder of a system, and of the purity of a quantum state. We define the dimensionless entropy of a quantum system as (Wehrl, 1978)

$$S = -Tr[\rho \ln \rho] \tag{42}$$

For a pure state this entropy vanishes, $S = 0$, whereas for a statistical mixture the entropy is nonzero, $S \neq 0$. The entropy of a system remains constant whenever the time-dependent Schrödinger equation governs the time evolution. Of more use, therefore, are the partial entropies of system components, such as the field and atom subsystems of the JCM. From the reduced density matrices of the atom and field we form the partial entropies (Phoenix and Knight, 1988; Barnett and Phoenix, 1989)

$$S_A(t) = -Tr_A[\rho^A(t) \ln \rho^A(t)], \quad S_F(t) = -Tr_F[\rho^F(t) \ln \rho^F(t)]. \tag{43}$$

Unlike the entropy of the complete system, these partial entropies may change with time. According to Araki and Lieb (1970; cf. Phoenix and Knight, 1988, 1990), if the combined system begins as a pure quantum state, then at all times the component entropies are equal, $S_A(t) = S_F(t)$. A decrease of partial entropy means that each subsystem evolves toward a pure quantum state, whereas a rise in partial entropy means that the two components tend to lose their individuality and become correlated or entangled. The individual components are in their purest state when their entropies are smallest. The components are most strongly correlated when their individual entropies are large. Thus a plot of $S_A(t)$ will reveal a history of atom-field entanglement.

Starting from initially pure states of atom and field, and hence from zero partial entropies, one expects, and finds, that the partial entropies increase during the interval of collapsing Rabi oscillations (see Figure 5). One might guess that the field and atom return most closely to their initial pure state at the peak of the revivals. Surprisingly, the peak of the revival does not correspond to the recreation of a pure atomic state, as measured by small entropy; instead the system returns most closely to a pure state of atom or field at the half-revival time $t = t_R/2$, during the collapse, when population appears static (Phoenix and Knight, 1988, 1991; Gea-Banacloche, 1990, 1991, 1992). The momentarily created nearly-pure atomic state is a coherent superposition of the two energy states of the atom. Because it occurs even with large mean-photon numbers, it provides an example of

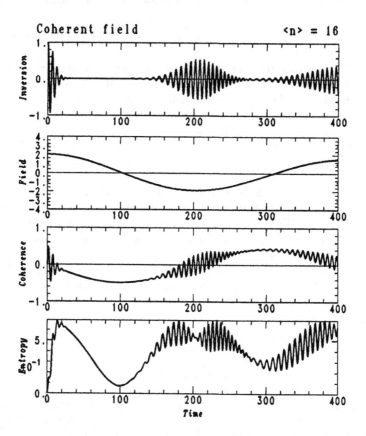

Fig. 5. Histories of inversion, field $\langle X \rangle$, coherence $\langle \hat{S}_{12} \rangle$ and partial entropy (for atom or field; they are equal), for the example of Fig. 3.

a macroscopic superposition state (Buzek et al, 1992). The occurrence of revivals proves the existence of the superposition states: cavity damping will introduce dephasing during the collapse interval, and thereby eliminate the interferences that appear as revivals.

14. Extending the JCM

Much effort has been directed toward generalizing the original model. In part the extensions have aimed at more realistic descriptions of experiments, such as the inclusion of photon loss. Other papers seek more general Hamiltonians that have analytic solutions. For example, Yoo and Eberly (1985) presented an extensive study of the three-level JCM. We shall mention only two of the many generalizations and extensions of the JCM.

The original JCM assumed that each photon change balanced an atomic change in such a way that the combination of photon energy plus atom energy remained constant. Jaynes and Cummings recognized that, just as with semiclassical theory, a more complete description of the radiation interaction should include additional terms, in the form

$$H = \hbar\omega\hat{a}^\dagger\hat{a} + E_1\hat{S}_{11} + E_2\hat{S}_{22} + \frac{\hbar}{2}\Omega_1[\hat{a}^\dagger + \hat{a}][\hat{S}_{12} + \hat{S}_{21}]. \tag{44}$$

The original JCM is a quantized version of the rotating wave approximation (RWA); the additional terms are counterrotating terms (Shirley 1965). Typically these are treated, if at all, as sources of slight frequency shifts (Bloch-Siegert shifts: Bloch and Siegert, 1940; Shirley 1965). Although these terms are small, they can profoundly affect the long-time behavior of the system. In particular, the semiclassical version of the JCM is perfectly periodic, whereas the inclusion of counter-rotating terms in the semiclassical equations leads to chaotic behavior (Kujawski, 1988; Milonni et al, 1983; Milonni, Shi, and Ackerhalt, 1987).

Apart from the collapse produced by the destructive interference of different Rabi frequencies associated with the distribution of photon numbers, two dissipative mechanisms affect atoms in real cavities: spontaneous emission into any continuum of field modes other than the JCM cavity mode, say at rate γ; and the loss of energy from the cavity (through walls or mirrors) at a rate ω/Q, parametrized by the quality factor Q (Haroche and Raimond, 1985). One can recognize the Markov stochastic nature of these dissipative processes by augmenting the Liouville equations to include both damping and fluctuations, as in the density matrix equation

$$\hbar\frac{\partial}{\partial t}\rho = -i[H,\rho] + \frac{\hbar\omega}{Q}(2\hat{a}\rho\hat{a}^\dagger - \rho\hat{a}^\dagger\hat{a}\rho - \rho\hat{a}^\dagger\hat{a})$$
$$+ \frac{\hbar\gamma}{2}(2\hat{S}_{12}\rho\hat{S}_{21} - \hat{S}_{11}\rho - \rho\hat{S}_{11}) \tag{45}$$

The two dissipative mechanisms have different effects upon the dynamics (Quang, Knight and Buzek, 1991): revivals of population oscillations are much more sensitive to the cavity damping than to spontaneous emission. Even slight cavity damping is sufficient to destroy revivals (Barnett and Knight, 1986).

15. Conclusions

The Jaynes-Cummings model continues today to fulfill the objectives of its originators to permit examination of basic properties of quantum electrodynamics: Rabi oscillations of population as energy shifted between field and matter; the collapse and subsequent revival of these oscillations; the appearance of chaos in an appropriate semiclassical limit; the nonclassical statistical properties of the cavity field; and the recent recognition of the formation of macroscopic quantum states. The extension to three-state atoms and pairs of cavity modes introduce additional properties, though none yet rival the surprises of the two-state model. We can hope that studies of still more general extensions, involving multiple excitations of multilevel atoms, may in time reveal still further surprises.

ACKNOWLEDGMENTS. We dedicate this paper to the memory of our good friend and colleague, Dr. Jay Ackerhalt (1947 - 1992), from whom we first learned of the beauty and complexity of the JCM. His many contributions to the study of the Jaynes-Cummings model and the Heisenberg equations have done much to further general understanding of quantum optics. The work of B.W.S. was supported under the auspices of the U.S. Department of Energy at Lawrence Livermore

National Laboratory under contract W-7405-Eng-48. That of P. L. K. was supported by the UK Science and Engineering Research Council.

REFERENCES

Ackerhalt, J. R.: 1974, Thesis, Univ. Rochester.

Ackerhalt, J. R. and Rzazewski, K.: 1975, Phys. Rev. A **12**, 2549.

Agarwal, G. S. and Puri, R. R.: 1988, JOSA B **5**, 1669.

Allen, L. and Eberly, J. H.: 1975, *Optical Resonance and Two Level Atoms*, Dover, N.Y.

Araki, H. and Lieb, E.: 1970, Commun. Math. Phys. **18**, 160.

Aravind, P. K. and Hu, G.: 1988, Physica C **150**, 427.

Bai, Y. S., Yodh, A. G. and Mossberg, T. W.: 1985, Phys. Rev. Lett. **55**, 1277.

Barnett, S. M., Filipowicz, P., Javanainen, J., Knight, P. L. and Meystre, P.: 1986, in E. R. Pike and S. Sarkar (ed.), *Frontiers in Quantum Optics* Adam Hilger, Bristol, pp. 485.

Barnett, S. M. and Knight, P. L.: 1986, Phys. Rev. A **33**, 2444.

Barnett, S. M. and Phoenix, S. J. D.: 1989, Phys. Rev. A **40**, 2404.

Bloch, F. and Siegert, A.: 1940, Phys. Rev. **57**, 522.

Brune, M., Raimond, J. M., Goy, P., Davidovich, L. and Haroche, S.: 1987, Phys. Rev. Lett. **59**, 1899.

Buzek, V., Moya-Cessa, H., Knight, P. L., and Phoenix, S. J. D.: 1992, Phys. Rev. A**45**, 8190.

Cummings, F. W.: 1965, Phys. Rev. **140**, A1051.

Eberly, J. H., Narozhny, N. B. and Sanchez-Mondragon, J. J.: 1980, Phys. Rev. Letts. **44**, 1323.

Eiselt, J. and Risken, H.: 1989, Opt. Comm. **72**, 351.

Eiselt, J. and Risken, H.: 1991, Phys. Rev. A **43**, 346.

Gea-Banacloche, J.: 1990, Phys. Rev. Lett. **65**, 3385.

Gea-Banacloche, J.: 1991, Phys. Rev. A **44**, 5913.

Gea-Banacloche, J.: 1992, Opt. Commun. **88**, 531.

Glauber, R. J.: 1963, Phys. Rev. **131**, 2766.

Haroche, S. and Raimond, J. M.: 1985, Adv. At. Mol. Phys. **20**, 347.

Hillery, M.: 1989, Phys. Rev. A **39**, 1556.

Jaynes, E. T.: 1958, Stanford Microwave Laboratory Report.

Jaynes, E. T. and Cummings, F. W.: 1963, Proc. IEEE **51**, 89.

Joshi, A. and Puri, R. R.: 1989, J. Mod. Opt. **36**, 557.

Kim, M. S., de Oliveira, F. A. M. and Knight, P. L.: 1989, Phys. Rev. A **40**, 2494.

Knight, P. L. and Milonni, P. W.: 1980, Phys. Repts. **66**, 21.

Knight, P. L. and Radmore, P. M.: 1982, Phys. Rev. A **26**, 676.

Krause, J., Scully, M. O. and Walther, H.: 1986, Phys. Rev. A **34**, 2032.

Kujawski, A. : 1988, Phys. Rev. A **37**, 1386.

Kuklinski, J. R., and Madajczyk, J. L.: 1988, Phys. Rev. A **37**, 317.

Loudon, R. and Knight, P. L.: 1987, J. Mod. Opt. **34**, 709.

Lu, N., Berman, P. R., Yodh, A. G., Bai, Y. S. and Mossberg, T. W.: 1986, Phys. Rev. A **33**, 3956.

Mandel, L.: 1976, Prog. Optics **13**, 29.

Meschede, D., Walther, H. and Muller, G.: 1985, Phys. Rev. Lett. **54**, 551.

Meystre, P., Geneux, E., Quattropani, A. and Faust, A.: 1975, Nuov. Cim. B **25**, 521.

Meystre, P. and Zubairy, M. S.: 1982, Phys. Lett. **89A**, 390.

Meystre, P. and Sargent, M., III: 1990, *Elements of Quantum Optics*, Springer-Verlag, N.Y.

Meystre, P.: 1992, in E. Wolf (ed.), *Progress in Optics XXX*, Elsevier, N.Y., pp. 263.

Milburn, G. J.: 1984, Opt. Acta **31**, 671.

Milonni, P. W.: 1976, Phys. Rep. **25c**, 1.

Milonni, P. W., Ackerhalt, J. R. and Galbraith, H. W.: 1983, Phys. Rev. Lett. **50**, 9660.

Milonni, P. W., Shih, M.-L. and Ackerhalt, J. R.: 1987, *Chaos in Laser-Matter Interactions*, World Scientific, Singapore.

Milonni, P. W. and Singh, S.: 1991, Adv. Atomic Mol. Opt. Physics **28**, 75.

Narozhny, N. B., Sanchez-Mondragon, J. J. and Eberly, J. H.: 1981, Phys. Rev. A **23**, 236.

Phoenix, S. J. D. and Knight, P. L.: 1988, Ann. Phys. **186**, 381.

Phoenix, S. J. D. and Knight, P. L.: 1990, JOSA B **7**, 116.

Phoenix, S. J. D. and Knight, P. L.: 1991, Phys. Rev. A **44**, 6023.

Puri, R. R. and Agarwal, G. S.: 1987, Phys. Rev. A **35**, 3433.

Quang, T., Knight, P. L., and Buzek, V.: 1991, Phys. Rev. A **44**, 6092.

Rabi, I. I.: 1937, Phys. Rev. **51**, 652.

Rempe, G., Walther, H. and Dobiasch, P.: 1986, in A. Kujawski and M. Lewenstein (ed.), *Quantum Optics* Polish Academy of Sciences, Wroclaw.

Rempe, G., Walther, H. and Klein, N. 1987, Phys. Rev. Lett. **58**, 353.

Shirley, J. H.: 1965, Phys. Rev. **138 B**, 979.

Shore, B. W.: 1990, *The Theory of Coherent Atomic Excitation*, Wiley, N.Y.

Stenholm, S.: 1973, Phys. Rep. **6C**, 1.

von Foerster, T.: 1975, J. Phys. A **8**, 95.

Wehrl, A.: 1978, Rev. Mod. Phys. **50**, 221.

Yoo, H.-I., Sanchez-Mondragon, J. J. and Eberly, J. H.: 1981, J. Phys. A **14**, 1383.

Yoo, H. I. and Eberly, J. H.: 1985, Phys. Rept. **118**, 239.

Zaheer, K., and Zubairy, M. S.: 1990, Adv. At. Mol. Opt. Phys. **28**, 143.

THE JAYNES-CUMMINGS MODEL AND THE ONE-ATOM-MASER

H. Walther
Sektion Physik der Universität München and
Max-Planck-Institut für Quantenoptik
8046 Garching
Germany

ABSTRACT. In this paper experiments performed with the one-atom maser are reviewed. Furthermore, possible experiments to test basic quantum physics are discussed.

1. Introduction, the One-Atom-Maser

The most promising avenue to study the generation process of radiation in lasers and masers is to drive a maser consisting of a single mode cavity by single atoms. This system, at first glance, seems to be another example of a Gedanken-experiment treated in the pioneering work of Jaynes and Cummings (1963), but such a one-atom maser (Meschede, Walther and Müller, 1985) really exists and can in addition be used to study the basic principles of radiation-atom interaction. The advantages of the system are:

(1) it is the first maser which sustains oscillations with less than one atom on the average in the cavity,
(2) this setup allows to study in detail the conditions necessary to obtain nonclassical radiation, especially radiation with sub-Poissonian photon statistics in a maser system directly in a Poissonian pumping process, and
(3) it is possible to study a variety of phenomena of a quantum field including the quantum measurement process.

What are the tools that make this device work: It was the enormous progress in constructing superconducting cavities together with the laser preparation of highly excited atoms – Rydberg atoms – that have made the realization of such a one-atom maser possible. Rydberg atoms have quite remarkable properties (Haroche and Raimond, 1985; Gallas, Leuchs, Walther and Figger, 1985) which make them ideal for such experiments: The probability of induced transitions between neighboring states of a Rydberg atom scales as n^4, where n denotes the principle quantum number. Consequently, a few photons are enough to saturate the transition between adjacent levels. Moreover, the spontaneous lifetime of a highly excited state is very large. We obtain a maser by injecting these Rydberg atoms into a superconducting cavity with a high quality factor. The injection rate is such that on the average there is less than one atom present inside the resonator at any time. A transition between two neighboring Rydberg levels is resonantly coupled to a single mode of the cavity field. Due to the high quality factor of the cavity, the radiation decay time is much larger than the characteristic time of the atom-field interaction, which is given by the inverse of the single-photon Rabi-frequency. Therefore it is possible to observe the dynamics (Jaynes and Cummings, 1963) of the energy exchange between atom and field mode leading to collapse and revivals in the Rabi oscillations (Eberly, Narozhny and Sanchez-Mondragon, 1980; Rempe, Walther and Klein, 1987). Moreover a field is built up inside the cavity when the mean time between the atoms injected into the cavity is shorter than the cavity decay time.

The detailed experimental setup of the one-atom maser is shown in Figure 1. A highly collimated beam of rubidium atoms passes through a Fizeau velocity selector. Before entering the superconducting cavity, the atoms are excited into the upper maser level $63p_{3/2}$ by the frequency-doubled light of a cw ring dye laser. The laser frequency is stabilized onto the atomic transition $5s_{1/2} \rightarrow 63p_{3/2}$, which has a width determined by the laser linewidth and the transit time broadening corresponding to a total of a few MHz. In this way, it is possible to prepare a very stable beam of excited atoms. The ultraviolet light is linearly polarized parallel to the electric field of the cavity. Therefore only $\Delta m = 0$ transitions are excited by both the laser beam and the microwave field. The superconducting niobium maser cavity is cooled down to a temperature of 0.5 K by means of a ^{3}He cryostat. At such a low temperature the number of thermal photons is reduced to about 0.15 at a frequency of 21.5 GHz. The cryostat is carefully designed to prevent room temperature microwave photons from leaking into the cavity. This would considerably increase the temperature of the radiation field above the temperature of the cavity walls. The quality factor of the cavity is 3×10^{10} corresponding to a photon storage time of about 0.2 s. The cavity is carefully shielded against magnetic fields by several layers of cryoperm. In addition, three pairs of Helmholtz coils are used to compensate the earth's magnetic field to a value of several mG in a volume of $10 \times 4 \times 4$ cm^3. This is necessary in order to achieve the high quality factor and prevent the different magnetic substates of the maser levels from mixing during the atom-field interaction time. Two maser transitions from the $63p_{3/2}$ level to the $61d_{3/2}$ and to the $61d_{5/2}$ level are studied.

The Rydberg atoms in the upper and lower maser levels are detected in two separate field ionization detectors. The field strength is adjusted so as to ensure that in the first detector the atoms in the upper level are ionized, but not those in the lower level, they are then ionized in the second field.

To demonstrate maser operation, the cavity is tuned over the $63p_{3/2} - 61d_{3/2}$ transition and the flux of atoms in the excited state is recorded simultaneously. Transitions from the initially prepared $63p_{3/2}$ state to the $61d_{3/2}$ level (21.50658 GHz) are detected by a reduction of the electron count rate.

In the case of measurements at a cavity temperature of 0.5 K, shown in Figure 2, a reduction of the $63p_{3/2}$ signal can be clearly seen for atomic fluxes as small as 1750 atoms/s. An increase in flux causes power broadening and a small shift. This shift is attributed to the ac Stark effect, caused predominantly by virtual transitions to neighboring Rydberg levels. Over the range from 1750 to 28000 atoms/s the field ionization signal at resonance is independent of the particle flux which indicates that the transition is saturated. This, and the observed power broadening show that there is a multiple exchange of photons between Rydberg atoms and the cavity field.

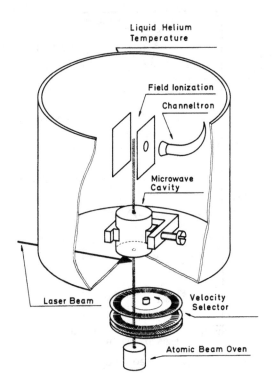

Fig. 1. Scheme of the one-atom maser. To suppress blackbody induced transitions to neighboring states, the Rydberg atoms are excited inside the liquid-Helium-cooled environment.

For an average transit time of the Rydberg atoms through the cavity of 50 μs and a flux of 1750 atoms/s we obtain that approximately 0.09 Rydberg atoms are in the cavity on the average. According to Poisson statistics this implies that more than 90% of the events are due to single atoms. This clearly demonstrates that single atoms are able to maintain a continuous oscillation of the cavity with a field corresponding to a mean number of photons between unity and several hundreds.

2. Single Atom Inside a Resonant Cavity - Oscillatory Regime

The experimental set-up described in section 1 is suitable to test the Jaynes-Cummings model. An important requirement is that the atoms of the beam have a homogeneous velocity so that it is possible to observe the Rabi-nutation induced by the cavity field directly. This is not possible with the broad Maxwellian velocity distribution. The Fizeau-type velocity selector is therefore used, so that a fixed atom-field interaction time is obtained (Rempe, Walther, Klein, 1987). Changing the selected velocity leads to a different interaction time and leaves the atom in another phase of the Rabi cycle when it reaches the detector.

The experimental results obtained for the $63p_{3/2}-61d_{5/2}$ transition are shown in Figure 3. In the figure the ratio between the field ionization signals on and off resonance are plotted versus the interaction time of the atoms in the cavity. The solid curve was calculated using

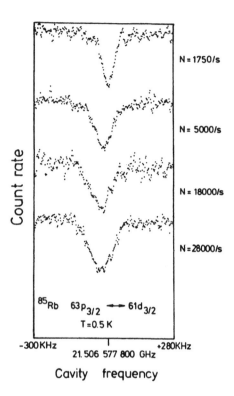

Fig. 2. A maser transition of the one-atom maser manifests itself in a decrease of atoms in the excited state. The flux of excited atoms N governs the pump intensity. Power broadening of the resonance line demonstrates the multiple exchange of a photon between the cavity field and the atom passing through the resonator. For these experiments no velocity selection of the atoms has been used.

the Jaynes-Cummings model (Jaynes, Cummings, 1963), which is in very good agreement with the experiment. The total uncertainty in the velocity of the atoms is 4% in this measurement. The error in the signal follows from the statistics of the ionization signal and amounts to 4%. The measurement is made with the cavity at 2.5K and $Q = 2.7 \cdot 10^8$ ($r_d = 2ms$). There are on the average 2 thermal photons in the cavity. The number of maser photons is small compared with the number of blackbody photons.

The experimental result shown in the upper part of Figure 3 is obtained with very low atom beam flux ($N = 3000s^{-1}$ and $n_m = 0.5$; n_m is the number of photons accumulated in the cavity). When the atomic beam flux is increased more photons are piled up in the cavity. Measurements with $N = 2000s^{-1}$ ($n_m = 2$) and $N = 3000s^{-1}$ ($n_m = 3$) are shown in the lower part of Figure 3. The maximum of $P_e(t)$ at $70\mu s$ flattens with increasing photon number, thus demonstrating the collapse of the Rabi nutation induced by the resonant maser field. Figure 3 (bottom) shows that for atom-field interaction times between $50\mu s$ and about $130\mu s$ $P_e(t)$ does not change as a function of time. Nevertheless, at about $150\mu s$, $P_e(t)$ starts to oscillate again, thus showing the revival predicted by the Jaynes-Cummings model. The variation of the Rabi nutation dynamics with increasing atomic beam fluxes and thus with increasing photon numbers in the cavity generated by stimulated emission is obvious from Figures 3.

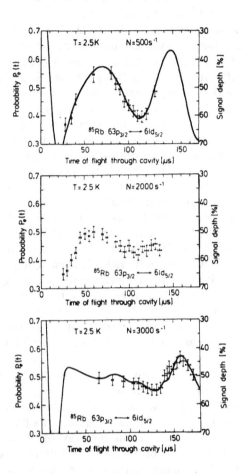

Fig. 3. Measured probability of finding the Rydberg atom in the upper maser level after it passed through the cavity. The cavity is tuned to the $63p_{3/2} - 61d_{5/2}$ transition for increasing flux of the atoms $N = 500s^{-1}$, $N = 2000s^{-1}$ and $N = 3000s^{-1}$. The solid lines represent the theoretical predictions of the Jaynes-Cummings model.

3. A New Source of Nonclassical Light

One of the most interesting questions in connection with the one- atom maser is the photon statistics of the electromagnetic field generated in the superconducting cavity. This problem will be discussed in this section.

Electromagnetic radiation can show nonclassical properties (Walls, 1979, 1983, 1986), that is properties that cannot be explained by classical probability theory. Loosely speaking we need to invoke "negative probabilities" to get deeper insight into these features. We know of essentially three phenomena which demonstrate the nonclassical character of light: photon antibunching (Kimble, Dagenais and Mandel, 1977; Cresser, Häger, Leuchs, Rateike and Walther, 1982), sub-Poissonian photon statistics (Short and Mandel, 1983) and squeezing (Slusher, Hollberg, Yurke, Mertz and Valley, 1985; Loudon and Knight, 1987). Mostly methods of nonlinear optics are employed to generate nonclassical radiation.

However, also the fluorescence light from a single atom caught in a trap exhibits nonclassical features (Carmichael and Walls, 1976; Diedrich and Walther, 1987).

Another nonclassical light generator is the one-atom maser. We recall that the Fizeau velocity selector preselects the velocity of the atoms: Hence the interaction time is well-defined which leads to conditions usually not achievable in standard masers (Filipowicz, Javanainen and Meystre, 1986; Lugiato, Scully and Walther, 1987; Krause, Scully and Walther, 1987; Krause, Scully, Walther and Walther, 1989; Meystre, 1987; Meystre, Rempe and Walther, 1988; Slosser, Meystre and Wright, 1990). This has a very important consequence when the intensity of the maser field grows as more and more atoms give their excitation energy to the field: Even in the absence of dissipation this increase in photon number is stopped when the increasing Rabi-frequency leads to a situation where the atoms reabsorb the photon and leave the cavity in the upper state. For any photon number, this can be achieved by appropriately adjusting the velocity of the atoms. In this case the maser field is not changed any more and the number distribution of the photons in the cavity is sub-Poissonian (Filipowicz, Javanainen and Meystre, 1986; Lugiato, Scully and Walther, 1987), that is narrower than a Poisson distribution. Even a number state that is a state of well-defined photon number can be generated (Krause, Scully and Walther, 1987; Krause, Scully, Walther and Walther, 1989; Meystre, 1987) using a cavity with a high enough quality factor. If there are no thermal photons in the cavity - a condition achievable by cooling the resonator to an extremely low temperature - very interesting features such as trapping states show up (Meystre, Rempe and Walther, 1988). In addition, steady-state macroscopic quantum superpositions can be generated in the field of the one-atom maser pumped by two-level atoms injected in a coherent superposition of their upper and lower states (Slosser, Meystre and Wright, 1990).

Unfortunately, the measurement of the nonclassical photon statistics in the cavity is not that straightforward. The measurement process of the field invokes the coupling to a measuring device, with losses leading inevitably to a destruction of the nonclassical properties. The ultimate technique to obtain information about the field employs the Rydberg atoms themselves: Measure the photon statistics via the dynamic behavior of the atoms in the radiation field, i.e. via the collapse and the revivals of the Rabi oscillations, that is one possibility. However, since the photon statistics depends on the interaction time which has to be changed when collapse and revivals are measured, it is much better to probe the population of the atoms in the upper and lower maser levels when they leave the cavity. In this case, the interaction time is kept constant. Moreover, this measurement is relatively easy since electric fields can be used to perform selective ionization of the atoms. The detection sensitivity is sufficient so that the atomic statistics can be investigated. This technique maps the photon statistics of the field inside the cavity via the atomic statistics.

In this way, the number of maser photons can be inferred from the number of atoms detected in the lower level (Meschede, Walther and Müller, 1985). In addition, the variance of the photon number distribution can be deduced from the number fluctuations of the lower-level atoms (Rempe and Walther, 1990). In the experiment, we are therefore mainly interested in the atoms in the lower maser level. Experiments carried out along these lines are described in the following section.

4. Experimental Results - A Beam of Atoms with Sub-Poissonian Statistics

Under steady state conditions, the photon statistics of the field is essentially determined by the dimensionless parameter $\Theta = (N_{ex} + 1)^{1/2} \Omega t_{int}$, which can be understood as a pump parameter for the one-atom maser (Filipowicz, Javanainen and Meystre, 1986). Here, N_{ex} is the average number of atoms that enter the cavity during the lifetime of the field T_c, t_{int} the time of flight of the atoms through the cavity, and Ω the atom-field coupling constant (one-photon Rabi frequency). The one-atom maser threshold is reached for $\Theta = 1$. At this value and also at $\Theta = 2\pi$ and integer multiples thereof, the photon statistics is super-Poissonian. At these points, the maser field undergoes first-order phase transitions (Filipowicz, Javanainen and Meystre, 1986). In the regions between these points sub-Poissonian statistics is expected. The experimental investigation of the photon number fluctuation is the subject of the following discussion.

In the experiments (Rempe, Schmidt-Kaler and Walther, 1990), the number N of atoms in the lower maser level is counted for a fixed time interval T roughly equal to the storage time T_c of the photons. By repeating this measurement many times the probability distribution p(N) of finding N atoms in the lower level is obtained. The normalized variance (Mandel, 1979) $Q_a = [\langle N^2 \rangle - \langle N \rangle^2 - \langle N \rangle]/\langle N \rangle$ is evaluated and is used to characterize the deviation from Poissonian statistics. A negative (positive) Q_a value indicates sub-Poissonian (super-Poissonian) statistics, while $Q_a = 0$ corresponds to a Poisson distribution with $\langle N^2 \rangle - \langle N \rangle^2 = \langle N \rangle$. The atomic Q_a is related to the normalized variance Q_f of the photon number by the formula

$$Q_a = \epsilon P Q_f (2 + Q_f), \tag{1}$$

which was derived by Rempe and Walther (1990) with P denoting the probability of finding an atom in the lower maser level. It follows from formula (1) that the nonclassical photon statistics can be observed via sub-Poissonian atomic statistics. The detection efficiency ϵ for the Rydberg atoms reduces the sub-Poissonian character of the experimental result. The detection efficiency was 10% in our experiment; this includes the natural decay of the Rydberg states between the cavity and field ionization. It was determined by both monitoring the power-broadened resonance line as a function of flux (Meschede, Walther and Müller, 1985) and observing the Rabi-oscillation for constant flux but different atom-field interaction times (Rempe, Walther and Klein, 1987). In addition, this result is consistent with all other measurements described in the following, especially with those on the second maser phase transition.

Experimental results for the transition $63p_{3/2} \leftrightarrow 61d_{3/2}$ are shown in Figure 4. The measured normalized variance Q_a is plotted as a function of the flux of atoms. The atom-field interaction time is fixed at $t_{int} = 50\mu s$. The atom-field coupling constant Ω is rather small for this transition, $\Omega = 10$kHz. A relatively high flux of atoms $N_{ex} > 10$ is therefore needed to drive the one-atom maser above threshold. The large positive Q_a observed in the experiment proves the large intensity fluctuations at the onset of maser oscillation at $\Theta = 1$. The solid line is plotted according to Eq.(1) using the theoretical predictions for Q_f of the photon statistics (Filipowicz, Javanainen and Meystre, 1986; Lugiato, Scully and Walther, 1987). The error in the signal follows from the statistics of the counting distribution p(N). About 2×10^4 measurement intervals are needed to keep the error of Q_a below 1%. The statistics of the atomic beam is measured with a detuned cavity. The result is a Poisson distribution. The error bars of the flux follow from this measurement. The agreement between theory and experiment is good.

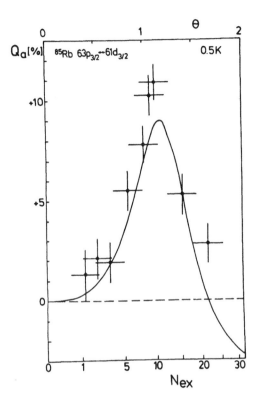

Fig. 4. Variance Q_a of the atoms in the lower maser level as a function of flux N_{ex} near the onset of maser oscillation for the $63p_{3/2} \leftrightarrow 61d_{3/2}$ transition (Rempe, Schmidt-Kaler and Walther 1990).

The nonclassical photon statistics of the one-atom maser is observed at a higher flux of atoms or a larger atom-field coupling constant. The $63p_{3/2} \leftrightarrow 61d_{5/2}$ maser transition with $\Omega=44$kHz is therefore studied. Experimental results are shown in Figure 5. Fast atoms with an atom-cavity interaction time of $t_{int}=35\mu s$ are used. A very low flux of atoms of $N_{ex} >1$ is already sufficient to generate a nonclassical maser field. This is the case since the vacuum field initiates a transition of the atom to the lower maser level, thus driving the maser above threshold.

The sub-Poissonian statistics can be understood from Figure 6, where the probability of finding the atom in the upper level is plotted as a function of the atomic flux. The oscillation observed is closely related to the Rabi nutation induced by the maser field. The solid curve was calculated according to the one-atom maser theory with a velocity dispersion of 4%. A higher flux generally leads to a higher photon number, but for $N_{ex} < 10$ the probability of finding the atom in the lower level decreases. An increase in the photon number is therefore counterbalanced by the fact that the probability of photon emission in the cavity is reduced. This negative feed-back leads to a stabilization of the photon number (Rempe and Walther, 1990). The feed-back changes sign at a flux $N_{ex} \approx 10$, where the second maser phase transition is observed at $\Theta = 2\pi$. This is again characterized by

Fig. 5. Same as Figure 3, but above
threshold for the $63p_{3/2} \leftrightarrow 61d_{5/2}$
transition (Rempe, Schmidt-Kaler
and Walther 1990).

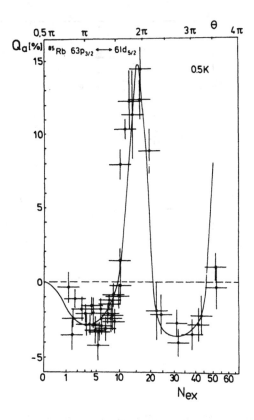

large fluctuations of the photon number. Here the probability of finding an atom in the lower level increases with increasing flux. For even higher fluxes, the state of the field is again highly nonclassical. The solid line in Figure 5 represents the result of the one-atom maser theory using Eq.(1) to calculate Q_a. The agreement with the experiment is very good. The sub-Poissonian statistics of atoms near $N_{ex}=30$, $Q_a=-4\%$ and $P=0.45$ (see Figure 6) is generated by a photon field with a variance $\langle n^2 \rangle - \langle n \rangle^2 = 0.3*\langle n \rangle$, which is 70% below the shot noise level. Again, this result agrees with the prediction of the theory (Filipowicz, Javanainen and Meystre, 1986; Lugiato, Scully and Walther, 1987). The mean number of photons in the cavity is about 2 and 13 in the regions $N_{ex} \approx 3$ and $N_{ex} \approx 30$, respectively. Near $N_{ex} \approx 15$, the photon number changes abruptly between these two values. The next maser phase transition with a super-Poissonian photon number distribution occurs above $N_{ex} \approx 50$.

Sub-Poissonian statistics is closely related to the phenomenon of antibunching, for which the probability of detecting a next event shows a minimum immediately after a triggering event. The duration of the time interval with reduced probability is of the order of the coherence time of the radiation field. In our case this time is determined by the storage time of the photons. The Q_a value therefore depends on the measuring interval T. The

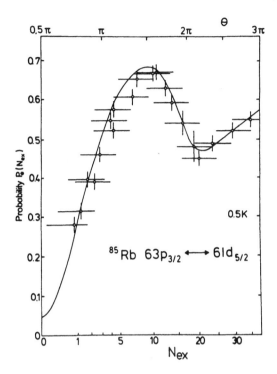

Fig. 6. Probability $P_e(N_{ex})$ of finding the atom in the upper maser level $63p_{3/2}$ for the $63p_{3/2} \leftrightarrow 61d_{5/2}$ transition as a function of the atomic flux.

measured Q_a value approaches a time-independent value for $T > T_c$. For very short sampling intervals, the statistics of atoms in the lower level shows a Poisson distribution. This means that the cavity cannot stabilize the flux of atoms in the lower level on a time scale which is short in relation to the intrinsic cavity damping time.

We want to emphasize that the reason for the sub-Poissonian atomic statistics is the following: A changing flux of atoms changes the Rabi-frequency via the stored photon number in the cavity. By adjusting the interaction time, the phase of the Rabi-nutation cycle can be chosen so that the probability for the atoms leaving the cavity in the upper maser level increases when the flux and therefore the photon number is enlarged or vice versa. We observe sub-Poissonian atomic statistics in the case where the number of atoms in the lower state is decreasing with increasing flux and photon number in the cavity. The same argument can be applied to understand the nonclassical photon statistics of the maser field: Any deviation of the number of light quanta from its mean value is counterbalanced by a correspondingly changed probability of photon emission for the atoms. This effect leads to a natural stabilization of the maser intensity by a feed-back loop incorporated into the dynamics of the coupled atom-field system.

The experimental results presented here clearly show the sub-Poissonian photon statistics of the one-atom maser field. An increase in the flux of atoms leads to the predicted second maser phase transition. In addition, the maser experiment leads to an atomic beam with atoms in the lower maser level showing number fluctuations which are up to 40% below those of a Poissonian distribution found in usual atomic beams. This is interesting, because atoms in the lower level have emitted a photon to compensate for cavity losses inevitably present under steady-state conditions. But this is a purely dissipative phenomenon giving rise to fluctuations. Nevertheless the atoms still obey sub-Poissonian statistics.

5. A New Probe of Complementarity in Quantum Mechanics

The preceding section discussed how to generate a nonclassical field inside the maser cavity. But this field is extremely fragile because any attenuation causes a considerable broadening of the photon number distribution. Therefore it is difficult to couple the field out of the cavity while preserving its nonclassical character. But what is the use of such a field? In the present section we want to propose a new series of experiments performed inside the maser cavity to test the "wave-particle" duality of nature, or better said "complementarity" in quantum mechanics.

Complementarity (Bohm, 1951; Jammer, 1974) lies at the heart of quantum mechanics: Matter sometimes displays wave-like properties manifesting themselves in interference phenomena, and at other times it displays particle-like behavior thus providing "which-path" information. No other experiment illustrates this wave-particle duality in a more striking way than the classic Young's double-slit experiment (Wootters and Zurek, 1979; Wheeler and Zurek, 1983). Here we find it impossible to tell which slit light went through while observing an interference pattern. In other words, any attempt to gain which-path information disturbs the light so as to wash out the interference fringes. This point has been emphasized by Bohr in his rebuttal to Einstein's ingenious proposal of using recoiling slits (Wheeler and Zurek, 1983) to obtain "which-path" information while still observing interference. The physical positions of the recoiling slits, Bohr argues, are only known to within the uncertainty principle. This error contributes a random phase shift to the light beams which destroys the interference pattern.

Such random-phase arguments illustrating in a vivid way how the "which-path" information destroys the coherent-wave-like interference aspects of a given experimental setup, are appealing. Unfortunately, they are incomplete: In principle, and in practice, it is possible to design experiments which provide "which-path" information via detectors which do not disturb the system in any noticeable way. Such "Welcher Weg"- (German for "which-path") detectors have been recently considered within the context of studies involving spin coherence (Englert, Schwinger and Scully, 1988). In the present section we describe a quantum optical experiment (Scully and Walther, 1989) which shows that the loss of coherence occasioned by "Welcher Weg" information, that is, by the presence of a "Welcher Weg"-detector, is due to the establishing of quantum correlations. It is in no way associated with large random-phase factors as in Einstein's recoiling slits.

The details of this application of the micromaser are discussed by Scully, Englert and Walther (1991). Here only the essential features are given. We consider an atomic interferometer where the two particle beams pass through two maser cavities before they reach the two slits of the Youngs interferometer. The interference pattern observed is then also determined by the state of the maser cavity. The interference term is given by:

$$\langle \Phi_1^{(f)}, \Phi_2^{(i)} \mid \Phi_1^{(i)}, \Phi_2^{(-f)} \rangle,$$

where $\mid \Phi_j^{(i)} \rangle$ and $\mid \Phi_j^{(f)} \rangle$ denote the initial and final states of the maser cavity.

Let us prepare, for example, both one-atom masers in coherent states $\mid \Phi_j^{(i)}\rangle = \mid \alpha_j\rangle$ of large average photon number $\langle n\rangle = \mid \alpha_j \mid^2 \gg 1$. The Poissonian photon number distribution of such a coherent state is very broad, $\Delta n \approx \alpha \gg 1$. Hence the two fields are not changed much by the addition of a single photon associated with the two corresponding transitions. We may therefore write

$$\mid \Phi_j^{(f)}\rangle \cong \mid \alpha_j\rangle$$

which to a very good approximation yields

$$\langle \Phi_1^{(f)}, \Phi_2^{(i)} \mid \Phi_1^{(i)}, \Phi_2^{(-f)}\rangle, \cong \langle \alpha_1, \alpha_2 \mid \alpha_1, \alpha_2\rangle = 1.$$

Thus there is an interference cross term different from zero. When we, however, prepare both maser fields in number states $\mid n_j\rangle$ (Krause, Scully and Walther, 1987; Krause, Scully, Walther and Walther, 1989; Meystre, 1987; Meystre, Rempe and Walther, 1988) the situation is quite different. After the transition of an atom to the d-state, that is after emitting a photon in the cavity the final states read

$$\mid \Phi_j^{(f)}\rangle = \mid n_j + 1\rangle$$

and hence

$$\langle \Phi_1^{(f)}, \Phi_2^{(i)} \mid \Phi_1^{(i)}, \Phi_2^{(f)}\rangle = \langle n_1, n_2 \mid n_1, n_2 + 1\rangle = 0,$$

that is the coherence cross term vanishes and no interferences are observed.

On first sight this result might seem a bit surprising when we recall that in the case of a coherent state the transition did not destroy the coherent cross term, i.e. did not affect the interference fringes. However, in the example of number states we can, by simply "looking" at the one-atom maser state, tell which "path" the atom took.

It should be pointed out that the beats disappear not only for a number state. For example, a thermal field leads to the same result. In this regard, we note that it is not enough to have an indeterminate photon number to ensure interferences. The state $\mid \Phi_j^{(f)}\rangle$ goes as $a_j^+ \mid \Phi_j^{(i)}\rangle$ where a_j^+ is the creation operator for the j-th maser. Hence the inner product

$$\langle \Phi_j^{(i)} \mid \Phi_j^{(f)}\rangle \rightarrow \langle \Phi_j^{(i)} \mid a_j^+ \mid \Phi_j^{(i)}\rangle,$$

and in terms of a more general density matrix formalism we have

$$\langle \Phi^{(i)} \mid \Phi^{(f)}\rangle \rightarrow \sum_n \sqrt{n+1}\rho_{n,n+1}^{(i)}.$$

Thus we see that an off-diagonal density matrix is needed for the production of beats. For example, a thermal field having indeterminate photon number would not lead to interferences since the photon number distribution is diagonal in this case.

The atomic interference experiment in connection with one-atom maser cavities is a rather complicated scheme for a "Welcher Weg"-detector. There is a much simpler possibility which we will discuss briefly in the following. This is based on the logic of the famous "Ramsey fringe" experiment. In this experiment two microwave fields are applied to the atoms one after the other. The interference occurs since the transition from an upper state to a lower state may either occur in the first or in the second interaction region. In order to

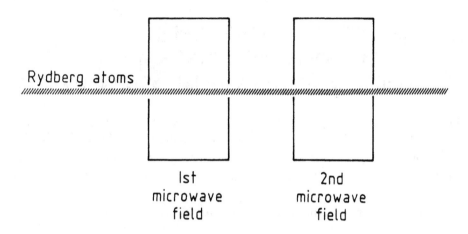

Rydberg atoms

1st
microwave
field

2nd
microwave
field

Fig. 7. Setup for the Ramsey experiment.

calculate the transition probability we must sum the two amplitudes and then square, thus leading to an interference term. We will show here only the principle of this experiment; a more detailed discussion is the subject of another paper (Englert, Walther and Scully, 1992). In the setup we discuss here, the two Ramsey fields are two one-atom maser cavities (see Figure 7). The atoms enter the first cavity in the upper state and are weakly driven into a lower state $| b \rangle$. That is, each microwave cavity induces a small transition amplitude governed by $m\tau$ where m is the atom-field coupling constant and τ is the time of flight across the cavity.

Now if the quantum state of the initial (final) field in the j-th cavity is given by $\Phi_j^{(i)}$ ($\Phi_j^{(f)}$) then the state of the atom + maser 1 + maser 2 at the various relevant times is given in terms of the coupling constant m_j and interaction times τ_j, initial $| \Phi_j^{(i)} \rangle$ and final states $| \Phi_j^{(f)} \rangle$ of the j-th maser by:

$$| \psi(0) \rangle = | a, \Phi_1^{(i)}, \Phi_2^{(i)} \rangle$$

$$| \psi(\tau_1) \rangle \cong | a, \Phi_1^{(i)}, \Phi_2^{(i)} \rangle - im_1\tau_1 \; | b, \Phi_1^{(f)}, \Phi_2^{(i)} \rangle$$

$$| \psi(\tau_1 + T) \rangle \cong | a, \Phi_1^{(i)}, \Phi_2^{(i)} \rangle - im_1\tau_1 \; | b, \Phi_1^{(f)}, \Phi_2^{(i)} \rangle e^{-i\Delta\omega T}$$

$$| \psi(\tau_1 + T + \tau_2) \rangle \cong | a, \Phi_1^{(i)}, \Phi_2^{(i)} \rangle - im_1\tau_1 \; | b, \Phi_1^{(f)}, \Phi_2^{(i)} \rangle e^{-i\Delta\omega T}$$

$$- im_2\tau_2 \; | b, \Phi_1^{(i)}, \Phi_2^{(f)} \rangle$$

where $\Delta\omega$ is the atom-cavity detuning and $T \gg \tau_j$ the time of flight between the two cavities. If we ask for P_b, the probability that the atom exits cavity 2 in the lower state $| b \rangle$, this is given by

$$P = \left[\langle \Phi_1^{(f)}, \Phi_2^{(i)} \mid m_1^* \tau_1 e^{i\Delta\omega T} + \langle \Phi_1^{(i)}, \Phi_2^{(f)} \mid m_2^* \tau_2 \right]$$
$$\times \left[\mid \Phi_1^{(f)}, \Phi_2^{(i)} \rangle m_1 \tau_1 e^{-i\Delta\omega T} + \mid \Phi_1^{(i)}, \Phi_2^{(f)} \rangle m_2 \tau_2 \right]$$
$$= m_1^* m_1 \tau_1^2 + m_2^* m_2 \tau_2^2 + (m_1^* m_2 \tau_1 \tau_2 e^{i\Delta\omega T} \langle \Phi_1^{(f)}, \Phi_2^{(i)} \mid \Phi_1^{(i)}, \Phi_2^{(f)} \rangle + c.c.).$$

Now in the usual Ramsey experiment $\mid \Phi_j^{(i)} \rangle = \mid \Phi_j^{(f)} \rangle = \mid \alpha_j \rangle$, where $\mid \alpha_j \rangle$ is the coherent state in the j-th maser, which is not changed by the addition of a single photon. Thus the "fringes" appear going as $\exp(i\Delta\omega T)$. However, consider the situation in which $\mid \Phi_j^{(i)} \rangle$ is a number state e.g. the state $\mid 0_j^{(i)} \rangle$ having no photons in the j-th cavity initially; now we have

$$P = m_1^* m_1 \tau_1^2 + m_2^* m_2 \tau_2^2 + (m_1^* m_2 \tau_1 \tau_2 e^{i\Delta\omega T} \langle 1_1, 0_2 \mid 0_1, 1_2 \rangle + c.c.).$$

In this case, the one-atom masers are now acting as "Welcher Weg"-detectors, and the interference term vanishes due to the atom-maser quantum correlation.

We note that the more usual Ramsey fringe experiment involves a strong field "$\frac{\pi}{2}$-pulse" interaction in the two regions. This treatment is more involved than necessary for the present purposes. A more detailed analysis of the one-atom maser Ramsey problem is given elsewhere (Englert, Walther and Scully, 1992).

We conclude this section by emphasizing again that this new and potentially experimental example of wave-particle duality and observation in quantum mechanics displays a feature which makes it distinctly different from the Bohr-Einstein recoiling-slit experiment. In the latter the coherence, that is the interference, is lost due to a phase disturbance of the light beams. In the present case, however, the loss of coherence is due to the correlation established between the system and the one-atom maser. Random-phase arguments never enter the discussion. We emphasize that the argument of the number state not having a well-defined phase is not relevant here; the important dynamics is due to the atomic transition. It is the fact that which-path information is made available which washes out the interference cross terms (Scully, Englert and Walther, 1991).

6. Conclusion

In this paper a review of the work with the one-atom-maser was given emphasizing the tests of the Jaynes-Cummings model. The experiments show nicely that this model, originally considered as a treatment of a purely academic problem, got nice confirmation after so many years.

REFERENCES

Bohm, D.: 1951, 'Quantum Theory', Prentice Hall, Englewood Cliffs.
Carmichael, H.J., and Walls, D.F.: 1976, 'Proposal for the Measurement of the Resonant Stark Effect by Photon Correlation Techniques', *J. Phys.* **B9**, L 43.
Carmichael, H.J., and Walls, D.F.: 1976, 'A Quantum-Mechanical Master Equation Treatment of the Dynamical Stark Effect', *J. Phys.* **B9**, 1199.

Cresser, J.D., Häger, J., Leuchs, G., Rateike, M., and Walther, H.: 1982, 'Resonance Fluorescence of Atoms in Strong Monochromatic Laser Fields', in R. Bonifacio (ed.), Dissipative Systems in Quantum Optics, Springer, Berlin. *Topics in Current Physics* **27**, 21.

Diedrich, F., and Walther, H.: 1987, 'Non-classical Radiation of a Single Stored Ion', *Phys. Rev. Lett.*, **58**, 203.

Eberly, J.H., Narozhny, N.B., and Sanchez-Mondragon, JJ.: 1980, 'Periodic Spontaneous Collapse and Revival in a Simple Quantum Model', *Phys. Rev. Lett.* **44**, 1323.

Englert, B.-G., Schwinger, J., and Scully, M.O.: 1988, 'Is Spin Coherence Like Humpty-Dumpty?', *Found. Phys.* **18**, 1045.

Englert, B.-G., Walther, H., and Scully, M.O.: 1992, 'Quantum Optical Ramsey Fringes and Complementarity', *Appl. Phys.* **B54**, 366.

Filipowicz, P., Javanainen, J., and Meystre, P.: 1986, 'The Microscopic Maser', *Opt. Comm.* **58**, 327.

Filipowicz, P., Javanainen, J., and Meystre, P.: 1986, 'Theory of a Microscopic Maser', *Phys. Rev.* **A34**, 3077.

Filipowicz, P., Javanainen, J., and Meystre, P.: 1986, 'Quantum and Semiclassical Steady States of a Kicked Cavity Mode', *J. Opt. Soc. Am.* **B3**, 906.

Gallas, J.A., Leuchs, G., Walther, H., and Figger, H.: 1985, 'Rydberg Atoms: High Resolution Spectroscopy and Radiation Interaction - Rydberg Molecules', in D.R. Bates, B. Bederson (eds.), *Adv. Atom. Molec. Phys.*, **20**, 413, Academic, New York.

Haroche, S., and Raimond, J.M.: 1985, 'Radiative Properties of Rydberg States in Resonant Cavities', in D.R. Bates, B. Bederson (eds.), *Adv. Atom. Molec. Phys.* **20**, 350, Academic, New York.

Jaynes, E.T., and Cummings, F.W.: 1963, *Proc. IEEE* **51**, 89.

Jammer, M.: 1974, 'The Philosophy of Quantum Mechanics', Wiley, New York.

Kimble, H.J., Dagenais, M., and Mandel, L.: 1977, 'Multiatom and Transit-Time Effects on Photon-Correlation Measurements in Resonance Fluorescence', *Phys. Rev. Lett.* **39**, 691.

Krause, J., Scully, M.O., and Walther, H.: 1987, 'State Reduction and $|n\rangle$-State Preparation in a High-Q Micromaser', *Phys. Rev.* **A36**, 4547.

Krause, J., Scully, M.O., Walther, T., and Walther, H.: 1989, 'Preparation of a Pure Number State and Measurement of the Photon Statistics in a High - Q Micromaser', *Phys. Rev.* **A39**, 1915.

Loudon, R., and Knight, P.L.: 1987, 'Squeezed Light', *J. Mod. Opt.* **34**, 709.

Lugiato, L., Scully, M.O., and Walther, H.: 1987, 'Connection between Microscopic and Macroscopic Maser Theory', *Phys. Rev.* **A36**, 740.

Mandel, L.: 1979, Opt. Lett. 4, 205.

Meschede, D., Walther, H., and Müller, G.: 1985, 'The One-Atom Maser', *Phys. Rev. Lett.* **54**, 551.

Meystre, P.: 1987, 'Repeated Quantum Measurements on a Single-Harmonic Oscillator', *Opt. Lett.* **12**, 669.

Meystre, P., Rempe, G., and Walther, H.: 1988, 'Very-Low Temperature Behaviour of a Micromaser', *Optics Lett.* **13**, 1078.

Rempe, G., Walther, H., and Klein, N.: 1987, 'Observation of Quantum Collapse and Revival in a One-Atom Maser' *Phys. Rev. Lett.*, **58**, 353.

Rempe, G., and Walther, H.: 1990, 'Sub-Poissonian Atomic Statistics in a Micromaser', *Phys. Rev.* **A42**, 1650.

Rempe, G., Schmidt-Kaler, F., and Walther, H.: 1990, 'Observation of Sub-Poissonian Photon Statistics in a Micromaser', *Phys. Rev. Lett.* **64**, 2783.

Scully, M.O., and Walther, H.: 1989, 'Quantum Optical Test of Observation and Complementarity in Quantum Mechanics', *Phys. Rev.* **A39**, 5229.

Scully, M.O., Englert, B.-G., and Walther, H.: 1991, 'Quantum Optical Tests of Complementarity', *Nature* **351**, 111.

Short, R., and Mandel, L.: 1983, 'Observation of Sub-Poissonian Photon Statistics', *Phys. Rev. Lett.* **51**, 384.

Slosser, J.J., Meystre, P., and Wright, E.M.: 1990, 'Generation of Macroscopic Superpositions in a Micromaser', *Opt. Lett.* **15**, 233.

Slusher, R.E., Hollberg, L.W., Yurke, B., Mertz, J.C., and Valley, J.F.: 1985, 'Observation of Squeezed States Generated by Four-Wave Mixing in a Optical Cavity', *Phys. Rev. Lett.* **55**, 2409.

Walls, D.F.: 1979, 'Evidence for the Quantum Nature of Light', *Nature* **280**, 451.

Walls, D.F.: 1983, *Nature* **306**, 141.

Walls, D.F.: 1986, 'Growing Expectations from Squeezed States of Light', *Nature* **324**, 210.

Wheeler, J.A., and Zurek, W.H.: 1983, 'Quantum Theory and Measurement', Princeton University, Princeton.

Wootters, W.K., and Zurek, W.H.: 1979, 'Complementarity in the Double-Slit Experiment: Quantum Nonseparability and a Quantative Statement of Bohr's Principle', *Phys. Rev.* **D19**, 473.

THE JAYNES-CUMMINGS MODEL IS ALIVE AND WELL

P. Meystre
Optical Sciences Center
University of Arizona
Tucson, Arizona 85721

1. Introduction

In 1963, E. T. Jaynes and F. W. Cummings published a paper entitled "Comparison of Quantum and Semiclassical Radiation Theories, with Application to the Beam Maser." This paper was written at a time when the quantum theory of the laser had not yet been worked out, and represented an attempt at discussing the new behaviour that could be expected from treating the field quantum mechanically instead of classically. Jaynes and Cummings began by considering the interaction between a single two-level molecule and a single field mode, a simple situation which has become known as the Jaynes-Cummings model. Treating then the maser medium as a beam of two-level systems initially in a mixture of their upper and lower states characterized by a temperature T, they found that the field mode reached a thermal equilibrium with temperature $T_f = \Omega T/\omega$, where ω is the transition frequency of the two-level systems and Ω is the field mode frequency, provided that $T > 0$. For $T < 0$, the steady state field was found to contain an infinite amount of energy, an unphysical result resulting from the neglect of losses in the model. Jaynes and Cummings then proceeded to develop an alternative semi-classical theory of the beam maser, introducing their famous neo-classical theory of spontaneous emission.

An important follow-up paper was published by F. W. Cummings, who studied the long-time behavior of a two-level system interacting with a quantized single-mode field initially in a coherent state. He found that the atomic upper state population undergoes non-exponentially damped oscillations so that the upper and lower state populations become constant (and equal for $\Omega = \omega$) after an interaction time on the order of the inverse atom-field dipole coupling constant. These results were later rediscovered and extended by Quattropani et al, Stenholm, and von Foerster.

The 1980's witnessed an explosion of work on the Jaynes-Cummings model. One major trigger to this activity was the work of Eberly and coworkers, who found that the "Cummings collapse" is followed by a sequence of "revivals" of the atomic inversion. They evaluated the revival times, and showed explicitly that in contrast to the collapse, which could also be obtained by describing the field as a fluctuating classical quantity, they are a true quantum feature due to the "granularity of the field". At about the same time, experimental groups were very busy learning to handle Rydberg atoms, i.e. atoms in states of high principal quantum number n. The study of these states become central to the

Jaynes-Cummings model for two reasons. First, they couple extremely strongly to the electromagnetic field, with a dipole coupling constant scaling as n^2, so that the transition probability between closely adjacent levels scales as n^4. This means that just one or a few photons are sufficient to saturate dipole transitions between adjacent Rydberg states. Their spontaneous lifetimes are also very large, scaling as n^3 and n^5 for low and high angular momentum states, respectively. Finally, these transitions can be made resonant with low-order modes of microwave cavities. Using superconducting materials, resonators of exceedingly high Q's can be constructed in this regime. They correspond to photon lifetimes that can be as long as hundreds of milliseconds, so that one can to an excellent degree of approximation neglect dissipation during the times, of the order of tens of microseconds, that atoms need to fly across the resonator. This allows one to experimentally realize conditions that approach the ideal Jaynes-Cummings model of a single two-level atom interacting with a single field mode.

The goal of this paper is to review some of these developments, at the occasion of Professor Jaynes' seventieth birthday. Section II recalls important aspects of the Jaynes-Cummings model. Section III discusses how micromasers operate under conditions that approach the ideal situation of this model. In Section IV, we show that single atom-single mode studies are also becoming reality in the optical domain. Finally, Section V illustrates how the Jaynes-Cummings model is taking center stage in the emerging field of atom optics.

2. The Jaynes-Cummings Model

The simplest form of interaction between a two-level atom and a single quantized mode of the electromagnetic field is described by the now ubiquitous Jaynes-Cummings Hamiltonian (Jaynes and Cummings [1963])

$$\mathcal{H} = \frac{1}{2}\hbar\omega\sigma_z + \hbar\Omega a^\dagger a + \hbar(ga^\dagger \sigma_- + adj.). \tag{1}$$

Here, ω is the atomic transition frequency, Ω the field frequency, a and $a^\dagger$ are the boson annihilation and creation operators of the field mode, with $[a, a^\dagger] = 1$, σ_z, σ_- and σ_+ are atomic pseudo-spin operators, with $[\sigma_+, \sigma_-] = \sigma_z$, and

$$g = \frac{\wp\mathcal{E}_\Omega}{2\hbar} \sin KZ \tag{2}$$

is the electric dipole matrix element at the location Z of the atom, where $\mathcal{E}_\Omega$ is the "electric field per photon" $\mathcal{E}_\Omega = [\hbar\Omega/\epsilon_0 V]^{1/2}$, V being the cavity volume (See e.g. Meystre and Sargent [1991].)

The Jaynes-Cummings model can be solved exactly, and as such provides non perturbative solutions that exhibit a number of truly quantum-mechanical dynamical features, such as a collapse followed by a series of revivals of the atomic inversion when the two-level atom interacts with a field initially in a coherent state (Eberly, Narozhny and Sanchez-Mondragon [1980].) It also gives the simplest description of quantum Rabi flopping, as well as the simplest illustration of spontaneous emission.

The eigenenergies of the Jaynes-Cummings Hamiltonian (1) are

$$E_{\pm n} = \hbar(n + 1/2)\Omega \pm \frac{1}{2}\hbar\mathcal{R}_n \tag{3}$$

where $\delta = \omega - \Omega$ is the atom-field frequency detuning and we have introduced the generalized n-photon Rabi flopping frequency

$$\mathcal{R}_n = \sqrt{\delta^2 + 4g^2(n+1)}. \tag{4}$$

The corresponding eigenvectors are

$$|+,n\rangle = \sin\theta_n|a,n\rangle + \cos\theta_n|b,n+1\rangle, \tag{5}$$

$$|-,n\rangle = \cos\theta_n|a,n\rangle - \sin\theta_n|b,n+1\rangle,$$

where the states $|a\rangle$ and $|b\rangle$ are the upper and lower atomic states, respectively, and $|n\rangle$ are number states of the field mode with $a^\dagger a|n\rangle = n|n\rangle$. The angle θ is defined by

$$\tan 2\theta_n = -\frac{2g\sqrt{n+1}}{\delta}.$$

For $n = 0$ and on resonance $\omega = \Omega$, the dressed levels $|+,0\rangle$ and $|-,0\rangle$ are separated by the frequency $\mathcal{R}_0 = 2g$, the so-called "vacuum Rabi splitting."

From these results, it is possible to compute all dynamical properties of the Jaynes-Cummings model. In particular, assuming that the atom is initially in its upper state $|a\rangle$ and the field is described by the photon statistics p_n, we find that on resonance $\omega = \Omega$, the probability $|C_a(t)|^2$ for the atom to be in the upper state at time t is given by

$$|C_a(t)|^2 = \sum_n p_n \cos^2(g\sqrt{n+1}\,t). \tag{6}$$

Similarly, if the atom is initially in its ground state,

$$|C_a(t)|^2 = \sum_n p_n \sin^2(g\sqrt{n}\,t). \tag{7}$$

When the field mode is initially in a coherent state, these probabilities undergo a collapse (Cummings [1965], Meystre, Quattropani, Faist and Geneux [1975], von Foerster [1975]) followed by a series of revivals (Eberly, Narozhny and Sanchez-Mondragon [1980]). Specifically, for the Poisson photon statistics of a coherent state with mean photon number $|\alpha|^2 \gg 1$ and for $t \ll |\alpha|/g$, one finds (Cummings [1965], Meystre, Quattropani and Baltes [1974])

$$|C_a(t)|^2 \simeq \frac{1}{2} + \frac{1}{2}\cos(2|\alpha|gt)\exp(-g^2t^2),$$

which predicts a collapse of the upper state population at a rate independent of the mean photon number. This collapse is due to the destructive interference of quantum Rabi floppings at different frequencies, and a similar effect would occur under the influence of a classical field with intensity fluctuations (Knight and Radmore [1982], Barnett, Filipowicz, Javanainen, Knight and Meystre [1986]). In contrast, the revivals are a purely quantum mechanical effect which finds its origin in the discreteness of the quantum states of the harmonic oscillator, the "granularity of the field" (Eberly, Narozhny and Sanchez-Mondragon [1980].)

An appealing interpretation of the revivals was given by Eiselt and Risken [1990] in terms of quasi-probability distributions. Starting from an initial coherent state of the field mode and the atom in its upper state, they found that the initially single-peaked Q-function of the field (Cahill and Glauber [1969]) splits into two single-peaked functions counter-rotating in the complex $(\mathrm{Re}(\alpha), \mathrm{Im}\,(\alpha))$ plane. Revivals in the upper state population occur when these two peaks collide and interfere.

Equations (6) and (7) yield the simplest form of "spontaneous emission". Considering a field initially in the vacuum state $|0\rangle$, we have

$$|C_a(t)|^2 = \cos^2(gt).$$

for an initially excited atom, while for an atom initially in its lower state we find

$$|C_a(t)|^2 = 0.$$

This single-mode version of spontaneous emission is of course unrealistic, except for cavity situations that we shall return to in the next Section. In general, spontaneous emission leads to an irreversible decay of the upper state population, rather than to a periodic exchange of energy between the atom and the cavity mode.

We mentioned that in masers where a sequence of inverted atoms are successively injected into the cavity, the intracavity field grows *ad infinitum*. For particular parameters of the system the field may however evolve to a steady state corresponding to a pure state of the field, in the simplest situation a number state (Filipowicz, Javanainen and Meystre [1986a], Slosser, Meystre and Braunstein [1989]). Such dynamics occur if there is a number state $|N\rangle$ of the field such that the atom undergoes a $2q\pi$ coherent interaction, where q is an integer, as it flies through the cavity in a time τ. Specifically, at resonance $\omega = \Omega$, the number state $|N\rangle$ with

$$g\sqrt{N+1}\tau = q\pi, \tag{8}$$

where q is an integer, is not coupled to the state $|N+1\rangle$ by the Jaynes-Cummings dynamics. Such a state is called a trapping state (Filipowicz, Javanainen and Meystre [1986a].) It is readily seen that the states $|m^2 N\rangle$, m integer, are also trapping states. They partition the Fock space into an infinite number of disconnected blocks of ever increasing size.

Physically, trapping states are such that initially inverted atoms undergo q complete Rabi oscillations and return to their initial state after interacting for a time τ with the field. They play an important role in the low-temperature dynamics of the micromaser (Meystre, Rempe and Walther [1988]), in the preparation of number states of the electromagnetic field (Filipowicz, Javanainen and Meystre [1986a]) and in the preparation and detection of "macroscopic quantum superpositions" (Slosser, Meystre and Braunstein [1989], Slosser, Meystre and Wright [1990]).

3. The Jaynes-Cummings model and the micromaser

In micromasers, a beam of Rydberg atoms is injected inside a single mode microwave cavity at such a low rate that at most one atom at a time is present inside the resonator. In many experiments, it is important to insure that the interaction time of the successive atoms with the cavity mode is constant. This is achieved by passing the atomic beam through a Fizeau velocity selector. The maser cavity is manufactured from pure Niobium and cooled down to a fraction of a degree Kelvin, thereby achieving quality factors Q up to $3 \cdot 10^{10}$. In the Max-Planck Institute single-photon micromaser experiments (Meschede, Walther and Müller [1984]), the $63P_{3/2}$ - $61D_{3/2}$, 21506.5 MHz transition and the $63P_{3/2}$ - $61D_{5/2}$, 21456.0 MHz transition of the Rubidium isotope 85 were investigated, the cavities being tuned to resonance by means of a piezoelectric transducer. In contrast, the Ecole Normale Superieure experiments (Brune, Raimond and Haroche [1987]) concentrated on two-photon micromasers, studying in particular the $40S$ - $39S$ transition of Rubidium at 2×68.41 GHZ, and taking advantage of the nearly equidistant intermediate $39P_{3/2}$ level, which is only 39 MHz away from the midpoint between the upper and lower maser levels.

The quantum theory of the micromaser (Filipowicz, Javanainen and Meystre [1986b]) relies on two important characteristics of this device. First, because the atom-field interaction takes place in a closed, single-mode cavity, spontaneous emission into free-space modes can be neglected. Second, due to the extremely high quality factors achieved in superconducting cavities, the photon lifetime is extremely long compared to the transit time τ of the successive atoms through the resonator. This means that cavity damping can practically be ignored in the rare instances when an atom interacts with the field (Barnett and Knight [1985], Nayak, Bullough and Thompson [1990]).

The strategy to theoretically describe the micromaser is then straightforward: while an atom is inside the cavity, the coupled atom-field system is described by the Jaynes-Cummings Hamiltonian (1), and during the intervals between atoms the evolution of the field density matrix ρ_f is governed by the master equation (Louisell [1990])

$$\frac{d\rho_f}{dt} \equiv L\rho_f(t_i) = (\kappa/2)(n_b + 1)(2a\rho_f a^\dagger - a^\dagger a\rho_f - \rho_f a^\dagger a)$$
$$+ (\kappa/2)n_b(2a^\dagger \rho_f a - aa^\dagger \rho_f - \rho_f aa^\dagger), \tag{9}$$

where n_b is the mean number of thermal photons present in the cavity mode.

At time t_i, the i-th atom enters the cavity containing the field described by the density operator $\rho_f(t_i)$. At this time, the density operator $\rho(t_i)$ of the combined atom-field system is simply the tensor product of $\rho_f(t_i)$ and the initial atomic density operator. After the interaction time τ, the atom exits the resonator and leaves the field in the state described by the reduced density operator

$$\rho_f(t_i + \tau) = Tr_a[U(\tau)\rho(t_i)U^\dagger(\tau)] \equiv F(\tau)\rho_f(t_i), \tag{10}$$

where $U(\tau)$ is the Jaynes-Cummings unitary evolution operator and Tr_a stands for trace over the atomic variables. Using Eq. (9) for the field evolution during the interval t_p until the time t_{i+1} when the next atom is injected, and noting that $t_p = t_{i+1} - t_i - \tau \simeq t_{i+1} - t_i$ we have then

$$\rho_f(t_{i+1}) = \exp(Lt_p)F(\tau)\rho_f(t_i). \tag{11}$$

This equation can be further simplified by assuming that the atoms enter the cavity according to a Poisson process with mean spacing $1/R$ between events, where R is the atomic flux. One then finds (Filipowicz, Javanainen and Meystre [1986b])

$$\bar{\rho}_f(t_{i+1}) = (1 - L/R)^{-1}F(\tau)\bar{\rho}_f(t_i).$$

In steady state $\bar{\rho}_f(t_{i+1}) = \bar{\rho}_f(t_i) = \bar{\rho}_{f,st}$, this equation yields the photon statistics

$$p_n = C \left[\frac{n_b}{1 + n_b} \right]^n \prod_{k=1}^{n} \left[1 + \frac{N_{ex} \sin^2 \mathcal{R}_k}{n_b k} \right]. \tag{12}$$

where $\mathcal{R}_k$ is given by Eq. (4), C is a normalization constant and

$$N_{ex} = R/\kappa$$

is the average number of atoms that traverse the cavity during the resonator damping time.

Figure 1 shows the normalized average photon number $\nu = \langle n \rangle / N_{ex}$ as a function of the dimensionless pump parameter

$$\Theta = \sqrt{N_{ex}} g\tau$$

for $N_{ex} = 200$ and for a mean thermal photon number $n_b = 0.1$. It is nearly zero for small Θ, but a finite value emerges at the threshold value $\Theta = 1$. For Θ increasing past this point, ν grows rapidly, but then decreases to reach a minimum at $\Theta \simeq 2\pi$, where the field jumps abruptly to a higher intensity. This general behaviour recurs roughly at integer multiples of 2π, but becomes less pronounced for increasing Θ.

As illustrated in Fig. 2, a similar behavior is apparent in the normalized standard deviation

$$\sigma = \left[\frac{\langle n^2 \rangle - \langle n \rangle^2}{\langle n \rangle} \right]^{1/2}. \tag{13}$$

Above $\Theta = 1$ the photon statistics are first strongly superpoissonian (poissonian photon statistics correspond to $\sigma = 1$), with further superpoissonian peaks occurring at the subsequent thresholds. In the remaining intervals of Θ, σ is typically of the order of 0.5, a signature of the subpoissonian nature of the field.

As it happens, the experimental verification of these predictions is not easy, since any information on the field statistics must be inferred from the state of the atoms as they leave the micromaser cavity. It is therefore necessary to find a correspondence between the atom statistics and the field statistics. Rempe and Walther [1990] have derived such a correspondence. For measurement times T longer than the cavity damping time κ^{-1}, they find that the atomic Mandel parameter Q_a is related to the field Mandel parameter Q_f by

$$Q_a = P(\langle N \rangle)Q_f(2 + Q_f). \tag{14}$$

This equation shows that subpoissonian field statistics ($Q_f < 0$) lead to subpoissonian statistics of the ground-state atoms. A maser field with reduced photon number fluctuations generates a stable flux of atoms in the lower level.

The experimental results of Rempe, Schmidt-Kaler and Walther [1990] are in remarkable agreement with theory, thus demonstrating that the micromaser is a ideal system to study in detail the subtle quantum effects present in this most fundamental form of atom-radiation interaction.

Micromaser experiments also permit to study the collapse and revivals predicted by the Jaynes-Cummings model (Rempe, Walther and Klein [1987]). One can readily see that

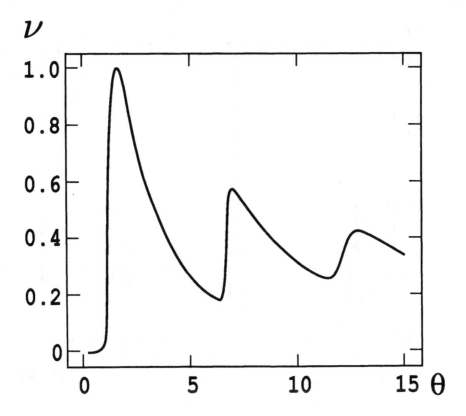

Fig. 1. Normalized steady-state mean photon number ν as a function of the pump parameter Θ
    for $N_{ex} = 200$ and $n_b = 0.1$. (After Filipowicz *et al* [1986b]).

the micromaser should lead to this kind of dynamics by comparing the probability (6)
for an atom to be in the upper state at time t in the Jaynes-Cummings model and the
corresponding probability

$$P_{up}(T) = \sum_n p_n(N_{ex})\cos^2(g\sqrt{n+1}\,T) \tag{15}$$

for the atom to exit the micromaser in the upper state.

It is however important to be careful when comparing these two situations. Equation
(15) is a steady-state result, with p_n the internally imposed steady-state micromaser photon
statistics, depending explicitly on N_{ex} and τ. In contrast, in the Jaynes-Cummings model
p_n is the externally imposed initial photon statistics. This leads quantitatively to the
difference that while in the Jaynes-Cummings, the collapse is followed by a series of revivals,
at most one collapse and one revival of $P_{up}(T)$ are possible in the micromaser (Wright and

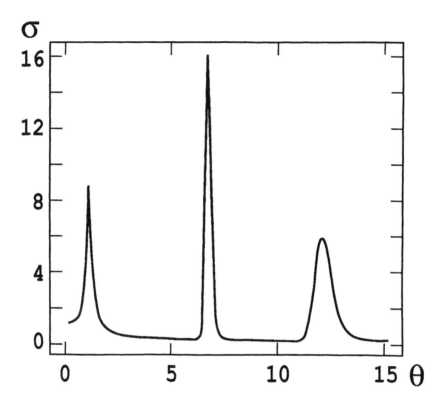

Fig. 2. Normalized standard deviation σ of the steady-state photon distribution for $N_{ex} = 200$
and $n_b = 0.1$. After Filipowicz *et al* [1986b]).

Meystre [1988]). In both the micromaser and in the Jaynes-Cummings model, however, the
revivals are an unambiguous signature of the "granularity of the field" (Eberly, Narozhny
and Sanchez-Mondragon [1980]).

4. The optical regime

For the closed cavities used in microwave experiments, two rates characterize the atom-
field dynamics. They are the cavity decay rate κ and the atom-field coupling constant g.
In contrast, the dynamics of an atom inside an open-sided optical cavity is characterized
in general by three coupling constants. In addition to g and κ, the rate γ of spontaneous
emission into free space plays an essential role. Most optical cavities encompass only a small
fraction of the free space 4π solid angle, and the coherent coupling rate g is usually exceeded
by κ and/or γ. Fortunately, the enhancement of $g \rightarrow g_{eff} = g\sqrt{N}$ due to the collective
response of N atoms inside the cavity (Haroche and Raimond [1985]) offers a way out of
this difficulty and permits to perform experiments in the so-called "strong coupling regime"

$g_{eff} \gg \gamma > \kappa$ (Raizen, Thompson, Brecha, Kimble and Carmichael [1989]). This multi-atom enhancement has been employed successfully to observe the vacuum Rabi splitting in the optical regime (Kaluzny et al [1983], Brecha et al [1986], Raizen et al [1989].) The strong coupling regime can also be reached by decreasing the volume of the cavity, thereby increasing the electric field per photon $\mathcal{E}_\Omega = [\hbar\Omega/\epsilon_0 V]^{1/2}$ and hence the dipole coupling constant (2). This technique has recently been used by Kimble's group to observe absorptive optical bistability in the optical regime with as few as 10 atoms or so at a time in the cavity (Rempe, Thompson, Brecha, Lee and Kimble [1991]).

The general situation where all three rates g, γ and κ are important has been investigated mostly by the groups of Carmichael, Kimble and Savage. Theoretically, one considers a single two-level atom coupled to a single cavity mode. The cavity mode is in turn coupled to a reservoir that accounts for cavity losses, while the atom is coupled to the free space modes not encompassed by the cavity, leading to a spontaneous decay rate γ. The interaction between the atom and a single cavity mode can be modeled by the master equation (Carmichael et al [1989], Savage [1990])

$$\dot{\rho} = \frac{1}{i\hbar}[\mathcal{V},\rho] + \frac{\gamma}{2}(2\sigma_-\rho\sigma_+ - \sigma_+\sigma_-\rho - \rho\sigma_+\sigma_-) + \kappa(2a\rho a^\dagger - a^\dagger a\rho - \rho a^\dagger a), \quad (16)$$

where ρ is the atom-cavity mode density matrix, 2κ is the photon decay rate of the cavity, and $\mathcal{V}$ is given by the Jaynes-Cummings interaction Hamiltonian.

To describe spontaneous emission for arbitrary values of γ, κ and g_{eff}, it is sufficient to solve the master equation (16) in the three-state basis $|a,0\rangle$, $|b,1\rangle$ and $|b,0\rangle$, where $|0\rangle$ and $|1\rangle$ are the zero- and one-photon Fock states of the field. In the strong coupling limit $g \gg \kappa, \gamma$, the spontaneous emission spectrum $S(\omega)$ splits into two lines which correspond to the normal-mode splitting of coupled harmonic oscillators,

$$2\pi S(\omega) = \frac{\kappa/2 + \gamma/4}{(\kappa/2 + \gamma/4)^2 + (\Omega - \omega - g)^2} + \frac{\kappa/2 + \gamma/4}{(\kappa/2 + \gamma/4)^2 + (\Omega - \omega + g)^2}.$$

The two peaks are separated by the vacuum Rabi frequency $2g$ and their half linewidth $(\kappa + \gamma)/2$ is less than the free space linewidth whenever $\kappa < \gamma/2$.

In the "bad cavity" limit $\kappa \gg g \gg \gamma/2$, in contrast, the spontaneous emission spectrum reduces to a conventional single-peak Lorentzian

$$2\pi S(\omega) = \frac{\gamma + 2g^2/\kappa}{(\gamma/2 + g^2/\kappa)^2 + (\omega - \Omega)^2}$$

which exhibits an increased linewidth corresponding to enhanced spontaneous emission (Carmichael, Brecha, Raizen, Kimble and Rice [1989]).

The experimental verification of the strong-coupling prediction was performed by Raizen, Thompson, Brecha, Kimble and Carmichael [1989], who used the cooperative response of N two-level atoms in a high-finesse cavity to reach the condition $g_{eff} = g\sqrt{N} \gg \gamma > \kappa$. These experiments verified that the spontaneous emission spectrum was split into a doublet separated by the vacuum Rabi frequency, and observed linewidth reductions of 25% relative to the free-space atomic decay.

5. The Jaynes-Cummings model in atom optics

Together with cavity QED, the manipulation of atomic trajectories by electromagnetic fields is one of the most exciting recent developments in quantum optics and laser spectroscopy.

Here, one exploits the fact that every time an atom exchanges energy with the field, the momentum of the absorbed or emitted light must be compensated by a mechanical motion of the atom. This leads to atomic trapping and cooling, state-selective atomic reflection and diffraction by light fields, atom interferometry, etc... In these situations, it is usually sufficient to describe the fields classically, while spontaneous emission is treated as a stochastic process. In cavity QED, in contrast, the mode structure of the field as well as its quantum nature are essential.

Recently, there has been a growing effort by several groups to marry these two areas of research (Meystre, Schumacher and Stenholm [1989], Englert, Schwinger, Barut and Scully [1991], Haroche, Brune and Raimond [1991].) A question of particular interest is related to the effects of the state of the field on the atomic motion. For instance, we can analyze the deflection of an atom in a quantized field by generalizing the Jaynes-Cummings Hamiltonian to account for the center of mass motion of the atom. For simplicity, we treat the motion of the atom in the direction $\hat{x}$ transverse to the field classically, that is, we ignore the velocity changes induced by the field in the $\hat{x}$-direction and describe the evolution of the atom-field system in a reference frame moving at the constant velocity P_x/M. We furthermore neglect the effects of spontaneous emission (see Wilkens, Schumacher and Meystre [1991] for a discussion of these effects). The atom-field Hamiltonian is then

$$\mathcal{H} = \frac{P_Z^2}{2M} + \hbar\Omega a^\dagger a + \frac{1}{2}\hbar\omega\sigma_z + \hbar g[\sigma_+ a + adj.]\cos KZ. \qquad (17)$$

with $[P_Z, Z] = -i\hbar$. The Hilbert space of the system is the direct sum of the Hilbert spaces for the center-of-mass motion of the atom, its internal degrees of freedom and the field mode, and the general state vector of the system is of the form

$$|\psi(t)\rangle = \int dP \sum_{i=a,b} \sum_{n=0}^{\infty} C_{P,i,n}(t)|P,i,n\rangle,$$

where we have dropped the index on the momentum variable for simplicity and $|P\rangle$ are the eigenstates of P_z. The analysis is considerably simplified by noting that as in the Jaynes-Cummings model, the state $|a,n\rangle$ is only coupled to $|b,n+1\rangle$. Hence, it is sufficient to solve the problem within one such manifold and to sum over the contributions of all manifolds in the end. Within each manifold, the quantized field problem is mathematically equivalent to the corresponding classical problem (Cook and Bernhardt [1978], Delone, Grinchuk, Kuzmichev, Nagavae, Kazantsev and Surdutovich [1980], Bernhardt and Shore [1981], Arimondo, Bambini and Stenholm [1981], Tanguy, Reynaud and Cohen-Tannoudji [1984], Kazantsev, Ryabenko, Surdutovich and Yakovlev [1985].) The quantum nature of the field appears only in the change in strength of the dipole coupling between the field and the atom from one manifold to the next.

As an example, Fig. 3 shows the probability $\mathcal{P}(p,t)$ for an atom to exit the interaction region with transverse momentum p for a field is initially either in a coherent state or in a thermal state. The three curves on each figure are for three values of the interaction time, all results being in the Raman-Nath regime where the kinetic energy gained by the atoms can be ignored. These results illustrate vividly the effects of photon statistics on atomic diffraction.

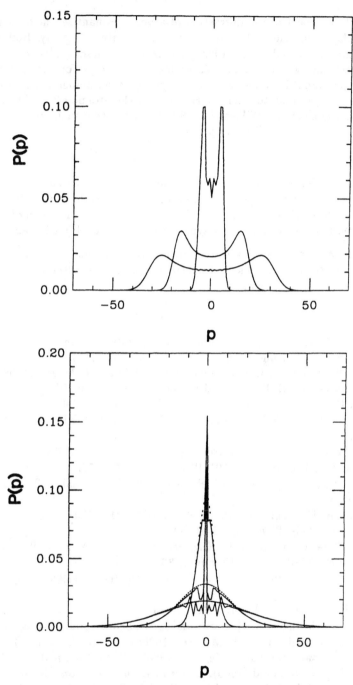

Fig. 3. (a) Momentum distribution $\mathcal{P}(p,t)$ for an atom to exit the field with momentum p, for a field initially in a coherent state with mean photon number $\langle n \rangle = 9$, and for three different interaction times; (b), same but for a field initially in thermal state with $\langle n \rangle = 9$. (After Meystre *et al* [1989].)

An important aspect of the diffraction of atoms by quantized fields is that the interaction leaves the atom-field system in an entangled state. Recently, Holland, Walls and Zoller [1992] and Marte and Zoller [1992] have exploited these results to propose quantum measurement schemes where atomic measurements can be performed to gain information on the field or prepare it in some well defined state (e.g. a number state). Conversely, field measurements can be made to infer information on the atomic state. Storey, Collett and Walls [1992] and Sleator and Wilkens [1992] have shown how such measurements can also be used to generate "virtual slits" for atoms.

6. Conclusion

The three examples discussed in this brief overview make it clear that the study of the subtle quantum effects that characterize the dynamics of a single two-level atom in interaction with a single mode of the electromagnetic field has now moved from the realm of theoretical "toy models" to experimental reality. The Jaynes-Cummings model is alive and well indeed!

ACKNOWLEDGMENTS. This work is supported by the U.S. Office of Naval Research contract N00014-91-J205, by the National Science Foundation Grants PHY-8902548 and INT-8712254, and by the Joint Services Optics Program.

REFERENCES

Arimondo, E. A. Bambini and S. Stenholm (1981), Optics Commun. **37**, 103.

Barnett, S. M. and P. L. Knight (1985), Phys. Rev. A **33**, 2444.

Barnett, S. M., P. Filipowicz, J. Javanainen, P. L. Knight and P. Meystre (1986), in *Frontiers in Quantum Optics*, E. R. Pike and S. Sarkar, eds., Adam Hilger, Bristol.

Bernhardt, A. F. and B. W. Shore (1981), Phys. Rev. A **23**, 1290.

Brecha, R. J., L. A. Orozco, M. G. Raizen, M. Xiao and H. J. Kimble (1986), J. Opt. Soc. Am. B **3**, 238.

Brune, M., J. M. Raimond and S. Haroche (1987), Phys. Rev. A **35**, 154.

Cahill, K. E. and R. J. Glauber (1969), Phys. Rev. **177**, 1857.

Carmichael, H. J., R. J. Brecha, M. G. Raizen, H. J. Kimble and P. R. Rice (1989), Phys. Rev. A **40**, 5516.

Cook, R. J. and A. F. Bernhardt (1978), Phys. Rev. A **18**, 2533.

Cummings, F. W. (1965) Phys. Rev. **140**, A1051.

Delone, G. A., V. A. Grinchuk, S. D. Kuzmichev, M. L. Nagavae, A. P. Kazantsev and G. I. Surdutovich (1980), Optics Commun. **33**, 149.

Eberly, J. H., N. B. Narozhny and J. J. Sanchez-Mondragon (1980), Phys. Rev. Lett.**44**, 1323.

Eiselt, J. and H. Risken (1990), Phys. Rev. A **43**, 346.

Englert, B. G., J. Schwinger, A. O. Barut and M. O. Scully (1991), Europhys. Lett. **14**, 25.

Filipowicz, P., J. Javanainen and P. Meystre (1986a), J. Opt. Soc. Am. B **3**, 906.

Filipowicz, P., J. Javanainen and P.Meystre (1986b), Phys. Rev. A **34**, 3077.

Haroche, S. and J. M. Raimond (1985), in *Advances in Atomic and Molecular Physics*, Vol. 20, D. Bates and B. Bederson, eds., Academic Press, New York.

Haroche, S., M. Brune and J. M. Raimond (1991), Europhys. Lett. **14**, 19.

Jaynes, E. T. and F. W. Cummings 1963, Proc. IEEE **51**, 89.

Holland, M. J., D. F. Walls and P. Zoller (1991), Phys. Rev. Lett. **67**, 1716.

Kaluzny, Y., P. Goy, M. Gross, J. M. Raimond and S. Haroche (1983), Phys. Rev. Lett. **51**, 1175.

Kazantsev, A. P., G. A. Ryabenko, G. I. Surdutovich and V. P. Yakovlev (1985), Physics Reports **129**, 77.

Knight, P. L. and Radmore, P. M. (1982), Phys. Lett. **90A**, 342.

Louisell, W. H. (1990), *Quantum Statistical Properties of Radiation*, Wiley, New York.

Marte, M. and P. Zoller (1992), Appl. Phys. B, in press.

Meschede, D., H. Walther and G. Müller (1984), Phys. Rev. Lett. **54**, 551.

Meystre, P. and M. Sargent III (1991), *Elements of Quantum Optics*, Springer, Berlini, Second Edition.

Meystre, P., A. Quattropani and H. P. Baltes (1974), Phys. Lett. **49A**, 85.

Meystre, P., A. Quattropani, A. Faist and E. Geneux (1975), Nuovo Cimento **25**, 521.

Meystre, P., G. Rempe and H. Walther (1988), Opt. Lett. **13**, 1078.

Meystre, P., E. Schumacher and S. Stenholm (1989), Opt. Commun. **73**, 443.

Nayak, N. R. K. Bullough and B. V. Thompson (1990), in *Coherence and Quantum Optics VI*, J. H. Eberly, L. Mandel and E. Wolf, eds., Plenum Press, New York.

Raizen, M. G., R. J. Thompson, R. J. Brecha, H. J. Kimble and H. J. Carmichael (1989), Phys. Rev. Lett. **63**, 240.

Rempe, G. and Walther H. (1990), Phys. Rev. A **42**, 1650.

Rempe, G., H. Walther and N. Klein (1987), Phys. Rev. Lett. **58**, 353.

Rempe, G., F. Schmidt-Kaler and H. Walther (1990), Phys. Rev. Lett. **64**, 2483.

Rempe, G., R. J. Thompson, R. J. Breche, W. D. Lee, and H. J. Kimball (1991), Phys. Rev. Lett. **67**, 1727.

Savage, C.M. (1990), Quantum Opt. **2**, 89.

Sleator, T. and M. Wilkens (1992), private communication.

Slosser, J. J., P. Meystre and S. L. Braunstein (1989), Phys. Rev. Lett. **63**, 934.

Slosser, J. J., P. Meystre and E. M. Wright (1990), Opt. Lett. **15**, 233.

Storey P., M. Collett and D. F. Walls (1992), Phys. Rev. lett. **68**, 472.

Tanguy, C., S. Reynaud and C. Cohen-Tannoudji (1984), J. Phys. B **17**, 4623.

von Foerster, T. (1975), J. Phys. A **8**, 95.

Wilkens, M, E. Schumacher and P. Meystre (1991), Optics Commun. **86**, 34.

Wright, E. M. and P. Meystre (1988), Opt. Lett. **14**, 177.

SELF-CONSISTENT RADIATION REACTION IN QUANTUM OPTICS – JAYNES' INFLUENCE AND A NEW EXAMPLE IN CAVITY QED

J. H. Eberly
Department of Physics and Astronomy
University of Rochester
Rochester, New York 14627

ABSTRACT. It is pointed out that the self-consistent treatment of radiation reaction is a common feature of many Ph.D. theses of E. T. Jaynes' students and "grandstudents" in the 30-year period 1962 - 1992. In this way they are all descended from Microwave Laboratory Report 502, written by Jaynes at Stanford in 1958. A number of examples are described briefly, and a new example that extends the framework laid down in M. L. 502 is presented.

1. Microwave Laboratory Report 502

In 1958 E. T. Jaynes wrote a paper entitled *Some Aspects of Maser Theory*[1] that was issued as Microwave Laboratory Report No. 502 at Stanford University. The cover of the report is shown in Fig. 1.

In the Abstract of M. L. 502 Jaynes identified the work as an examination of the relation between quantum electrodynamics and the semiclassical theory of radiation. M. L. 502 was intended to serve as a preliminary step in a study of the ultimate limitations on noise figure and frequency stability in molecular-beam masers. The Abstract says, among other things, that "the semiclassical theory, as extended here, is a far more reliable means of calculating radiation processes than usually supposed." In the present author's view, M. L. 502 reported the first examination of dynamical questions that are now associated with the field of cavity quantum electrodynamics.

I was a first-year graduate student at Stanford when M. L. 502 was being written, and I didn't see it until a year or two later, and then didn't understand its various messages for considerably longer than that. In this note I want to draw attention to only one of those messages. This is the message that the self-consistent treatment of radiation reaction is a powerful dynamical principle that can be used widely in quantum radiation theory. Of course I can't guarantee that the author of M. L. 502 will recognize the message I received as exactly the one that he intended to transmit, but that's the risk one takes in sending a message.

The specific term radiation reaction probably appears nowhere in the text of M. L. 502. Nevertheless it was a central concept, and it came into the picture when Jaynes insisted that even in the "semiclassical" treatment of a field-molecule interaction the reaction of the molecule back on the field should be taken into account. This insistence was not, and

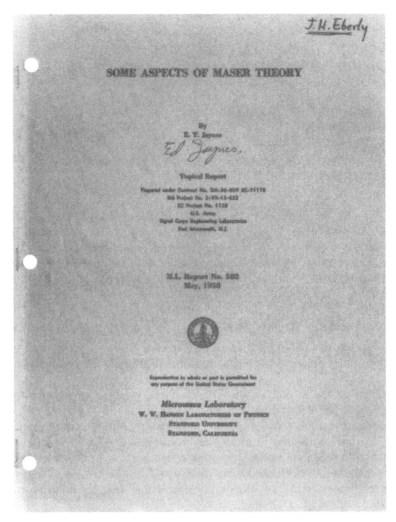

Fig. 1. Cover of the author's copy of M. L. 502

is not, compatible with the semiclassical theory that is conventionally taught to graduate students. In the conventional theory an external or incident radiation field is assumed to be strong enough that (i) it may be treated completely classically without field quantization and (ii) it may be considered to be the total field since any extra field due to radiation by the molecules is so weak by comparison that it can be safely ignored.

Jaynes understood that not only the strength of the field, but also more subtle correlation effects, are involved in the semiclassical assumptions. If the semiclassical theory is extended to take into account the radiation by the molecules one must immediately allow for the possibility that the molecular field will to some degree modify the total field and thus modify both radiation-matter correlations and also the molecular currents that determine the radiation by the molecule, and then a second loop of re-modifications of correlations and fields and currents must be considered, and so on. The inclusion of the molecular radiation field in semiclassical theory introduces both the principle of radiation reaction and the requirement of self-consistency and leads directly to a theoretical structure that is sufficiently novel to justify a distinctive name, and (also in 1958 but not in M. L. 502) for this purpose Jaynes devised the term "neoclassical" theory.

2. Radiation Reaction in Neoclassical Thesis Studies

Of course, self-consistent theories of the interaction of matter and radiation didn't begin with neoclassical theory. They are well-known in classical physics. For example, optical dispersion theory is based on the Lorentz equation for molecular oscillators self-consistently coupled to the Maxwell wave equation for the electric field. However, since both the Lorentz equation and the Maxwell wave equation are linear, their self-consistent solution is easily obtained in the Fourier domain and it only yields the familiar dispersion relation $k^2(\omega) = (\omega/c)^2[1 + N\alpha(\omega)/\epsilon_0]$. This leads to the usual formula for the refractive index, etc. However, M. L. 502 notes that a quantum mechanical treatment of the molecules changes the coupling constant of the Lorentz equation into a time-dependent parameter that is proportional to the molecular inversion, a parameter missing from classical theory, and this makes the corresponding quantum equations nonlinear and not trivially solvable by Fourier methods.

Molecular inversion is important in laser operation, and laser theory requires a self-consistent treatment of molecular radiation. Such a treatment was featured in several early laser theories with non-quantized radiation fields. In this sense the theories[2,3] of L. W. Davis (a Jaynes student at Stanford whose thesis project was in statistical mechanics, but who understood what was going on among Jaynes' electrodynamics students) and W. E. Lamb, Jr., were both really neoclassical laser theories, although the term neoclassical was not used by either author.

In the period 1958-63 the Ph.D. thesis projects at Stanford of F. W. Cummings[4] and M. J. Duggan,[5] as well as my own[6] all dealt with aspects of the nonlinear self-consistent coupling of matter and unquantized radiation. We developed neoclassical theories of the stability of maser operation, the generation of significant power at the third harmonic of the maser field, and the dynamical origin of thermal equilibrium (blackbody) radiation.

Obviously the need to take radiation by molecules into account is most critical when there is only a comparatively weak external field present, or no external field at all. Then even a small amount of molecular radiation may be dominant, as is certainly true in the case of spontaneous emission, where the molecular radiation is the only radiation present. An indirect examination of spontaneous emission was contained in my own thesis, but the theses at Washington University of M. D. Crisp[7] and C. R. Stroud[8] around 1968 made deliberate and direct applications of neoclassical theory to spontaneous emission and radiative level shifts. These applications were hotly controversial and they stimulated a wide range of other work. An early example was Stroud's attention to the question of spontaneous emission lineshapes in resonance fluorescence, which led to the first observation[9] of three-peak lineshapes in the thesis work at Rochester of F. Schuda in 1974.

Earlier, in 1967, McCall and Hahn proposed and confirmed experimentally (McCall's thesis work[10] on self-induced transparency at UC Berkeley) a novel theory of coherent light propagation in dielectrics obtained by coupling the Maxwell wave equation to the semiclassical Bloch equations in a two-level resonant absorber. Coupled Maxwell-Bloch theory was exactly the neoclassical theory of M. L. 502, but applied to radiation in free space instead of in a cavity. This was recognized by McCall who had learned informally about neoclassical theory from Davis, and also by Crisp[11] and Eberly[12] who applied the multi-pulse solutions developed for neoclassical cavity radiation (in M. L. 502 and in the theses of Cummings and Eberly) and found a wider variety of self-induced transparency pulse shapes. This work was further expanded in the Rochester thesis of L. Matulic,[13] the Brooklyn Polytechnic thesis of R. A. Marth,[14] and a decade later in the Rochester theses of M. Konopnicki, who introduced the multi-frequency optical soliton called a simulton,[15] and B. J. Herman, who extended the neoclassical Maxwell-Bloch theory to cover the multi-frequency propagating pulses involved in resonant Raman amplification.[16]

3. Radiation Reaction in Quantum Theory

Also in 1969 the attention of a Rochester student interested in Dicke's concept of super-radiance[17] was directed to the neoclassical principle of self-consistency. This led to the Ph.D. thesis of N. E. Rehler.[18] Self-consistent radiation reaction played a key role but field quantization was partially retained so his theory was not neoclassical. Rehler obtained a simplified equation for the evolution of the N-atom inversion applicable to superradiant samples much larger than a cubic wavelength. The price paid was a partial decorrelation of field and atomic dynamics. Slightly later Stroud, Eberly, Lama and Mandel[19] developed a fully decorrelated and thus fully neoclassical picture of superradiance and this played a role in Lama's thesis.

The connection between self-consistent radiation reaction and spontaneous emission was clarified more successfully in J. R. Ackerhalt's thesis project,[20] which considered a single-atom version of Rehler's two-level superradiance theory, but a version in which quantization of the radiation was fully retained along with self-consistency. This became known as "source-field" QED. As Ackerhalt, Knight and Eberly[21] first demonstrated, a proper single-atom source-field theory of spontaneous emission does not use any decorrelations. It was questioned whether this was a valid approach to quantum radiation theory, but Ackerhalt's operator equations were soon understood to be normal QED but written in the Heisenberg picture. For spontaneous emission Ackerhalt obtained pure exponential decay of the excited state population and exactly the conventional values of the Einstein A coefficient and Lamb shift appropriate to a single two-level transition. He also found the conventional divergences contained in the standard quantum theory.

Ackerhalt's work was the first in a series of operator source-field calculations. P. W. Milonni's thesis focused on the aspects of spontaneous emission requiring field quantization, and compared QED to the corresponding neoclassical and semiclassical radiation theories.[22] S. B. Lai's thesis made a non-relativistic evaluation of $g - 2$ for the electron,[23] showing how to obtain a positive value. This work was later completed by Grotch[24] and his colleagues at Penn State and was greatly elaborated by Cohen-Tannoudji's group[25] in Paris. R. Kornblith's thesis[26] dealt with the source-field of optical coherent transients, including an operator theory of quantum beats. Milonni has subsequently demonstrated in a variety of contexts the complete correspondence between two contrasting (and sometimes thought to be conflicting) views of radiative corrections – the views that they arise on the one hand from vacuum fluctuations or on the other hand from radiation reaction.[27]

4. Radiation Reaction in Cavity QED

Returning to cavity radiation, in 1980 the Rochester thesis investigation of Sanchez-Mondragon directed attention to unexpectedly complex and interesting behavior in the QED solutions of M. L. 502 when applied to coherent field states. As Eberly, Narozhny and Sanchez-Mondragon pointed out,[28] the coherent-state expectation values of the atomic inversion and dipole moment both develop coherently in time from spontaneous emission initial conditions, but only for a limited period, after which their Rabi oscillations collapse to zero. This collapse remains in effect only for a limited time after which a "revival" occurs. This revival of coherence was predicted to collapse and then be followed by an infinite and regularly spaced sequence of subsequent revivals. The revivals are evidence of the quantized radiation field's graininess. The thesis of H.-I. Yoo extended revival studies to lambda, vee and cascade three-level atoms, and introduced three-level dressed or "trapping" states into cavity QED.[29]

Shortly afterward in 1985, unaware of the direct relation of their work to M. L. 502, Meystre, Javanainen and Filipowicz[30] found another interesting extension of it. They found an independent solution to eq. (4-5) in M. L. 502 for the development of the field-atom density matrix under a succession of single-atom passages through the cavity. This led directly to the development of "micromaser" theory.[31] Semi-quantitative measurements by Walther's group of revival phenomena[32] in a micromaser were reported as early as 1987 and Lamb shift measurements[33] were completed a few years later.

These developments in the 1980's led to a very rapid expansion of interest in the physics of M. L. 502, what has been called since about 1975 the "Jaynes-Cummings model". This interest continues to grow. The JC model is ideal for theoretical investigations of non-classical radiation and its interaction with atoms, and Jaynes had already pointed out photon-atom tangled-state properties in M. L. 502 and made a brief examination of Einstein-Podolsky-Rosen correlations. Exact solutions of many variants of the JC model are now routinely used for understanding radiation-matter interactions non-perturbatively on a wider and wider scope within cavity QED.

5. Two-Channel Cavity QED

The JC model and all of its known variants deal only with "one-channel" interactions; there is only one photon-emission process connecting the two active levels of the atom in the cavity. Naturally many examples exist of interesting atom-field processes that involve "two-channel" interactions with more than one pathway between a pair of atomic levels. Among these are pump-probe interactions and quantum beats, wave-mixing processes of nonlinear optics, and optical pumping. They necessarily involve more than one field mode, and the Raman scattering example shown in Fig. 2 is possibly the simplest. The existence of a second independent channel causes a surprising amount of difficulty in the theory. This difficulty has been overcome by R. R. Puri and Liwei Wang and is described in detail in the recent thesis study of Wang.[34] Even a classical-pump version of the two-channel interaction has been shown by C. K. Law[35] to retain surprising quantum properties. These very recent studies may open a new "sector" in cavity QED, and a sketch of this new two-channel theory follows.

As is well known, the original JC interaction Hamiltonian involves mixed products of Bose-type and Fermi-type operators in the rotating wave approximation (RWA):

$$H_{int}^{JC} = g(a^\dagger \sigma_{-+} + a\sigma_{+-}) \tag{1}$$

Here the a's are the creation and destruction operators for cavity mode photons, and the σ's are the atomic raising and lowering operators appropriate to the single active transition

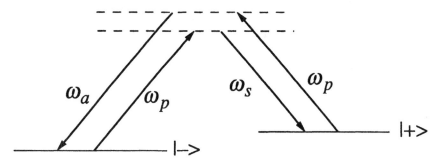

Fig. 2. Raman scattering example of two-channel cavity QED model

between the two atomic levels (labelled $-$ and $+$). They have the commutation properties $[\sigma_{km}, \sigma_{pq}] = \sigma_{kq}\delta_{mp} - \sigma_{pm}\delta_{qk}$. In contrast, the coupled-channel Raman Hamiltonian appropriate to Fig. 2 can be written

$$H_{int} = g[(a_p^\dagger a_s + a_a^\dagger a_p)\sigma_{-+} + \sigma_{+-}(a_s^\dagger a_p + a_p^\dagger a_a)] \qquad (2)$$

where the a's are labelled according to the mode they represent (pump, Stokes or anti-Stokes).

The exact two-photon resonance that is characteristic of all Raman processes leads to the requirements: $E_{+-} = (\omega_p - \omega_s) = (\omega_a - \omega_p)$, or $2\omega_p = \omega_s + \omega_a$. As in the JC model, both H_{int} and the "excitation number" (here the total photon number) are constants of the motion, and there is a new two-photon constant C in the two-channel model:

$$C = (a_a^\dagger a_a - a_s^\dagger a_s) + (1/2)(\sigma_{++} - \sigma_{--}) \qquad (3)$$

There are 10 nonlinearly coupled dynamical variables in the two-channel model: the creation and destruction operators for each of the three modes, the population variables for levels $|-\rangle$ and $|+\rangle$, and the two off-diagonal atomic coherences.

It is surprising that Ackerhalt's procedure[36] for obtaining coupled linear Heisenberg equations for operators in the JC model can also be applied to the two-channel situation. The difference is that a *two-photon operator* $B \equiv (a_p a_s^+ + a_a a_p^+)$ *involving all three modes* must be used in place of the one-mode destruction operator in the original Ackerhalt derivation. It is easy to see that B is the operator that "flips" the photon state accompanying an atomic state flip. Thus B and the atomic state-flipping operator $S_- \equiv \sigma_{-+}$ obey the same operator equation:

$$\left(i\frac{d}{dt} - E_{+-}\right)\left(i\frac{d}{dt} - E_{+-} + 2H_{int}\right)\left\{\begin{matrix} S_- \\ B \end{matrix}\right\} = -g^2\left(C + \frac{1}{2}\right)\left\{\begin{matrix} S_- \\ B \end{matrix}\right\}. \qquad (4)$$

where C is as given in (3).

A new property of the two-channel model is that it supports mode-mixing states for which the atom serves a catalytic role in a natural way. To illustrate the "chain" character of these mode-mixing states we can start in the bare atomic ground state $|-\rangle$ with a supply of pump photons. In an obvious notation this initial bare state is $|n_p, 0_s, 0_a; -\rangle$. The interaction Hamiltonian (2) couples this state to only one other bare state, namely $|(n-1)_p, 1_s, 0_a; +\rangle$. This is coupled back to the initial state but it is also coupled forward to one other bare state, namely $|(n-2)_p, 1_s, 1_a; -\rangle$. We note that after these two steps the atom has acted like a mode-mixing catalyst – it has returned to its initial bare state $|-\rangle$ but the distribution of photons among the three modes has been altered. After two more steps in the same direction we reach $|(n-4)_p, 2_s, 2_a; -\rangle$, which again has atomic state $|-\rangle$, and so on. Clearly this chain of bare states defines a quantized-pump-driven process that can be summarized by the mode-mixing relation

$$2\omega_p \rightarrow \omega_s + \omega_a. \tag{5}$$

This type of catalytic mode mixing is new within cavity QED.

All chains in the model are unbranched. One can use this fact to look for the model's eigensolutions. R. R. Puri and Liwei Wang have shown[34] how to solve $H_{int}|\Psi_\mu\rangle = E_\mu|\Psi_\mu\rangle$ for eigenstates that are compatible with photon-number conservation and C conservation. If one introduces the integer constants of motion $K = C + 1/2 \geq 0$, and $M = (N-K)/2$, then one can find the fundamental recursion relation of the model:

$$\mu^2 f_n = (2n+1)(2n+2)f_{n+1} + [2n(2M+K-2n+1) + M + K - n]f_n$$
$$+(M-n+1)(M+K-n+1)f_{n-1} \tag{6}$$

where $\mu = E_\mu/g$ is the desired eigenenergy in coupling-constant units. This recursion relation can be solved [for details see ref. 34], and the associated exact eigenvalues μ of H_{int} are given by the remarkably compact formula

$$\mu = \pm[2m(m+|C|)]^{1/2}, \tag{7}$$

where m ranges from its maximum value $(1/2)(N - |C| + 1/2)$ to either $1/2$ or 0 in integer steps.

The scaled eigenvalues of H_{int} for the lowest chain are given by $\mu = -\sqrt{3}, 0$, and $+\sqrt{3}$, and the corresponding dressed states, which we write $|N, 2C, \mu\rangle$, are:

$$|2, -1, -\sqrt{3}\rangle = [(1/\sqrt{3})|2_p, 0_s, 0_a; -\rangle - (1/\sqrt{2})|1_p, 1_s, 0_a; +\rangle$$
$$+ (1/\sqrt{6})|0_p, 1_s, 1_a; -\rangle], \tag{8a}$$

$$|2, -1, 0\rangle = [(1/\sqrt{3})|2_p, 0_s, 0_a; -\rangle - \sqrt{(2/3)}|0_p, 1_s, 1_a; -\rangle], \tag{8b}$$

$$|2, -1, +\sqrt{3}\rangle = [(1/\sqrt{3})|2_p, 0_s, 0_a; -\rangle + (1/\sqrt{2})|1_p, 1_s, 0_a; +\rangle$$
$$+ (1/\sqrt{6})|0_p, 1_s, 1_a; -\rangle], \tag{8c}$$

Appropriate combinations of such states can produce three-particle GHZ correlations.

The two independent pathways between the atomic states $|-\rangle$ and $|+\rangle$ produce interferences that are non-classical. The simplest example involves the "total field" $A(t) \equiv a_p(t)e^{-i\phi} + a_s(t) + a_a(t)$, where the pump phase has been shifted by ϕ. One can compute the average field $\langle A(t)\rangle$ as well as higher order field intensities: $\langle I(t)\rangle \equiv \langle A^\dagger(t)A(t)\rangle$ and $\langle I^2(t)\rangle \equiv \langle A^\dagger(t)A^\dagger(t)A(t)A(t)\rangle$. If computed in the middle Fock state of the chains shown in (8a) and (8c), i.e., the one in which the atom is in state $+$, one finds $\langle A(t)\rangle = 0$ because each mode has a random phase. There is also no ordinary phase interference among the three modes, so one finds $\langle I(t)\rangle = 2$, for all values of the shift ϕ.

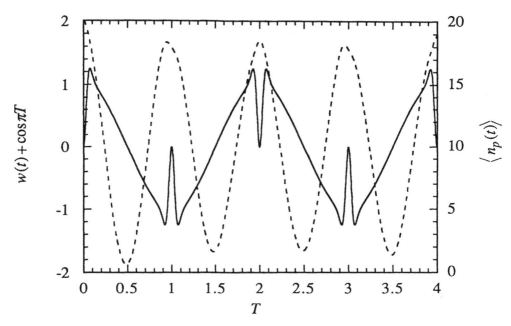

Fig. 3. Inversion (solid line and left scale) and pump expectation value (dashed line and right scale) for coherent pump initial conditions

Despite these results phase interference can still exist. If it does it must be non-classical. In the present case it appears in fourth order. For example, one can calculate the normalized degree of coherence $g^{(2)}(t;\tau)$ for zero time delay, i.e., $g^{(2)}(t;0) = \langle I^2(t)\rangle / \langle I(t)\rangle^2$, and one finds the results:[34]

$$g^{(2)}(t;0) = 1 + (1/3)\sin^2 \sqrt{3}gt, \quad \text{for} \quad \phi = 0 \tag{9a}$$

$$g^{(2)}(t;0) = \cos^2 \sqrt{3}gt, \quad \text{for} \quad \phi = \pi/2. \tag{9b}$$

The normalized degree of coherence is appropriate in the first case to super-Poisson statistics $[2 > g^{(2)} \geq 1]$, but to sub-Poisson statistics in the second $[1 \geq g^{(2)} \geq 0]$.

The average inversion, $w(t) \equiv \langle(\sigma_{++} - \sigma_{--})\rangle$, also shows two-channel interferences and they produce a style of revivals that is quite different from that found previously. There is an analytic expression, covering many revivals, that is an excellent approximation to the exact inversion:

$$w(t) \approx -\cos \pi T + 2\bar{n}\sin \pi T \sin 2\pi T \int_0^1 da \ \exp[-2\bar{n}(1 - a^2)\sin^2 \pi T] \tag{10}$$

where $2\pi T \equiv \sqrt{2}gt$ and $\bar{n}$ is assumed to be large. We show the second term of $w(t)$ in Fig. 3, where we also show the average pump photon number for comparison. Note that the value of n_p evolves smoothly to $n_p \approx 0$, far beyond the point ($n_p = 10$ in this case) where the sum of Stoke and anti-Stokes photons begins to exceed n_p, where the pump "should" stop acting like a pump. This is another indication of three-mode coherences in the model.

6. Conclusion

Detailed results on two-channel QED will appear elsewhere.[37] They will contribute to the long story of M. L. 502, and they will be part of the demonstration of the continuing influence of Ed Jaynes for thirty years in the field of electrodynamics. This is not really necessary to demonstrate because it is already evident in the careers of physicists who have had the good luck to be associated directly or indirectly with him. But later students may like to recognize the connections that exist between his work and theirs and for that this informal pocket history might be helpful.

ACKNOWLEDGMENTS. Liwei Wang supplied figures 2 and 3 and provided other assistance in preparing the manuscript. The research reported in Sec. 5 was supported by the National Science Foundation.

REFERENCES

1. E. T. Jaynes, *Some Aspects of Maser Theory*, Microwave Laboratory Report 502, Stanford University, 1958.
2. L. W. Davis, Proc. IEEE **51**, 76 (1963).
3. W. E. Lamb, Jr., Phys. Rev. **134**, A1429 (1964).
4. F. W. Cummings, *Comparison of Quantum and Semiclassical Radiation Theories with Applications to the Beam Maser*, Ph.D. Dissertation, Stanford University, 1962. Portions are published in E. T. Jaynes and F. W. Cummings, Proc. IEEE **51**, 89 (1963).
5. M. J. Duggan, *A Semiclassical Treatment of a Multiple Quantum Transition*, Ph.D. Dissertation, Stanford University, 1963.
6. J. H. Eberly, *Black-body Distribution Law in Semi-Classical Radiation Theory*, Ph.D. Dissertation, Stanford University, 1962.
7. M. D. Crisp, *Radiative Effects in Semi-Classical Theory*, Ph.D. Dissertation, Washington University 1968. See also M. D. Crisp and E. T. Jaynes, Phys. Rev. **179**, 1253 (1969).
8. C. R. Stroud, Jr., *Quantum and Semi-Classical Radiation Theories*, Ph.D. Dissertation, Washington University 1969. See also C. R. Stroud and E. T. Jaynes, Phys. Rev. A **1**, 106 (1970).
9. F. J. Schuda, *High Resolution Spectroscopy Using a CW Stabilized Dye Laser*, Ph.D. Dissertation, University of Rochester 1974. See also F. Schuda, C. R. Stroud, Jr., and M. Hercher, J. Phys. B **7**, L198 (1974).
10. S. L. McCall, *Self-Induced Transparency by Pulsed Coherent Light*, Ph.D. Dissertation, University of California, Berkeley 1968. See also S. L. McCall and E. L. Hahn, Phys. Rev. Lett. **18**, 108 (1967).
11. M. D. Crisp, Phys. Rev. Lett. **22**, 820 (1969).
12. J. H. Eberly, Phys. Rev. Lett. **22**, 760 (1969).
13. L. Matulic, *Chirping in Self-Induced Transparency*, Ph.D. Dissertation, University of Rochester 1971. See also L. Matulic and J. H. Eberly, Phys. Rev. A **6**, 822 (1972).
14. R. A. Marth, *A Theory of Chirped Optical Pulse Propagation in Resonant Absorbers*, Ph.D. Dissertation, Polytechnic Institute of Brooklyn 1972; and summarized in R. A. Marth, D. A. Holmes and J. H. Eberly, Phys. Rev. A **9**, 2733 (1974).
15. M. J. Konopnicki, *Theory of Coherent Propagation of Short Differential-Wavelength Optical Pulses*, Ph.D. Dissertation, University of Rochester 1980. See also M. J. Konopnicki and J. H. Eberly, Phys. Rev. A **24**, 2567 (1981).

16. B. J. Herman, *Coherent Three-Pulse Optical Propagation in a Collisionally Broadened Raman Amplifier*, Ph.D. Dissertation, University of Rochester 1986. See also B. J. Herman, J. H. Eberly and M. G. Raymer, Phys. Rev. A **39** 3447 (1989).
17. R. H. Dicke, Phys. Rev. **93**, 99 (1954).
18. N. E. Rehler, *A Theory of Cooperative Emission from Two-Level Atoms*, Ph.D. Dissertation, University of Rochester 1971. See also N. E. Rehler and J. H. Eberly, Phys. Rev. A **3**, 1735 (1971).
19. C. R. Stroud, Jr., J. H. Eberly, W. Lama and L. Mandel, Phys. Rev A **5**, 1094 (1972).
20. See J. R. Ackerhalt and J. H. Eberly, Phys. Rev. D **10**, 3350 (1974).
21. J. R. Ackerhalt, P. L. Knight and J. H. Eberly, Phys. Rev. Lett. **30**, 456 (1973).
22. P. W. Milonni, *Theoretical Aspects of Spontaneous Photon Emission from Atoms*, Ph.D. Dissertation, University of Rochester, 1974.
23. S. B. Lai, *The Heisenberg Equation of Motion Applied to the Free Electron, and to the Two-Level Atom Driven by a Finite-Bandwidth Laser Field*, Ph.D. Dissertation, University of Rochester, 1976.
24. H. Grotch and E. Kazes, Phys. Rev. Lett. **35**, 124 (1975).
25. See J. Dupont-Roc, C. Fabre and C. Cohen-Tannoudji, J. Phys. B **11**, 563 (1978).
26. R. Kornblith, *Transient Coherent Optical Effects in Atoms*, Ph.D. Dissertation, University of Rochester, 1976. See also R. Kornblith and J. H. Eberly, J. Phys. B **11**, 1545 (1978).
27. P. W. Milonni, Phys. Reports **25**, 1 (1976). See also P. W. Milonni and W. A. Smith, Phys. Rev. A **11**, 814 (1975).
28. J. J. Sanchez-Mondragon, *Dynamical and Spectral Properties of the Coherent State Jaynes-Cummings Model*, Ph.D. Dissertation, University of Rochester 1980. See also J. H. Eberly, N. B. Narozhny and J. J. Sanchez-Mondragon, Phys. Rev. Lett. **44**, 1323 (1980); and N. B. Narozhny, J. J. Sanchez-Mondragon and J. H. Eberly, Phys. Rev. A **23**, 236 (1981).
29. H. I. Yoo, *On-Resonance Properties of a Three-Level Atom with Quantized Field Modes*, Ph.D. Dissertation, University of Rochester 1983. See also H. I. Yoo and J. H. Eberly, Phys. Reports **118**, 239 (1985).
30. P. Filipowicz, J. Javanainen and P. Meystre, J. Opt. Soc. Am. B **3**, 906 (1986).
31. P. Filipowicz, J. Javanainen and P. Meystre, Optics Comm. **58**, 327 (1986), and Phys. Rev. A **34**, 3077 (1986). See also J. Krause, M. O. Scully and H. Walther *ibid.* **34**, 2032 (1986); and M. O. Scully and M. S. Zubairy, *ibid.* **35**, 752 (1987).
32. G. Rempe, H. Walther and N. Klein, Phys. Rev. Lett. **58**, 353 (1987).
33. H. Walther, private communication, 1992.
34. See Liwei Wang, R. R. Puri and J. H. Eberly, Phys. Rev. A (in press, 1992).
35. C. K. Law, Liwei Wang and J. H. Eberly, Phys. Rev. A **45**, 5089 (1992).
36. J. R. Ackerhalt, *Quantum electrodynamics Source-Field Method: Frequency Shifts and Decay Rates in Single-Atom Spontaneous Emission*, Ph.D. Dissertation, University of Rochester 1974. See also J. R. Ackerhalt and K. Rzazewski, Phys. Rev. A **12**, 2549 (1975).
37. One can consult refs. 34 and 35, for example, and also wait for the still-untitled theses of Liwei Wang and C. K. Law to be finished.

ENHANCING THE INDEX OF REFRACTION IN A NONABSORBING MEDIUM: PHASEONIUM VERSUS A MIXTURE OF TWO-LEVEL ATOMS

M. O. Scully and T. W. Hänsch
Max-Planck-Institut für Quantenoptik
Garching, Germany W-8046

M. Fleischhauer, C. H. Keitel, and Shi-Yao Zhu
Center for Advanced Studies and
Department of Physics and Astronomy
University of New Mexico
Albuquerque, New Mexico 87131.

ABSTRACT. We investigate the possibility of enhancing the refractive properties in a nonabsorbing medium via two fundamentally different schemes. First there is the coherent preparation of three-level atoms where absorption is cancelled due to destructive interference while the refractivity is not hampered in the same way. There also is the possibility of cancelling absorption via a mixture of absorbing and emitting two-level atoms without the need of a coherent preparation. One drawback here, however, is high sensitivity to Doppler broadening, collisions and number fluctuations which makes this scheme practically infeasible.

1. Introduction

The various application of atomic coherence in laser physics and quantum optics has recently attracted considerable interest. It has been shown that atomic coherence can lead to absorption cancellation (Alzetta, et al., 1976; and Gray, et al., 1979) and quenching of spontaneous emission noise (Scully, 1985). More recently, the notion of noninversion lasing has received attention, and it was shown that atomic coherence leading to cancellation of absorption does not necessarily influence emission (Harris, 1989; Scully, et al., 1989; and Kocharovskaya and Khanin, 1988). There has been extensive research on many schemes that involve coherence between an upper or lower level laser doublet due to various means: microwave or Raman coherent coupling and spontaneous and incoherent pumping coupling just to name the most important. Furthermore, it has been shown that absorption cancellation can also be connected with a high index of refraction (Scully, 1991; Scully and Zhu, 1992; Fleischhauer, et al., 1992a; and Fleischhauer, et al., 1992b). Such an ensemble of phase coherent atoms is a new state of matter with many new outstanding properties and thus deserves a new name. We call it "Phaseonium."

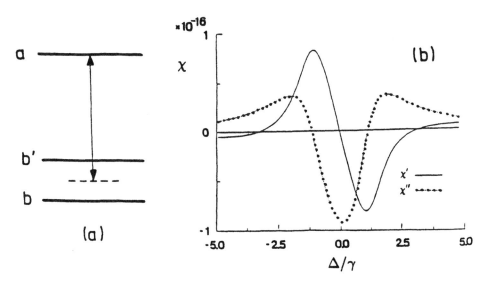

Fig. 1.(a) The Λ-quantum beat laser: The basic scheme for cancelling absorption and enhancing
the index of refraction via atomic coherence. (b) The index of refraction (solid) and the
absorption (dashed) for a medium consisting of atoms in the Λ-quantum beat configuration
with initial preparation of the atoms in a coherent superposition of the two lower levels and
density 1 atom per cm^3. The decay parameters and the phase of the initial state are chosen
such that the cancellation of the absorption and the maximum of the index of refraction
coincide.

In the following, we first outline the main properties of Phaseonium and then compare
it with a mixture of absorbing and emitting two-level atoms, that can also provide vanishing
absorption and a large index of refraction. We, however, show that Phaseonium can be
far less sensitive to Doppler broadening, collisions and certainly number fluctuations of the
atoms.

2. Phaseonium

In Fig. 1a we display the Λ-quantum beat configuration, the basic scheme for Phaseonium
atoms, which consists of one upper level and a lower level doublet. Absorption from a
coherent superposition of lower levels then obviously leads to interference terms since the
absorption reads

$$\begin{aligned}
|\langle a|H_{int}|(b|b\rangle + b'|b'\rangle)|^2 &= |b|^2|\langle a|H_{int}|b\rangle| \\
&+ |b'|^2|\langle a|H_{int}|b'\rangle|^2 \\
&+ 2\text{Re}[bb'\langle a|H_{int}|b\rangle\langle b'|H_{int}|a\rangle].
\end{aligned} \tag{1}$$

The interference term can cancel the two first terms for $|\langle a|H_{int}|b\rangle| \approx |\langle a|H_{int}|b'\rangle|$ and
$b = -b'$. We, however, do not have such interference terms for the emission process which
involves two orthogonal final states. Thus, the Λ-quantum beat laser gives us the necessary
asymmetry between emission and absorption for noninversion lasing.

In Fig. 2a we sketch the scheme that we suggest for an experimental implementation where the low frequency coherence is achieved via a coherent Raman field to an auxiliary level c. From the equation of motion

$$\dot{\rho}_{ab} = -(\gamma_{ab} + i\Delta_{ab})\rho_{ab}$$
$$+ i\left(\frac{\wp}{\hbar}E(\rho_{aa} - \rho_{bb}) - \frac{\wp'}{\hbar}E\rho_{b'b} + \Omega_R\rho_{ac}\right), \tag{2}$$

we realize that the Raman field of Rabi frequency Ω_R can hamper the ρ_{bb} absorption term both via the lower level doublet coherence $\rho_{b'b}$ and via the upper level doublet coherence ρ_{ac}.

The absorption and dispersion of the schemes displayed in Figs. 1a and 2a are governed by the imaginary and real part of the susceptibility:

$$\chi = -\left(\frac{\wp N}{\epsilon_0 E}\rho_{ab} + \frac{\wp' N}{\epsilon_0 E}\rho_{ab'}\right), \tag{3}$$

where ρ_{ab} and $\rho_{ab'}$, $\wp$ and $\wp'$ denote the coherences and dipole matrix elements of the $a-b$ and $a-b'$ transitions, respectively, and N describes the number of atoms and E the probe field on the $a-b$ and $a-b'$ transitions.

The equations of motion for ρ_{ab} and $\rho_{ab'}$ can be solved easily in steady state and to linear order in the probe field amplitude. For the Λ-quantum beat laser (Fig. 1a) with initial coherence between the lower two levels we have plotted the real and imaginary part of the susceptibility in Fig. 1b, where we learn that the index of refraction can be maximal at a frequency of vanishing absorption. In Fig. 2b we see that the Raman scheme can even provide vanishing and very small absorption on a very wide range of frequencies while the index of refraction is high. Moreover, the Raman scheme has the advantage of being insensitive to Doppler broadening, as can be seen from Fig. 2c and d. Since the Raman scheme is also insensitive to collisions, it seems most appropriate for an experimental implementation.

3. Mixture of Two-Level Atoms

It is, in principle, possible to generate a large index of refraction and absorption cancellation without the need of establishing atomic coherence. As displayed in Fig. 3a we assume a mixture of two-level atoms with slightly different transition frequencies. The first kind of atoms are completely in the lower level whereas the other is strongly pumped to its upper lasing level.

The total polarization of the mixture is given by

$$P = N_1\wp_1\rho_{ab} + N_2\wp_2\rho_{a'b'}, \tag{4}$$

where $\wp_{1/2}$ are the dipole matrix moments of the corresponding optical transitions, and ρ_{ab} and $\rho_{a'b'}$ are the off-diagonal matrix elements of the atomic density matrix for the corresponding transitions. The off-diagonal elements ρ_{ab} and $\rho_{a'b'}$ can be easily obtained from the steady state solution of the equation of motion. We find for the susceptibility, $\chi = \chi' + i\chi''$:

$$\chi' = \frac{N_1'(\Delta + \omega_0)\wp_1^2/\epsilon_0}{[\gamma_{ab}^2 + (\Delta + \omega_0)^2] + \frac{\gamma_{ab}G|\wp_1\mathcal{E}|^2}{D}} + \frac{N_2\wp_2^2(\omega_0 - \Delta)/\epsilon_0}{[\gamma_{ab}^2 + (\Delta + \omega_0)^2] + 2\gamma_{ab}'|\wp_2\mathcal{E}|^2\Gamma_{a'b'}} \tag{5a}$$

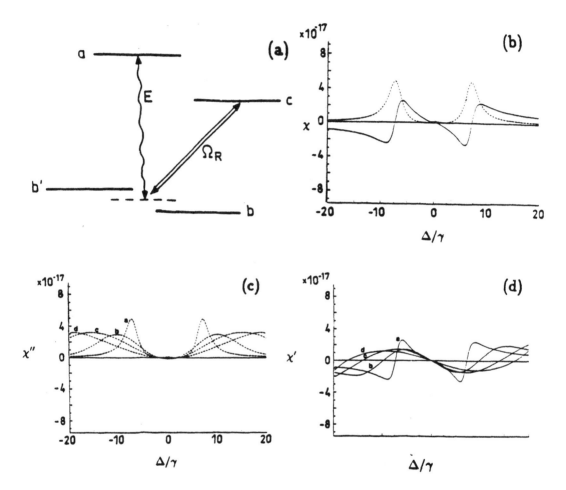

Fig. 2.(a) The Raman configuration: The experimentally most feasible scheme for cancelling absorption and enhancing the index of refraction via atomic coherence. (b) The index of refraction (solid) and the absorption (dashed) for a medium consisting of atoms in the Raman configuration and of density 1 atom per cm^3. One realizes that absorption cancellation is possible for a far range of frequency detuning while the index of refraction is relatively high. From Fig. 2 (c) (index of refraction) and (d) (absorption), it becomes obvious that the effect is fairly Doppler insensitive, where (a), (b), (c), (d) in the figures correspond to a Doppler width of $\Delta_{\text{Doppler}} = 0$, 5γ, 10γ and $20\ \gamma$, respectively.

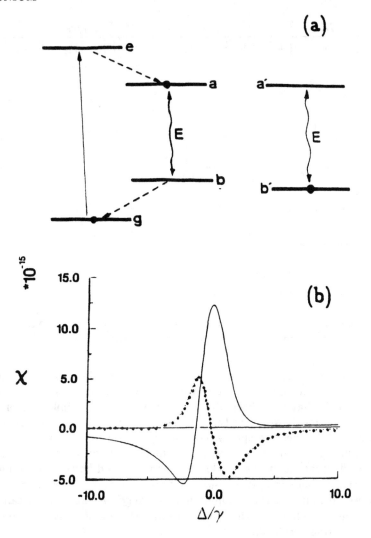

Fig. 3.(a) A mixture of absorbing and emitting two-level atoms without coherent superposition can also lead to simultaneous vanishing absorption (dashed) and high index of refraction (solid) as displayed in (b). Here, 80% of the atoms are assumed to be in the 4-level structure where strong pumping from g to e and a strong decay from b to g lead to inversion between a and b. The remaining 20% of the atoms are in the lower level b' of the two-level atoms are in the lower level of atoms in the scheme $a' - b'$. The frequency differences ω_{ab} and $\omega_{a'b'}$ on the lasing transitions differ by twice the decay from a to b and a' to b', defined to be γ.

$$\chi'' = \frac{N_1'\gamma_{ab}\wp_1^2/\epsilon_0}{[\gamma_{ab}^2 + (\Delta + \omega_0)^2] + \frac{\gamma_{ab}G|\wp_1\mathcal{E}|^2}{D}} - \frac{N_2\wp_2^2\gamma_{a'b'}/\epsilon_0}{[\gamma_{ab}^2 + (\Delta + \omega_0)^2] + 2\gamma_{ab}'|\wp_2\mathcal{E}|^2\Gamma_{a'b'}} \qquad (5b)$$

with

$$N_1' = \frac{\Gamma_{ea}(\Gamma_{bg} - \Gamma_{ab})}{D}, \qquad (6a)$$

$$G = 2\Gamma_{ea} + \Gamma_{bg}\left(1 + \frac{\Gamma_{ea} + \Gamma_{eg}}{R}\right), \qquad (6b)$$

$$D = \Gamma_{ea}\Gamma_{ab} + \Gamma_{ea}\Gamma_{bg} + \Gamma_{bg}\Gamma_{ab}\left(1 + \frac{\Gamma_{ea} + \Gamma_{eg}}{R}\right), \qquad (6c)$$

$$\omega_0 = \frac{1}{2}(\omega_{a'b'} - \omega_{ab}), \qquad (6d)$$

$$\Delta = \frac{1}{2}(\omega_{ab} + \omega_{a'b'}) - \nu, \qquad (6e)$$

where $\gamma_{\alpha,\beta}$ is the longitudinal decay rate from level α to level β; R is the pump rate from g to e; $\gamma_{\alpha\beta}$ is the transversal decay rate between α and β; and $|\mathcal{E}|^2$ is the intensity of the probe field.

For a proper choice of parameters, one can render the absorption vanishing and simultaneously the index of refraction for the total medium large, as can be seen from Fig. 3b. The problem, however, is that this effect does not persist under the influence of Doppler broadening. In Fig. 4a and b we show that the effect totally washes out for a modest Doppler broadening of 20 times the spontaneous decay rate γ. An atomic beam experiment to avoid Doppler broadening , moreover, is out of question for this scheme, since unavoidable number fluctuations in the two kinds of atoms would have a fatal influence on the effect. A further drawback is that collisions between the two species with close transition frequencies would induce a rapid population exchange. As absorption cancellation is very sensitive to the ratio between the number of absorbing and emitting atoms, these collisions represent a serious problem.

Finally, we note that there is a much larger percentage of excited atoms in the "2-level" incoherent mixture than is the case in a phased gas of "3-level" atoms.

In conclusion we note that the idea of mixing two kinds of absorbing and emitting two-level atoms with slightly different transition frequencies is an interesting possibility for obtaining a nonabsorbing medium with a large index of refraction, but for practical purposes, the Doppler and collisionally insensitive Raman scheme is preferred. Furthermore, in the Raman scheme, absorption is low over a wide range of frequencies and gain is quenched at all frequencies.

ACKNOWLEDGMENTS. This work has been supported by the Office of Naval Research.

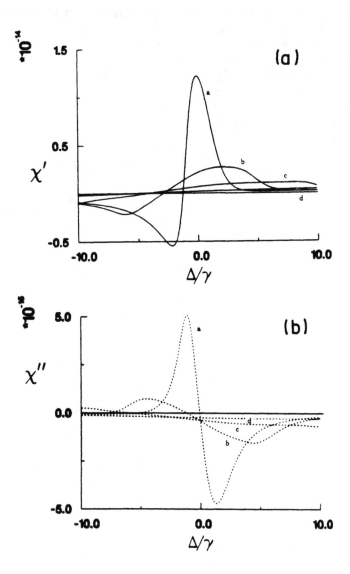

Fig. 4. The index of refraction (a) and the absorption (b) under the influence of Doppler broadening for the mixture of two-level atoms as displayed in the previous figure. The letters (a), (b), (c) and (d) in the figure refer to a Doppler width of $\Delta_{\text{Doppler}} = 0, 5\gamma, 10\gamma$ and 20γ, respectively. It shows that this scheme is very Doppler sensitive, a major drawback, which the Raman driven scheme does not suffer from.

REFERENCES

Alzetta, G., A. Gozzini, L. Moi and G. Orriols, Nuovo Cimento **36 B**, 5 (1976).

Fleischhauer, M., C. H. Keitel, M. O. Scully and Chang Su, Opt. Commun. 87, 109 (1992a)

Fleischhauer, M., C. H. Keitel, M. O. Scully, Chang Su, B. T. Ulrich and Shi-Yao Zhu, to be published in Phys. Rev. (1992b).

Gray, H., R. Whitley and C. Stroud, Opt. Lett. **3**, 218 (1979).

Harris, S., Phys. Rev. Lett. **62**, 1022 (1989).

Kocharovskaya, O., and Ya. I. Khanin, JETP Lett. **48**, 630 (1988).

Scully, M. O., Phys. Rev. Lett. **55**, 2802 (1985).

Scully, M. O., Phys. Rev. Lett. **67**, 1855 (1991).

Scully, M. O., S. Y. Zhu, Opt. Commun. **87**, 134 (1992).

Scully, M. O., S. Y. Zhu, and A. Gavrielides, Phys. Rev. Lett. **62**, 2813 (1989).

ED JAYNES' STEAK DINNER PROBLEM II

Michael D. Crisp
9113 Fairview Road
Silver Spring, MD 20910

ABSTRACT. During the Spring of 1966, Ed Jaynes presented a seminar course on quantum electronics that included the now famous "Jaynes-Cummings Model" and his Neoclassical Theory (NCT). As part of this seminar series, the NCT description of a two-level atom in an applied field was formulated as a formidable set of coupled nonlinear differential equations. Undaunted, Ed posted the equations on the Washington University Physics Department bulletin board and offered a prize of "a steak dinner for two" at a restaurant of the choice of the person who solves the equations. Within days, Bill Mitchell was able to present an elegant solution at one of the quantum electronics seminars. This early success of a new approach to doing theoretical physics encouraged Ed to challenge the knowledge hungry Physics Department with Steak Dinner Problem II. This problem was a specific mathematical formulation of the exact (i.e. without the Rotating Wave Approximation) description of the interaction of a two-level atom with a single quantized electromagnetic field mode. Jaynes' formulation of the problem appears to have anticipated the use of Bargmann Hilbert space in QED. This problem has remained unsolved for 26 years in spite of the efforts of numerous researchers, most of whom were probably unaware of Jaynes' offered prize. Recent efforts to solve this problem will be described.

> "If it were easy, it would already have been done."
> ... E.T. Jaynes, 1966.

1. Introduction

Ed Jaynes seems to enjoy pointing out that there are only a few problems in quantum electrodynamics (QED) for which an exact solution can be found. A well known example is the Jaynes-Cummings Model[1] (JCM) which is a solvable QED problem that is of fundamental importance in quantum optics. The JCM has become a standard part of text books on laser physics and quantum electronics.[2,3,4] As originally formulated, the model consists of a non-relativistic, two-level atomic system interacting with a single mode of a quantized electromagnetic field. The wave length of the field mode is assumed to be so long, compared with the atomic dimensions, that the dipole approximation can be made and effects of the diamagnetic term of the Hamiltonian are neglected. When Jaynes and Cummings originally applied their model to study the QED predictions for an ammonia beam maser, they found that even with the above simplifying assumptions, they could not find an exact solution for this application. Instead they had to make the additional approximation of neglecting transitions that correspond to processes that do not conserve energy.

The neglect of non-energy conserving terms, which was found necessary in the original derivation of the JCM, has become a common approximation in quantum optics and is usually referred to as the Rotating Wave Approximation (RWA). This terminology is chosen in analogy with work originally done with unquantized fields in magnetic resonance.[5] However, the necessity of the RWA in the unquantized field models for magnetic resonance experiments can be completely obviated by choosing a rotating magnetic field mode for the perturbing field. With such a choice of field polarization, the problem can be solved exactly without invoking the RWA.

This observation raises the natural question of whether the RWA can similarly be obviated in QED through the right choice of polarization of the perturbing field. A detailed analysis[6] of this question reveals that for a system that obeys the selection rule $\Delta m = \pm 1$, the non-energy conserving terms can be made insignificant for most experimental conditions with a judicious choice of polarization. For the $\Delta m = \pm 1$ case, the RWA terms can be made negligible for a sufficiently intense applied circularly polarized field and then the only approximation necessary to obtain a solution of the QED problem is to neglect the spontaneous emission of photons with the sense of polarization opposite that of the perturbing field. The matrix elements corresponding to this "virtual" process are independent of the number of photons in the applied field, n, and become relatively small, compared with those that correspond to energy conserving processes, as n becomes large.

However, for a system that obeys the selection rule $\Delta m = 0$, the emission of the photons that do not conserve energy is an induced process which corresponds to matrix elements that increase as $\sqrt{n+1}$. Thus neglecting the RWA terms for this case becomes less valid as the intensity of the perturbing field is increased.

It is of interest to note that the $\Delta m = 0$ case is the one that applies to the ammonia maser which was Jaynes and Cummings'[1] first application of their model. According to reference 6, this is the type of transition for which the RWA cannot be obviated through judicious choice of polarization of the perturbing field.

2. Atom-Field Hamiltonian

The perturbing field will be assumed to be linearly polarized, since it follows from reference 6 that for $\Delta m = 0$, the choice of a circularly polarized field does not simplify the problem. In the electric dipole approximation for a single frequency and a single linear polarization in the x-direction, the vector potential is given by the expression:

$$\boldsymbol{A}(\boldsymbol{r},t) = \sqrt{(2\pi\hbar c/Vk)}\boldsymbol{e}_x[a + a^+] \tag{1}$$

For this case, assume that quantum number m corresponds to the x- component of angular momentum then for $\Delta m = 0$ the momentum operator, $\boldsymbol{p}$, can be written in terms of Pauli matrices in the form:

$$\boldsymbol{p} = \boldsymbol{e}_x[p\sigma^+ + p^*\sigma^-] \tag{2a}$$

where the Pauli raising and lowering operators are given by:

$$\sigma^+ = (\sigma_x + i\sigma_y)/2 = \begin{pmatrix} 0 & 1 \\ 0 & 0 \end{pmatrix} \tag{2b}$$

and

$$\sigma^- = (\sigma_x - i\sigma_y)/2 = \begin{pmatrix} 0 & 0 \\ 1 & 0 \end{pmatrix} \tag{2c}$$

and the dipole moment matrix element is given by:

$$p = (e|\boldsymbol{p}.\boldsymbol{e}|g) = |p|e^{i\theta} \tag{2d}$$

The interaction Hamiltonian consists of two parts. The $\boldsymbol{A}.\boldsymbol{p}$ term which is:

$$-e\boldsymbol{A}.\boldsymbol{p}/mc = -(e/m)[2\pi\hbar/ckV]^{1/2}[p\sigma^+ + p^*\sigma^-][a + a^+] \tag{3a}$$

and the diamagnetic term which is given by:

$$e^2\boldsymbol{A}^2/2mc^2 = \hbar\gamma[a^2 + a^{+2} + 2aa^+ + 1] \tag{3b}$$

where:

$$\gamma = e^2\pi/mckV \tag{3c}$$

3. The Exact and JCM Hamiltonians

The "exact" QED Hamiltonian, including the diamagnetic term, is then:

$$\begin{aligned} H = &(E_e + E_g)/2 + (E_e - E_g)\sigma_z/2 + \hbar[\omega + 2\gamma][a^+a + 1/2] \\ &- [\alpha\sigma^+ + \alpha^*\sigma^-][a + a^+] \\ &+ \gamma[a^2 + a^{+2}] \end{aligned} \tag{4a}$$

where E_e and E_g are the unperturbed energy eigenvalues of the atom's excited and ground state respectively and where:

$$\alpha = (e/m)[2\pi\hbar/ckV]^{1/2}p = (e/m)\sqrt{(2\pi\hbar/\nu V)}p \tag{4b}$$

Two approximations are almost always made in the analyses of this Hamiltonian. The diamagnetic and the RWA terms are usually dropped. Both approximations involve the neglect of terms that are of the same order of magnitude so that consistency demands that an analysis which includes the RWA must also include the diamagnetic term. It has been demonstrated that, in the dipole approximation, a minor modification of the usual application of QED to <u>any atomic system</u> can exactly incorporate the effects of the diamagnetic term into a change of the electromagnetic cavity mode frequency, ν, relative to the frequency of the empty cavity, ω. The modified frequency can be expressed in terms of the empty cavity frequency according to:

$$\nu = \sqrt{\omega(\omega + 4\gamma)} \tag{5}$$

The JCM Hamiltonian which can be obtained by applying the RWA, which neglects the terms σ^+a^+ and σ^-a of the "exact" Hamiltonian of Eq.(4a), is:

$$H_{JCM} = (E_e + E_g)/2 + (E_e - E_g)\sigma_z/2 + \hbar\nu[a^+a + 1/2] - [\alpha\sigma^+a + \alpha^*\sigma^-a^+] \tag{6}$$

The effect of the diamagnetic term is <u>exactly</u> included in Eq.(6) by replacing the empty cavity frequency, ω, by the frequency ν, defined in Eq.(5).

The virtue of the JCM is that it is exactly solvable without any further approximations. The Hamiltonian of Eq.(6) can be represented by an infinite matrix with 2×2 submatrices along the diagonal, each of which can be easily diagonalized. Multiphoton absorption and the Bloch-Siegert shift[7] are examples of real physical effects which are lost when the RWA is applied to Eq.(4a).

4. Some Past Approaches to Solution

Swain[9] has expressed the eigenvalues of the Hamiltonian of Eq.(4a), without the diamagnetic term (c = 0), in terms of infinitely continued fractions. If the continued fraction converges, this procedure gives a formally exact solution of the problem. However, the final analysis of Swain's approach has to be done numerically by truncating the infinite fractions. The Hamiltonian matrix corresponding to Eq.(4a) is infinite. Graham and Höhnerbach[10,11,12,13] and Krus'[14] have studied the solution of Eq.(4a), again neglecting the diamagnetic terms, by numerically diagonalizing the truncated Hamiltonian matrix to include one or two hundred states. Schweber has studied this problem by using the Hilbert space of entire functions (Bargmann Hilbert space) which is reached by mapping the photon creation and annihilation operators into the variable x of an entire function, f(x), and its derivative d/dx respectively. He derives differential equations similar to the ones obtained from generating functions which appear below.

5. Heisenberg Representation

Three operators have been found that commute with the Hamiltonian of Eq.(4a). Thus the Heisenberg equations of motion for the five operators appearing in the Hamiltonian of Eq.(4) have three constants of motion. First the total energy, H, of this system is independent of time:

$$H(t) = H(0) \tag{7a}$$

$$2(\sigma^+\sigma^- + \sigma^-\sigma^+) + \sigma_z^2 = 3 \tag{7b}$$

and the parity operator[8]:

$$\exp[i\pi(a^+a + \omega_z/2)] = 1$$
$$= -\omega_z \exp[i\pi a^+ a] \tag{7c}$$

All three of these constants of motion also apply to the JCM Hamiltonian of Eq.(6). In addition, another operator is found to commute with the JCM Hamiltonian and provide a condition more restrictive than the parity conservation expressed in of Eq.(7c):

$$a^+a + \omega^+\omega^- - 1/2 = a^+a + \omega_z/2 = \text{constant} \tag{8}$$

which implies that Eq.(7c) will also be satisfied. Equation (8) states that in the RWA, the total number of excitations (number of photons plus number of excited atoms – there is only one atom of course) is constant, while Eq.(7c) is true if, and only if:

$$N = a^+a + \omega_z/2 = \text{constant} + 2n(t) \tag{9a}$$

where n(t) is a function of time that can only take on the values:

$$n(t) = 0, \pm 1, \pm 2, \ldots \tag{9b}$$

Without the RWA, the constant of Eq.(8) has a time dependence which is given by:

$$d[a^+a + \omega_z/2]/dt = (1/i\hbar)[a^+a + \omega_z/2,\ H]$$
$$= -2[\alpha\omega^+a^+ + a^*\omega^-a] - 2\gamma[a^2 - a^{+2}] \tag{10}$$

When the RWA is made and the diamagnetic term is neglected, the right hand side of Eq.(10) is set equal to zero and Eq.(8) is satisfied.

6. Schrödinger Representation

Any state of the combined two-level atom and field system may be expressed in the Schrödinger representation as a linear combination of the product states $|r, n >$ each of which corresponds to the atom being in state r and n photons in the field mode:

$$|\Psi\rangle = \sum_{n=0}^{\infty}[a_{e,n}|e, n\rangle + a_{g,n}|g, n\rangle] \tag{11}$$

For which the Schrödinger Equation may be written out as:

$$i\hbar da_{e,n}/dt = (E_e + n\hbar\nu)a_{e,n} + [\beta\sqrt{n}a_{g,n-1} + \alpha\sqrt{n+1}a_{g,n+1}] \tag{12a}$$

$$i\hbar da_{g,n}/dt = (E_g + n\hbar\nu)a_{g,n} + [\alpha^*\sqrt{n}a_{e,n-1} + \beta^*\sqrt{n+1}a_{e,n+1}] \tag{12b}$$

where the coupling parameters α and β are defined by:

$$\hbar\alpha = -(e/mc)[2\pi\hbar/kV]^{1/2}(e|e^{i\boldsymbol{k}.\boldsymbol{r}}\boldsymbol{e}.\boldsymbol{p}|g) = \hbar\nu a \tag{12c}$$

$$\hbar\beta = -(e/mc)[2\pi\hbar/kV]^{1/2}(e|e^{-i\boldsymbol{k}.\boldsymbol{r}}\boldsymbol{e}.\boldsymbol{p}|g) = \hbar\nu b \tag{12d}$$

The new parameters a and b are dimensionless expressions for the atom-field coupling matrix elements. In the dipole approximation limit, k → 0 and a = b. However, it is still useful to keep the two parameters distinct because all the effects of the non-energy conserving terms involve the parameter b. Thus setting b = 0 is equivalent to the RWA.

The Hamiltonian for this problem was written out by Ed Jaynes during his 1966 seminar series as a formidable matrix:

$$H/\hbar\nu = \tag{13a}$$

	$|g,0\rangle$	$|e,0\rangle$	$|g,1\rangle$	$|e,1\rangle$	$|g,2\rangle$	$|e,2\rangle$	$|g,3\rangle$	$|e,3\rangle$	
$|g,0\rangle$	ϵ_g	0	0	b^*	0	0	0	0	$\cdots$
$|e,0\rangle$	0	ϵ_e	a	0	0	0	0	0	$\cdots$
$|g,1\rangle$	0	a^*	ϵ_g+1	0	0	$\sqrt{2}b*$	0	0	$\cdots$
$|e,1\rangle$	b	0	0	ϵ_e+1	$\sqrt{2}a$	0	0	0	$\cdots$
$|g,2\rangle$	0	0	0	$\sqrt{2}a*$	ϵ_g+2	0	0	$\sqrt{3}b*$	$\cdots$
$|e,2\rangle$	0	0	$\sqrt{2}b$	0	0	ϵ_e+2	$\sqrt{3}a$	0	$\cdots$
$|g,3\rangle$	0	0	0	0	0	$\sqrt{3}a*$	ϵ_g+3	0	$\cdots$
$|e,3\rangle$	0	0	0	0	$\sqrt{3}b$	0	0	ϵ_e+3	$\cdots$
$|g,4\rangle$	0	0	0	0	0	0	0	$\sqrt{4}a*$	$\cdots$
	$\vdots$	$\vdots$	$\vdots$	$\vdots$	$\vdots$	$\vdots$	$\vdots$	$\vdots$	$\vdots$

where the unperturbed energy eigenvalues are expressed in dimensionless parameters given by:

$$\epsilon_g = E_g/\hbar\nu \tag{13b}$$

$$\epsilon_e = E_e/\hbar\nu \tag{13c}$$

It can be readily seen that setting b = 0 reduces this Hamiltonian to an infinite matrix with zero elements except for 2 × 2 matrices along the diagonal which is the JCM matrix of Hamiltonian of Eq.(6). Setting b = 0 is equivalent to application of the RWA.

The equation for the energy eigenvalues, that are denoted by q, is:

$$b\sqrt{n}a_{g,n-1} + (p+n-q)a_{e,n} + a\sqrt{n+1}a_{g,n+1} = 0 \tag{14a}$$

$$a^*\sqrt{n}a_{e,n-1} - (p-n+q)a_{g,n} + b^*\sqrt{n+1}a_{e,n+1} = 0 \tag{14b}$$

where the parameter:

$$p = E_e/\hbar\nu = -E_g/\hbar\nu \tag{14c}$$

assumes that the reference point for measuring energy is chosen so that $E_e + E_g = 0$. Equation (14) is a formulation of the problem in terms of two coupled three term difference equations.

Graham and Höhnerbach[10,11,12,13] and Krus'[14] have shown that the conservation of parity condition which is expressed in Eq.(7c) can be used to simplify the matrix form of this problem. Conservation of parity restricts the states that can be connected by a transition among elements of two independent sets. The atom-field states that have even parity, $|e, 2k + 1\rangle$ and $|g, 2k\rangle$, are not connected with the states that have odd parity, $|g, 2k + 1\rangle$ and $|e, 2k\rangle$.

The system's Hamiltonian can be rearranged so that even parity states label its upper right hand part and odd parity states label its lower left hand part. Then the Hamiltonian may be written:

$$H = \begin{bmatrix} H_{even} & 0 \\ 0 & H_{odd} \end{bmatrix} \tag{15}$$

where the even part is in the form of a continuant matrix:

$$H_{even}/\hbar\nu = \tag{16}$$

$$
\begin{array}{c}
 \\
|g,0\rangle \\
|e,1\rangle \\
|g,2\rangle \\
|e,3\rangle \\
|g,4\rangle \\
|e,5\rangle \\
\end{array}
\begin{array}{c}
\begin{array}{cccccc}
|g,0\rangle & |e,1\rangle & |g,2\rangle & |e,3\rangle & |g,4\rangle & |e,5\rangle
\end{array} \\
\left(\begin{array}{cccccccc}
\epsilon_g & b^* & 0 & 0 & 0 & 0 & \cdots \\
b & \epsilon_e+1 & \sqrt{2}a & 0 & 0 & 0 & \cdots \\
0 & \sqrt{2}a^* & \epsilon_g+2 & \sqrt{3}b^* & 0 & 0 & \cdots \\
0 & 0 & \sqrt{3}b & \epsilon_e+3 & \sqrt{4}a & 0 & \cdots \\
0 & 0 & 0 & \sqrt{4}a^* & \epsilon_g+4 & \sqrt{5}b^* & \cdots \\
0 & 0 & 0 & 0 & \sqrt{5}b & \epsilon_e+5 & \cdots \\
\vdots & \vdots & \vdots & \vdots & \vdots & \vdots &
\end{array}\right)
\end{array}
$$

The odd parity submatrix, H_{odd}, can be obtained from the matrix of Eq.(16) by interchanging subscripts e and g and interchanging parameters a and b*.

7. Continuants and Continued Fractions

A matrix of the form shown in Eq.(16), which has non-zero elements along the principal diagonal and along the diagonals on either side of the principal diagonal, is called a continuant (or tridiagonal) matrix. The secular equation for the matrix shown above is of a special type of determinant called a continuant. A determinant of a continuant matrix is called a continuant because it is intimately connected to a continued fraction.[15]

Continuant matrices arise in various physics problems. For example, in the mathematical analysis of a linear chain of particles interacting with only their nearest neighbors[17] and in the computation of the impedance of Ladder networks in electrical engineering[18]. The continuant matrix Hamiltonian that describes the two-level atom interacting with a single quantized field mode, is more complicated than these two examples. It has a closer similarity with the continuant matrix that describes a diatomic linear chain such as Born's model[17] for a sodium chloride lattice. For this application, Born found that the frequency spectrum had two branches, separated by a "no pass" zone where there were no eigenfrequencies. The lower branch is called the acoustical branch and corresponds to motion of particles such that in each short section of the line all particles move in the same direction at any instant. The upper branch is called the optical branch and corresponds to one or more types of particles moving in the direction opposite to that of the others at any given instant.

The analogy between the linear diatomic lattice and the QED two-level problem is not complete because the QED interaction matrix elements increase as $\sqrt{n}$ in contrast to the fixed mechanical spring constants in the corresponding lattice problem. In addition to this, the QED problem has diagonal matrix elements that are the sum of a term that increases as n and a term that alternates between the two unperturbed energy eigenvalues. In contrast the diagonal matrix elements of the diatomic lattice problem just alternate between two values. Graham and Höhnerbach[11] predict a band structure of the eigenvalue spectrum, similar to the two branches of the diatomic lattice, in the limit that $\nu \to 0$. In this limit the diagonal matrix elements of the two-level atom Hamiltonian no longer vary with photon number n. Thus this limit is more closely related to the diatomic linear chain.

8. Introduction of Generating Functions

In his original formulation of the Steak Dinner Problem II, Jaynes introduced generator functions which are defined by:

$$f_e(x) = \sum_{n=0}^{\infty} a_{e,n} x^n / \sqrt{n!} \tag{17a}$$

$$f_g(x) = \sum_{n=0}^{\infty} a_{g,n} x^n / \sqrt{n!} \tag{17b}$$

Then the formulation of Eq.(14) is equivalent to:

$$x f_e' + (p - q)f_e + a\, f_g' + b\, x f_g = 0 \tag{18a}$$

$$b^* f_e' + a^* x f_e + x f_g' - (p + q)f_g = 0 \tag{18b}$$

It is not essential[19], but Jaynes simplified the equations by assuming that the atomic wave functions are chosen so that the matrices are real and that:

$$a = a^* = b. \tag{19a}$$

With this assumption made, Ed introduced new variables:

$$f = f_e + f_g \tag{19b}$$

$$g = f_e - f_g \tag{19c}$$

which allowed him to write:

$$(x + a)f' + [a(x + a) - (q + a^2)]f = -pg \tag{20a}$$

$$(x - a)g' - [a(x - a) + (q + a^2)]g = -pf \tag{20b}$$

This is the form in which Jaynes originally proposed his Steak Dinner Problem Number II. Jaynes must have formulated this QED problem in terms of generator functions at some time prior to his Spring seminar series in 1966. Recently I discovered a similar formulation of this problem which was published a year after Jaynes' 1966 seminar. Schweber[16] used a Bargmann Hilbert space formulation which is mathematically equivalent to the generator function approach used here, to study the dynamics of a two-level system described by the Hamiltonian of Eq.(4a).

The two coupled first order differential equations of Eq.(20) can be rewritten as individual second order differential equations of the Fuchs type:

$$(x^2 - a^2)f'' + [(1 - 2a^2 - 2q)x - a]f' - (a^2x^2 - ax + p^2 - q^2 + a^2)f = 0 \tag{21a}$$

$$(x^2 - a^2)g'' + [(1 - 2a^2 - 2q)x + a]g' - (a^2x^2 + ax + p^2 - q^2 + a^2)g = 0 \tag{21b}$$

The theory of ordinary differential equations guarantees that a solution to these equations exists. A general approach is to express the solution as a power series about the ordinary singular points x = ± a. This ansatz results in a formidable four term recurrence relation for the series coefficients. However, the following transformations involving the exponential function $e^{\pm ax}$ can be used to "factor out" the solution's behavior at infinity:

$$f(x) = e^{ax}F(x) = e^{ax}\sum_{n=0}^{\infty} C_{+,n}(x - a)^{\mu+n} \tag{22a}$$

$$g(x) = e^{-ax}G(x) = e^{-ax}\sum_{n=0}^{\infty} C_{-,n}(x + a)^{\mu+n} \tag{22b}$$

and reduces the problem to a three term recurrence relation in terms of the coefficients of the power series. Substituting Eq.(22) into Eq.(21) yields an indicial equation that has roots $\mu = 0$ and q+a^2+1. For the μ=0 root, the three term recurrence relation can be written as:

$$(n - q - a^2)\{2aT_{n-1} + (n + 3a^2 - q)T_n + 2a(n + 1)T_{n+1}\} - p^2T_n = 0 \tag{23a}$$

where the new variable T_n is defined by:

$$T_n = (\pm 1)^n C_{\pm,n} \qquad (23b)$$

The simplification obtained in going from $C_{\pm,n}$ to T_n is a consequence of conservation of parity.

Although a three term recurrence relation still represents a difficult mathematical problem, it offers the hope of providing useful information about the solution. In principle, all the coefficients T_n can be expressed in terms of the two possible initial values of T_0 (which are $B_{+,0}$ and $B_{-,0}$) through repeated application of Eq.(23). For example, the Mathieu function[20] has been quite thoroughly explored from this point of view. However, the physical interpretation of the results of a mathematical analysis would be complicated by the transformation involved in Eq.(17) and Eq.(22). However, it has been shown[21] that the coefficients $C_{\pm,n}$ can be interpreted as the coefficients of the system's wave function when it is expanded in a basis that consists of the product of displaced harmonic oscillator states for the field and a momentum eigenstate for the atomic system.

9. Conclusion

In conclusion, I have eliminated the diamagnetic term from the problem without approximation and formulated it as a difference equation, a special form of a secular determinant equation known as a continuant, two coupled first order differential equations and a single second order ordinary differential equation. The second order differential equation has been solved in the sense that a three term recurrence relation has been provided that would enable one to write down the coefficients of its power series. I invite the reader to take up the gauntlet and continue the effort to find a closed form solution of this problem.

I would like to say that this problem illustrates the way in which Ed Jaynes could start his students off in a direction that could produce valuable research for much of their professional careers. Even though a general solution has eluded me to date, I have learned a lot about QED and mathematics from my attempts to solve this problem.

REFERENCES

1. E.T. Jaynes and F.W. Cummings. Proc. IEEE **51**, 89 (1963).
2. William Louisell, *Radiation and Noise in Quantum Electronics*, (McGraw-Hill, New York, 1964) p. 212ff.
3. William H. Louisell, *Quantum Statistical Properties of Radiation*, (John Wiley & Sons, New York, 1973) p. 323ff.
4. M. Sargent III, M.O. Scully and W.E. Lamb, Jr., *Laser Physics* (Addison-Wesley, Reading, MA, 1974).
5. I.I. Rabi, Phys. Rev. **49**, (1936) 324; Phys. Rev. **51** (1937) 652.
6. M.D. Crisp, Phys. Rev. **43**, 2430 (1991).
7. F. Bloch and A. Siegert, Phys. Rev. **57**, 522 (1940).
8. M.D. Crisp, Phys. Rev. **44**, 563 (1991).
9. S. Swain, J. Phys. **A6**, 192 (1973) and S. Swain, J. Phys. **A6**, 1919 (1972).
10. R. Graham and M. Höhnerbach, Phys. Lett. **101A** (2) 61 (1984).
11. R. Graham and M. Höhnerbach, Zeitschrift für Physik B **58** (1984).
12. R. Graham and M. Höhnerbach, Acta Physica Austriaca **56**, 45-56 (1984).

13. R. Graham and M. Höhnerbach, Phys. Rev. Lett. **57**, 1378 (1986).

14. M. Krus', Phys. Rev. Lett. **54**, 1343 (1985).

15. A.C. Aitken, *Determinants and Matrices*, (Oliver and Boyd, Edinburgh and London, 1958) See page 127.

16. S. Schweber, Ann. of Phys. **41**, 205 (1967).

17. L. Brillouin, *Wave Propagation in Periodic Structures*, 1953 (Dover Publications Inc. New York) Note especially page 44ff for a discussion of Born's model for a sodium chloride lattice.

18. Bath Van der Pol and H. Bremmer, *Operational Calculus based on the Two-Sided Laplace Integral* 1950 (Cambridge at the University Press) p. 185ff.

19. If $a = b = |a| \exp[i\theta]$ then the transformation of Eq.(19b) and Eq.(19c) would take the form $f = f_e + f_g \exp(i\theta)$ and $g = f_e - f_g \exp(i\theta)$ respectively. Under this change in variables, Eq.(20) would be valid with $|a|$ replacing a.

20. Mathews and Walker, *Mathematical Methods of Physics* (W.A. Benjamin Inc., New York, 1965) page 189ff.

21. M.D. Crisp, Phys. Rev. A, to be published.

SOURCE THEORY OF VACUUM FIELD EFFECTS

Peter W. Milonni
Theoretical Division
Los Alamos National Laboratory
Los Alamos, New Mexico 87545

When you follow two separate chains of thought, Watson, you will find some point of intersection which should approximate the truth.

—Sir Arthur Conan Doyle, *The Disappearance of Lady Frances Carfax*

ABSTRACT. It is now well established that effects traditionally associated with vacuum electromagnetic field fluctuations can be described equally well in terms of source fields (radiation reaction). This remarkable reconciliation of two previously unconnected points of view, which was stimulated by Jaynes' neoclassical theory, is reviewed and explained in a general and simple way.

1. Background

The Jaynes-Cummings paper seems to have been the first to employ dressed states of two-state atoms in fields (Jaynes 1963). One result found in that paper is that semiclassical radiation theory could serve as an excellent approximation even under certain conditions where the average number of photons in the field is small. The accuracy of semiclassical theory paved the way to Jaynes' "neoclassical theory," where even spontaneous emission was treated without field quantization (Jaynes 1972, Milonni 1976).

The neoclassical theory was a subject of much debate and misunderstanding in the 1970s. It turned out not to be so easy to dismiss it on *experimental* grounds. The most compelling evidence against it, and classical electromagnetic theory in general, was provided by the results of photon polarization correlation experiments of the type that would later be used to test Bell inequalities (Clauser 1972). Neoclassical theory was no substitute for QED, but it sharpened the limits of semiclassical theory *vis-à-vis* QED.

In neoclassical theory there is no nontrivial vacuum electromagnetic field. Spontaneous emission in this theory is attributed to the more classically familiar concept of radiation reaction, the field produced by the atom acting back on the same atom. The fact that the correct Einstein A coefficient was obtained by neoclassical theory led Ackerhalt *et al* to develop the theory of spontaneous emission in the Heisenberg picture, where one could most easily interpret things physically, and they showed that radiation reaction in QED could be regarded as the basis for spontaneous emission and the nonrelativistic portion of the Lamb shift associated with the Bethe logarithm, just as neoclassical theory suggested (Ackerhalt 1973).

Early in 1973 Jay Ackerhalt gave a seminar on this work at the University of Rochester, where he and I were graduate students. After the talk Professor Mandel raised an interesting question: how did this work, where the level shift was obviously attributed to the atom's source field, relate to the well known physical interpretation of the same shift in terms of vacuum field fluctuations (Welton 1948)? The latter interpretation, after all, is textbook material (Bjorken 1964).

That evening I went home and did a calculation similar to that of Ackerhalt, Knight, and Eberly, except that I used a non-normal ordering of annihilation and creation operators for the field instead of the standard normal ordering. This is permissible because of the commutativity of equal-time atom and field operators. With a non-normal ordering it is obvious that there will be an explicit contribution from the vacuum field, because a photon creation operator acting on the vacuum state of the field does not give zero. The following day I discussed this at some length with Jay, with whom I shared an office, and Professor Eberly. About a week later, as I recall, we learned that a similar Heisenberg-picture calculation had been done by Wally Smith, a postdoc with Melvin Lax at City College of New York. Shortly thereafter Smith came to Rochester to present his work, and shortly after that he, Jay, and I submitted a paper to Physical Review Letters (Milonni 1973; 1975). A similar and completely independent paper by I.R. Senitzky was received two days before ours, but the editor was generous enough to publish the two papers together.

To me the most interesting aspect of the work during that "neoclassical period" is that it produced, through the freedom to choose different operator orderings, a much better understanding of all the standard "vacuum fluctuation effects" — spontaneous emission, the Lamb shift, the anomalous moment of the electron, the van der Waals interaction, Casimir effects, and the laser linewidth, among others. Authoritative pronouncements over many years about vacuum field fluctuations have been shown to be oversimplified and, in some instances, even incorrect. Edwin Jaynes deserves a lot of credit for this clarification, not only because it was spawned by his own work, but also because of his incessant challenging of othodoxy that has awakened and inspired so many of us.

The basic ideas I will discuss below have been published in various papers and reviews, and are also discussed in some detail in a book on *The Quantum Vacuum* (Milonni 1993). Similar ideas, with a somewhat different emphasis, have also been discussed by many others, especially Dalibard *et al* (Dalibard 1982; 1984). Aside from the fact that the subject still gives me pleasure, I think it may be useful to present it once more, in a way I believe may be the simplest possible.

2. Energy of Induced Polarization

Recall that an induced dipole $\mathbf{p}$ in an electric field $\mathbf{E}$ has energy $-\frac{1}{2}\mathbf{p} \cdot \mathbf{E}$ classically, and that for N dipoles per unit volume defining a polarization density $\mathbf{P} = N\mathbf{p}$, the expectation value of the energy in quantum theory is similarly

$$\langle E \rangle = -\frac{1}{2} \int d^3r \langle \mathbf{P} \cdot \mathbf{E} \rangle . \tag{1}$$

Now we can write $\mathbf{E}(\mathbf{r}, t) = \mathbf{E}^{(+)}(\mathbf{r}, t) + \mathbf{E}^{(-)}(\mathbf{r}, t)$, where $\mathbf{E}^{(+)}$ and $\mathbf{E}^{(-)}$ involve (time-dependent) photon annihilation and creation operators, respectively. We also have

$$\mathbf{E}^{(\pm)}(\mathbf{r}, t) = \mathbf{E}_o^{(\pm)}(\mathbf{r}, t) + \mathbf{E}_s^{(\pm)}(\mathbf{r}, t), \tag{2}$$

where the subscripts o and s denote *free* and *source* parts of the field, respectively. The free part is always present, regardless of whether there are sources of radiation. It is the homogeneous solution of the Maxwell equation for the electric field, and has the property

$$\mathbf{E}_o^{(+)}(\mathbf{r},t)|\text{vac}\rangle = \langle\text{vac}|\mathbf{E}_o^{(-)}(\mathbf{r},t) = 0, \tag{3}$$

where $|\text{vac}\rangle$ is the vacuum state of the field.

Now the equal-time commutators $[\mathbf{P},\cdot\mathbf{E}^{(\pm)}] = 0$ allow us to write (1) in various equivalent ways. For instance, we can normally order the field operators and write

$$\langle E\rangle = -\frac{1}{2}\int d^3r\langle\mathbf{E}^{(-)}\cdot\mathbf{P} + \mathbf{P}\cdot\mathbf{E}^{(+)}\rangle . \tag{4}$$

From equations (3) we see that this implies

$$\langle E\rangle = -\frac{1}{2}\int d^3r\langle\mathbf{E}_s^{(-)}\cdot\mathbf{P} + \mathbf{P}\cdot\mathbf{E}_s^{(+)}\rangle . \tag{5}$$

We can also choose, for instance, a symmetric ordering of field operators (Milonni 1975) in equation (1):

$$\begin{aligned}
\langle E\rangle &= -\frac{1}{2}\int d^3r\langle\frac{1}{2}\mathbf{P}\cdot[\mathbf{E}^{(+)} + \mathbf{E}^{(-)}] + \frac{1}{2}[\mathbf{E}^{(+)} + \mathbf{E}^{(-)}]\cdot\mathbf{P}\rangle \\
&= \langle E\rangle_{VF} + \langle E\rangle_s ,
\end{aligned} \tag{6}$$

where, assuming the vacuum state for the field,

$$\begin{aligned}
\langle E\rangle_{VF} &\equiv -\frac{1}{2}\int d^3r\langle\frac{1}{2}\mathbf{P}\cdot[\mathbf{E}_o^{(+)} + \mathbf{E}_o^{(-)}] + \frac{1}{2}[\mathbf{E}_o^{(+)} + \mathbf{E}_o^{(-)}]\cdot\mathbf{P}\rangle \\
&= -\frac{1}{2}\int d^3r\langle\frac{1}{2}\mathbf{P}\cdot\mathbf{E}_o^{(-)} + \frac{1}{2}\mathbf{E}_o^{(+)}\cdot\mathbf{P}\rangle ,
\end{aligned} \tag{7}$$

$$\begin{aligned}
\langle E\rangle_s &\equiv -\frac{1}{2}\int d^3r\langle\frac{1}{2}\mathbf{P}\cdot[\mathbf{E}_s^{(+)} + \mathbf{E}_s^{(-)}] + \frac{1}{2}[\mathbf{E}_s^{(+)} + \mathbf{E}_s^{(-)}]\cdot\mathbf{P}\rangle \\
&= -\frac{1}{2}\int d^3r\langle\frac{1}{2}\mathbf{P}\cdot\mathbf{E}_s + \frac{1}{2}\mathbf{E}_s\cdot\mathbf{P}\rangle ,
\end{aligned} \tag{8}$$

and $\mathbf{E} = \mathbf{E}_s^{(+)} + \mathbf{E}_s^{(-)}$ is the full electric field of the source. It may be shown, within a Markov approximation appropriate to everything here, that $\langle E\rangle_s = 0$ (Milonni 1992a; 1993), so that

$$\langle E\rangle = \langle E\rangle_{VF} = -\frac{1}{4}\int d^3r\langle\mathbf{P}\cdot\mathbf{E}_o^{(-)} + \mathbf{E}_o^{(+)}\cdot\mathbf{P}\rangle \tag{9}$$

when a symmetric ordering of field operators is used. The subscript VF denotes a vacuum field contribution; the sense in which (9) represents a contribution due solely to the vacuum field will be clarified below.

Thus the freedom to choose different operator orderings has led to two distinct expressions for the energy $\langle E\rangle$ associated with induced dipoles. (Of course there are infinitely many such expressions!) In expression (5) only the source field operators $\mathbf{E}_s^{(\pm)}$ appear explicitly. This leads naturally to an interpretation in terms of source fields, or radiation reaction. The *equivalent* expression (9), on the other hand, has an interpretation in terms of the vacuum field and zero-point electromagnetic energy, as we shall see.

3. Examples

I will now describe a few examples where calculations based on (5) and (9) may be shown explicitly to produce the same results.

RADIATIVE LEVEL SHIFTS

Consider a point dipole system with isotropic polarizability $\alpha(\omega)$. Then the polarization $\mathbf{P} = \alpha(\omega)[\mathbf{E}_o^{(+)}(\omega) + \mathbf{E}_o^{(-)}(\omega)]\delta^3(\mathbf{r})$ is induced by the vacuum field (or more generally any "external" field not due to the dipole itself), and (9) becomes

$$\langle E \rangle = -\frac{1}{2} \int d\omega \alpha(\omega) \langle \mathbf{E}_o^{(+)}(\omega) \cdot \mathbf{E}_o^{(-)}(\omega) \rangle = -\frac{1}{2} \int d\omega \alpha(\omega) \langle \mathbf{E}_o^2(\omega) \rangle \qquad (10)$$

when we use the property (3) of the vacuum field state. $\langle E \rangle$ here is evidently the Stark shift, *due to the vacuum field*, of the state of the dipole system with polarizability $\alpha(\omega)$. In fact if we subtract from $\langle E \rangle$ the free-electron energy $\langle e^2 \mathbf{A}_o^2/2mc^2 \rangle$, and introduce a high-frequency cutoff $mc^2/\hbar$, we obtain the famous "Bethe logarithm" for the nonrelativistic contribution to the Lamb shift (Milonni 1988a).

We can relate this to another interpretation of the Lamb shift (Feynman 1961; Power 1966) by using the relation $n(\omega) \cong 1 + 2\pi N\alpha(\omega)$ for the refractive index of N particles per unit volume, each of polarizability $\alpha(\omega)$, to write (10) as*

$$\langle E \rangle \cong -\frac{1}{4\pi N} \int d\omega [n(\omega) - 1] \langle \mathbf{E}_o^2(\omega) \rangle = -\frac{1}{N} \int d\omega [n(\omega) - 1] \rho_o(\omega)$$

$$= -\frac{1}{N} \int d\omega [n(\omega) - 1] \frac{\hbar \omega^3}{2\pi^2 c^3} \rightarrow -\frac{1}{NV} \sum_k [n(\omega_k) - 1] \frac{1}{2} \hbar \omega_k \qquad (11)$$

$$\cong \sum_k \left[\frac{\frac{1}{2}\hbar\omega_k}{n(\omega_k)} - \frac{1}{2}\hbar\omega_k \right],$$

which is just the *change in the zero-point energy of the field due to the presence of one polarizable particle (NV = 1)*. That is, $\frac{1}{2}\hbar\omega_k/n(\omega_k)$ is the zero-point energy of mode k in the presence of the particles, and $\frac{1}{2}\hbar\omega_k$ is the zero-point energy in their absence. The equivalence of this interpretation to the "vacuum Stark shift" interpretation following from (10) obviously results from the relation between the refractive index and the polarizability.

How to relate this to radiation reaction? Suppose we take (5) as our starting point instead of (9). The source (radiation reaction) electric field for a point dipole $\mathbf{p}$ is

$$\mathbf{E}_s(t) = \frac{2}{3c^3} \dddot{\mathbf{p}}(t) - \frac{\delta m}{e^2} \ddot{\mathbf{p}}(t)$$

$$= -\frac{4\pi}{V} \sum_k \mathbf{e}_k \mathbf{e}_k \cdot \int_0^t dt' \dot{\mathbf{p}}(t') \cos \omega_k(t' - t), \qquad (12)$$

---

* Here $\rho_o(\omega)$ is the spectral energy density of the vacuum field ($\omega^2/\pi^2 c^3$ modes per unit volume in the frequency interval $[\omega, \omega + d\omega]$, each mode having an energy $\frac{1}{2}\hbar\omega$), and we have employed the usual relation $\sum_k \rightarrow (V/4\pi^2 c^3) \sum_{\lambda=1}^2 \int d\omega \omega^2$ for free space.

where $\mathbf{e}_k$ is a unit polarization vector for mode k and δm is the electromagnetic mass, which is linearly divergent in the nonrelativistic theory. Identifying from this

$$\mathbf{E}_s^{(+)}(t) \cong -\frac{2\pi}{V} \sum_k \mathbf{e}_k \mathbf{e}_k \cdot \int_0^t \dot{\mathbf{p}}(t') e^{i\omega_k(t'-t)} \tag{13}$$

and $\mathbf{P} = \mathbf{p}(t)\delta^3(\mathbf{r})$, we have, from (5),

$$\langle E \rangle = -\mathrm{Re}\langle \mathbf{p}(t) \cdot \mathbf{E}_s^{(+)}(t) \rangle = -\frac{2\pi}{V}\mathrm{Re}\sum_k e_{km}e_{kn}\int_0^t dt' \langle p_m(t)\dot{p}_n(t')\rangle e^{i\omega_k(t'-t)} . \tag{14}$$

Now in the approximation of nearly unperturbed motion (Milonni 1982)

$$\langle p_m(t)\dot{p}_n(t')\rangle \cong \langle j|p_m(t)\dot{p}_n(t')|j\rangle = i\sum_i \omega_{ji}p_{jim}p_{ijn}e^{i\omega_{ji}(t-t')} , \tag{15}$$

for an atom in state j, where the sum is over a complete set of states i and the $\mathbf{p}_{ji}$ and ω_{ji} are the $j \leftrightarrow i$ transition dipole matrix elements and Bohr transition frequencies, respectively. Then (14) becomes

$$\langle E \rangle = -\frac{2}{3\pi c^3}\sum_i \omega_{ji}p_{ji}^2 \int \frac{d\omega\omega^2}{\omega - \omega_{ji}} , \tag{16}$$

or

$$\langle E \rangle - \langle E \rangle_{\mathrm{free}} = -\frac{2}{3\pi c^2}\sum_i \omega_{ji}^2 p_{ji}^2 \int \frac{d\omega\omega}{\omega - \omega_{ji}} \tag{17}$$

when we subtract the free-electron energy as above. Finally - since this calculation is based on radiation reaction - we must renormalize the mass in order to avoid double-counting the electromagnetic mass δm. That is, we must subtract $-(\delta m/m)\langle j|p^2/2m|j\rangle$ from (17). When this is done we obtain exactly the Bethe logarithm for the nonrelativistic Lamb shift. In other words, calculations based on expressions (5) and (9) lead to exactly the same result: *the Bethe log can be derived from considerations based on either the vacuum field or the source field.* A similar conclusion applies to the radiative decay rate (Milonni 1976; 1988a).

Van der Waals forces

What if there are *two* (ground-state) atoms? In this case equations (5) and (9) are still applicable (and equivalent), but now the field mode functions are not simply plane waves. The atom at $\mathbf{r}_A$ scatters a plane wave to produce a new mode function that is the sum of the plane wave plus a scattered dipole field:

$$\begin{aligned}\mathbf{e}_{k\lambda}e^{i\mathbf{k}\cdot\mathbf{r}} \rightarrow \mathbf{e}_{k\lambda}e^{i\mathbf{k}\cdot\mathbf{r}} &+ \alpha_A(\omega_k)k^3 e^{i\mathbf{k}\cdot\mathbf{r}_A}e^{ikR}\left[\mathbf{e}_{k\lambda}\left(\frac{1}{kR} + \frac{1}{k^2R^2} - \frac{1}{k^3R^3}\right)\right.\\ &\left. - \frac{1}{R^2}(\mathbf{e}_{k\lambda}\cdot\mathbf{R})\mathbf{R}\left(\frac{1}{kR} + \frac{3i}{k^2R^2} - \frac{3}{k^3R^3}\right)\right],\end{aligned} \tag{18}$$

where $\mathbf{R} = \mathbf{r} - \mathbf{r}_A$ and $\alpha_A(\omega_k)$ is the polarizability of atom A at frequency $\omega_k = kc$. Using these modified mode functions in the expression for the vacuum field $\mathbf{E}_o$, and $\mathbf{P} =$

$\alpha_B(\omega)\mathbf{E}_o(\omega)\delta^3(\mathbf{r}_B)$ for the polarizability of atom B, we obtain from (9) a contribution to the level shift of atom B (or atom A) depending on the interatomic distance $r = |\mathbf{r}_A - \mathbf{r}_B|$. This turns out to be exactly the van der Waals interaction, varying as r^{-6} and r^{-7} at small and large distances, respectively, exactly as found by Casimir and Polder (Milonni 1982; 1992a).

Physically, this calculation based on (9) gives the following explanation for the van der Waals interaction. The vacuum field induces dipole moments in each atom and, because the vacuum field is correlated with itself over finite distances (Milonni 1993), these fluctuating dipole moments are themselves correlated, and this correlation leads to a nonvanishing expectation value of the dipole-dipole interaction. This expectation value is the van der Waals interaction.

We can also calculate the van der Waals interaction starting from equation (5). In this case the source field of each atom is modified by the presence of the other atom, i.e., *the radiation reaction field depends on the modes of the field*, and each atom modifies these modes according to (18). The calculation is very similar to that summarized in (12)-(16), and leads again to the van der Waals interaction (Milonni 1982). So here again we have an interpretation in terms of either vacuum or source fields.

CAVITY QED

Exactly the same considerations apply in cavity QED, where one considers radiative processes for atoms contained within cavities. As in the case of van der Waals forces the level shifts, for instance, are affected by the presence of all the other atoms, in this case the atoms comprising the cavity walls. For effective virtual photon wavelengths much larger than the wall interatomic spacings we can simply treat the cavity according to *classical*, macroscopic boundary conditions (Milonni 1992b). Then the modifications of both the vacuum field and the source field from their free-space forms are determined straightforwardly by the cavity mode functions and, again depending upon the operator ordering we choose, we can attribute cavity QED effects to vacuum fields or to radiation reaction.

CASIMIR FORCES

Consider now the famous Casimir force between conducting plates. With few exceptions (Milonni 1982; 1992a), this effect is nearly always described in terms of changes of vacuum field energy, or slight variations thereof (Milonni 1988b).

The simplest way, perhaps, to see that the Casimir force can be obtained from either vacuum or source fields is to begin with the case of two dielectric plates separated by a distance d and characterized by dielectric constants $\epsilon_1(\omega)$ and $\epsilon_2(\omega)$. The force between the plates is then the macroscopic manifestation of the interatomic van der Waals forces between the atoms of the two plates. Since this interatomic force can be gotten from either vacuum or source fields, so too can the net force between the plates. When the limits $\epsilon_1(\omega), \epsilon_2(\omega) \to \infty$ are taken in order to describe the case of conductors, the result is precisely the Casimir force $F(d) = -\pi^2 \hbar c / 240 d^4$ between perfectly conducting plates. We conclude then that the Casimir force, a macroscopic manifestation of van der Waals forces, can be regarded alternatively as a consequence of source fields (radiation reaction) or vacuum field fluctuations (Milonni 1992a; 1993).

A technical difficulty with any approach to macroscopic van der Waals forces is that these forces are not simply additive. The nonadditivity is closely related to the multiple scattering processes that determine the field appearing in equation (1) (Milonni 1992b). An *approximate* way of handling this difficulty is to assume that all the virtual photon wavelengths are large compared with the interatomic distances. Then, as mentioned above, we can employ the standard boundary conditions to calculate mode functions, and the spatial variations of these mode functions then determine the forces between dielectric bodies. In essence, this macroscopic approach assumes that the field acting on any one atom propagates in a *continuum* medium with a simple dielectric constant associated with all the other atoms. It can be derived via the Ewald-Oseen extinction theorem, and underlies the Lifshitz theory of van der Waals forces between dielectrics (Lifshitz 1956, Milonni 1992a; 1992b).

To derive the Casimir force from the source-field perspective we use in equation (5) the expressions (Milonni 1992a)

$$\mathbf{E}_s^{(+)}(\mathbf{r},t) = 8\pi \int d^3r' \int_0^t dt' \overset{\leftrightarrow}{G}^{(+)}(\mathbf{r},\mathbf{r}';t,t') \cdot \mathbf{P}(\mathbf{r}',t'), \tag{19}$$

$$\overset{\leftrightarrow}{G}(\mathbf{r},\mathbf{r}';t,t') = \frac{1}{2\pi} \int_{-\infty}^{\infty} d\omega \, \overset{\leftrightarrow}{\Gamma}(\mathbf{r},\mathbf{r}',\omega) e^{-i\omega(t-t')}, \tag{20}$$

where $\overset{\leftrightarrow}{G}^{(+)}$ is a dyadic Green function whose Fourier transform satisfies

$$-\nabla \times \nabla \times \overset{\leftrightarrow}{\Gamma} + \frac{\omega^2}{c^2}\epsilon(\omega)\overset{\leftrightarrow}{\Gamma} = -\frac{\omega^2}{c^2}\overset{\leftrightarrow}{1}\delta^3(\mathbf{r}-\mathbf{r}'). \tag{21}$$

Equation (5) then leads to the following expression for the source-induced dipole energy (Milonni 1992a):

$$\langle \delta E \rangle = -4\text{Re} \int d^3r \int d^3r' \int_0^t dt' \int_0^{\infty} d\omega \Gamma_{ij}(\mathbf{r},\mathbf{r}',\omega)\langle P_i'(\mathbf{r},t)P_j'(\mathbf{r}',t')\rangle e^{-i\omega(t-t')}, \tag{22}$$

with

$$\langle P_i'(\mathbf{r},t)P_j'(\mathbf{r}',t')\rangle = \langle p_j(t)p_i(t')\rangle N(\mathbf{r})\delta^3(\mathbf{r}-\mathbf{r}'). \tag{23}$$

Here $N(\mathbf{r})$ is the number density of (identical) atoms and $\mathbf{p}(t)$ is the dipole moment operator associated with a single atom. For the case of dielectric half-spaces with dielectric constants $\epsilon_1(\omega)$ and $\epsilon_2(\omega)$, and separated by a distance d, equation (22) leads to the Lifshitz formula for the force, and the limits $\epsilon_1, \epsilon_2 \to \infty$ then give the Casimir force between perfect conductors (Milonni 1992a).

The derivation just outlined shows why Schwinger, *et al*, from the perspective of ordinary QED, are able to obtain the Casimir force using Schwinger's source theory, where there is no nontrivial vacuum field (Schwinger, *et al*, 1978, Milonni 1992a). This theory leads to an expression of the same general form as (22), and involves the same (classical) dyadic Green function. On the other hand the freedom to choose different operator orderings, and in particular to start with equation (9) instead of (5), explains why the Casimir force can also be derived in terms of the change in the vacuum field energy due to the presence of the plates. The equivalence of these two perspectives is enforced by the proportionality

$$\overset{\leftrightarrow}{\Gamma}(\mathbf{r},\mathbf{r}',\omega) \propto \langle \mathbf{E}_o^{(+)}(\mathbf{r},\omega)\mathbf{E}_o^{(-)}(\mathbf{r}',\omega)\rangle \tag{24}$$

between the Green dyadid and the vacuum field correlation function (Milonni 1992a).

LASER LINEWIDTH

Very similar considerations apply to other vacuum fluctuation effects, and here we will consider just one more example, namely the fundamental, quantum-mechanical limit to the laser linewidth. A standard theory leads to the conclusion that the linewidth "depends on *both* photon and atomic noise sources" (Lax 1966; see also Sargent 1974).

Here we will simply outline the calculation in a largely symbolic fashion. (For details see Goldberg 1991.) With a the photon annihilation operator associated with a single laser mode of frequency ω_o, we calculate the correlation function

$$\langle a^\dagger(t)a(t+\tau)\rangle = \overline{n}e^{-(C/2\overline{n})\tau}e^{-i\omega_o\tau} , \tag{25}$$

where C is the rate of spontaneous emission into the single mode and $\overline{n}$ is the average intracavity photon number. This implies the linewidth

$$\delta\nu = \frac{C}{2\pi\overline{n}} = \frac{\overline{N}_2}{\Delta N_t}\frac{cg_t}{2\pi\overline{n}} , \tag{26}$$

where g_t is the threshold gain coefficient, $\overline{N}_2$ ($\overline{N}_1$) is the steady-state population of the upper (lower) level of the lasing transition, and $\Delta N_t = \overline{N}_2 - \overline{N}_1$ is the threshold population difference. Using $P_{\text{out}} = cg_t h\nu\overline{n}$, we recover the Schawlow-Townes formula for the case in which the gain and the intracavity intensity are assumed to be spatially uniform.

No vacuum field fluctuations enter into this calculation; this is because we have chosen a normal ordering of the field annihilation and creation operators. We have, symbolically,

$$a = a_o + \Sigma/(\Delta N_t)^{1/2} , \tag{27}$$

where a_o is the source-free (vacuum) field operator and Σ represents a collective atomic lowering operator such that the steady-state expectation values of interest are

$$\langle\Sigma^\dagger\Sigma\rangle = \overline{N}_2, \quad \langle\Sigma\Sigma^\dagger\rangle = \overline{N}_1 . \tag{28}$$

In the calculation outlined above, with a normal ordering of field operators, we used

$$\langle a^\dagger a\rangle = \frac{\langle\Sigma^\dagger\Sigma\rangle}{\Delta N_t} = \frac{\overline{N}_2}{\Delta N_t} . \tag{29}$$

There is no vacuum contribution because $\langle a_o^\dagger a_o\rangle = 0$. If we use instead a symmetric ordering, however, we obtain

$$\begin{aligned}
\frac{1}{2}\langle aa^\dagger + a^\dagger a\rangle &= \frac{1}{2}\langle a_o a_o^\dagger + a_o^\dagger a_o\rangle + \frac{1}{2}\langle\Sigma\Sigma^\dagger + \Sigma^\dagger\Sigma\rangle/\Delta N_t \\
&= \frac{1}{2}\left[1 + \frac{\overline{N}_1 + \overline{N}_2}{\overline{N}_2 - \overline{N}_1}\right] = \frac{\overline{N}_2}{\Delta N_t} ,
\end{aligned} \tag{30}$$

since $\langle a_o a_o^\dagger\rangle = 1$. Or we can use, for instance, an anti-normal ordering:

$$\langle aa^\dagger\rangle = \langle a_o a_o^\dagger\rangle + \frac{\langle\Sigma\Sigma^\dagger\rangle}{\Delta N_t} = 1 + \frac{\overline{N}_1}{\overline{N}_2 - \overline{N}_1} = \frac{\overline{N}_2}{\Delta N_t} . \tag{31}$$

Obviously these different operator orderings lead to the same result for the laser linewidth. But obvious also is the different *interpretation* of the final, equivalent expressions. In particular, (30) is perfectly consistent with the idea that both photon and atomic noise sources contribute to the linewidth, while (29) suggests an interpretation in terms of only atomic noise sources. We see again the intimate, fluctuation-dissipation connection between vacuum and source fields.

4. Remarks

These considerations of operator orderings can be applied to other problems, such as amplifier noise or the buildup of laser oscillation from noise. They can also be used to explain, for instance, why there is no spontaneous *absorption* from the vacuum field (Milonni 1973; 1976; 1984).

Dalibard, *et al* have advocated the symmetric ordering as a means of removing the "ambiguity" between source and vacuum field effects. They argue that the symmetric ordering is the preferred one in the sense that the source and vacuum field contributions are separately Hermitian. The symmetric ordering certainly facilitates physical interpretation along classical lines in the sense that the field annihilation and creation operators appear in the symmetric combination defining the vector potential or electric and magnetic fields (Milonni 1975; 1976). But of course a preference for symmetric ordering is subjective, since nothing in the quantum formalism requires the source and vacuum field contributions to be *separately* Hermitian.

ACKNOWLEDGMENTS. I am indebted to Edwin Jaynes not only through my fortunate association with his students, colleagues, and other grandstudents, but also because he has encouraged me to think, through his work and writings, that it may still be possible to work in a lighthouse.

REFERENCES

Ackerhalt, J.R.; Knight, P.L.; Eberly, J.H.: 1973, Phys. Rev. Lett. **30**, 456.
Bjorken, J.D.; Drell, S.D.: 1964, *Relativistic Quantum Mechanics*, McGraw-Hill, New York.
Clauser, J.F.: 1972, Phys. Rev. A**6**, 49.
Dalibard, J.; Dupont-Roc, J.; Cohen-Tannoudji, C.: 1982, J. Physique **43**, 1617.
Dalibard, J.; Dupont-Roc, J.; Cohen-Tannoudji, C.: 1984, J. Physique **45**, 637.
Feynman, R.P.: 1961, in R. Stoops (ed.), *The Quantum Theory of Fields*,
Wiley Interscience, New York.
Goldberg, P.; Milonni, P.W.; Sundaram, B.: 1991, Phys. Rev. A**44**, 1969.
Jaynes, E.T.; Cummings, F.W.: 1963, Proc. IEEE J. **51**, 89.
Jaynes, E.T.: 1972, in L. Mandel and E. Wolf (eds.), *Coherence and Quantum Optics*,
    Plenum, N.Y.
Lax, M.: 1966, in P.L. Kelley, B. Lax, and P.E. Tannenwald (eds.), *Physics of Quantum Electronics*, McGraw-Hill, N.Y.
Lifshitz, E.M.: 1956, Sov. Phys. JETP **2**, 73.
Milonni, P.W.; Ackerhalt, J.R.; Smith, W.A.: 1973, Phys. Rev. Lett. **31**, 958.
Milonni, P.W.; Smith, W.A.: 1975, Phys. Rev. A**11**, 814.
Milonni, P.W.: 1976, Phys. Rep. **25**, 1.
Milonni, P.W.: 1982, Phys. Rev. A**25**, 1315.
Milonni, P.W.: 1984, Am. J. Phys. **52**, 340.

Milonni, P.W.: 1988a, Phys. Scr. **T21**, 102.

Milonni, P.W.; Cook, R.J.; Goggin, M.: 1988b, Phys. Rev. **A30**, 1621.

Milonni, P.W.; Shih, M.-L.: 1992a, Phys. Rev. **A45**, 4241.

Milonni, P.W.; Lerner, P.B.: 1992b, Phys. Rev. **A46**, 1185.

Milonni, P.W.: 1993, *The Quantum Vacuum*, Academic Press, Boston.

Power, E.A.: 1966, Am. J. Phys. **34**, 516.

Sargent, M., III; Scully, M.O.; Lamb, W.E., Jr.: 1974, *Laser Physics*, Addison-Wesley, Reading, MA.

Schwinger, J.; DeRaad, L.L., Jr.; Milton, K.A.: 1978, Ann. Phys. (N.Y.) **115**, 1.

Welton, T.A.: 1948, Phys. Rev. **74**, 1157.

THE NATURAL LINE SHAPE

Edwin A. Power*
Center for Nonlinear Studies
Los Alamos National Laboratory
Los Alamos, New Mexico 87545

All these fifty years of conscious brooding have brought me no nearer to the answer to the question "what are light quanta?" Nowadays every Tom, Dick and Harry thinks he knows it, but he is mistaken.

— Albert Einstein [1]

ABSTRACT. It is suggested that the apparent paradox, which arose from a variety of predictions for the distortion from a Lorentzian shape of the natural spontaneous emission line, is resolved by noting that the concept of "photon" is different for the different theories. In this paper it is shown that, taking into account the small differences in the meaning of a photon, there is no contradiction between the various theoretical predictions. Furthermore, if a precise definition of line shape is made which involves the Maxwell fields themselves, a definition which does not involve the idea of a photon, there is a unique shape to the line. To the lowest order in perturbation theory and for a straightforward detector theory this shape is Lorentzian.

1. Introduction

The contrasting predictions for the shape of a spontaneous emission line arise, not from assumptions of differing dynamics (they all use some aspect of "Wigner - Weisskopf" theory), but from the well-known question as to the form of interaction between the atoms and their radiation field. Most quantum opticians use the familiar dipole coupling $-\vec{\mu}\cdot\vec{E}$ while particle theorists prefer the so called minimal coupling interaction $e\vec{p}\cdot\vec{A}/mc + e^2\vec{A}^2/2mc^2$.

There has been an explosive literature on the topic of this difference. Ed Jaynes in 1976 [2] put the problem succinctly in this quote:

... a whole generation of physicists has stumbled on this problem and lived, not only under the shadow of the immediate difficulty: "How can I ever know whether a practical calculation has been done right?" but the deeper mystery: "How is it possible that a theory, for which formal invariance is proved easily once and for all, can lead to grossly non-invariant results as soon as we try to apply it to the simplest real problem?"

I would like to answer Ed's questions, at least partially in the context of electrodynamics, by showing that Tom's photons are not the same as Dick's photons — and as for

---

* On leave from University College, London.

Harry's, ...! The different results remarked upon in the quotation from Ed Jaynes occur if the results are the predicted answers to questions referring to "photons". The differences disappear if the questions posed refer to the electromagnetic fields per se.

In the present paper our working hypothesis is conventional quantum electrodynamics and we do not consider semiclassical theory. It is true that the original problem involving the differences in the form of coupling was raised by Lamb in his 1952 paper [3]. In that paper he pointed out the two differing predictions, not for spontaneous emission, but for the induced quenching rate for the $2S_{1/2}$ state of atomic hydrogen in his Lamb shift experiments. His analysis, and many subsequent papers, refer to semiclassical theory. The questions raised in semiclassical theory are different to those in full electrodynamics — classical as well as quantum. This is clear since in electrodynamics not only do the charges obey the dynamical equations but the electromagnetic fields must satisfy Maxwell's equations. In addition to not encompassing semiclassical use of the two couplings this work does not consider any gauge transformations. The analysis is made entirely within the Coulomb gauge: $\mathrm{div}\,\vec{A} = 0$. For a well defined propagation direction $\vec{k}$ we have $\vec{k} \cdot \hat{e} = 0$ where $\hat{e}$ is the direction of polarization. Gauge transformations change only the longitudinal and scalar potentials and here we want the scalar potential to be just the Coulomb potential that binds the atomic or molecular system[1].

The only process analyzed here is spontaneous emission so that the initial state is one in which the atom/molecule is excited, say in the state $|p>$ with energy E_p and the electromagnetic field is in its ground state. Note we do not say the radiation field is the no-photon state. This is because we don't know if there is none of Tom's photons or none of Dick's! There is a Schrödinger wave-functional that represents the ground state of the radiation field in the "configuration space" labeled by the vector potential $\vec{A}(\vec{r})$. It is

$$\Psi_0[\vec{A}] = N \exp\left\{ -\frac{1}{16\pi^3\hbar c} \int \frac{\mathrm{curl}\,\vec{A}(\vec{r}) \cdot \mathrm{curl}\,\vec{A}(\vec{r'})}{|\vec{r} - \vec{r'}|^2} d^3r\, d^3r' \right\} \qquad (1)$$

It is not necessary to use Fock space to specify the state. To simplify the calculations we ignore recoil and assume the atoms are fixed.

2. The Textbook Theory

In the Schrödinger picture the initial wave function is assumed to be

$$\Psi(t = 0) = |p\rangle\, |\,\mathrm{vacuum}\rangle \qquad (2)$$

which develops to the state

$$\Psi(t) = C_0(t)\, |p\rangle\, |\,\mathrm{vacuum}\rangle + \sum_{\mathrm{modes}} c_{\mathrm{mode}}(t)\, |g\rangle\, |1; \mathrm{mode}\rangle + ... \qquad (3)$$

---

[1]Some authors write that the difference between the interactions is a gauge change. This can just squeeze through what would otherwise be an abuse of language if, within semiclassical theory, no spatial dependence of the fields is considered, i.e. the electric dipole approximation is made and only a single atom is considered. Then $\mathrm{div}\,\vec{A} = 0$ is an empty equation and, on the other hand, the Coulomb scalar potential can be transformed by adding $-\vec{r} \cdot \vec{E}(t)$.

The line shape for the transition g← p with energy E_{po} is found by considering

$$\lim_{t\to\infty} \sum_{\text{'modes'}} |c_{\text{mode}}(t)|^2 \tag{4}$$

as a function of mode frequency $\omega = ck$. The sum 'modes' is restricted to those modes $(\vec{k}, \lambda)$ with $\hbar ck \approx E_{po}$. The equivalent computation in the Heisenberg picture is to consider the number operator sum

$$\sum_{\text{'modes'}} n(\vec{k}, \lambda; t) = \sum_{\text{'modes'}} a^\dagger(t)a(t) \tag{5}$$

as a function of ω for the stationary distribution that develops at large t. A restricted choice of states out of the full expansion, equation (3), makes the calculation feasible and these chosen states are often called essential states.

Clearly, to determine (5) we need the equation of motion for the $a(t)$ operators. This, for a system Hamiltonian H, is

$$\dot{a}(t) = -\frac{i}{\hbar}[a(t), H] \tag{6}$$

Now

$$H = H_{\text{EMITTER}} + H_{\text{RADIATION}} + H_{\text{EMIT-RAD}}^{\text{INT}} \tag{7}$$

and it is here that the bête noir, the precise form of the interaction Hamiltonian, surfaces. It has been known for 35 years [4] that if

$$H_{\text{EMIT-RAD}}^{\text{INT}} = -\vec{\mu} \cdot \sum_{\text{modes}} \sqrt{\frac{2\pi\hbar ck}{V}} i(\vec{e}ae^{i\vec{k}\cdot\vec{R}} - \vec{e}^* a^\dagger e^{-i\vec{k}\cdot\vec{R}}) \tag{8}$$

then

$$\sum_{\text{'modes'}} n(\vec{k}, \lambda; t) \to \frac{(2|\mu|^2/3\hbar c^3)\omega^3 d\omega}{\pi[(\omega_E - \omega)^2 + (\gamma/2)^2]}; \tag{9}$$

but, if

$$H_{\text{EMIT-RAD}}^{\text{INT}} = \frac{e\vec{p}\cdot\vec{A}(\vec{R})}{mc} + \frac{e^2\vec{A}^2(\vec{R})}{2mc^2} \tag{10}$$

then

$$\sum_{\text{'modes'}} n(\vec{k}, \lambda; t) \to \frac{(2|\mu|^2/3\hbar c^3)\omega\omega_E^2 d\omega}{\pi[(\omega_E - \omega)^2 + (\gamma/2)^2]} \tag{11}$$

In both (9) and (11) the mode sums are over those modes where $ck = \omega$ is close to $E_{po}/\hbar = \omega_E$ by an amount $d\omega$. The predicted lines are (differently) skewed Lorentzians.

3. Paradox Resolved

How can it be that the two number distributions (9) and (11) are different? If after doing some counting we have different numbers it must be because we are counting different things. There is no paradox if we have different kinds of photons and the results (9) and (11) are predictions for Dick's photons and Tom's photons respectively. In the next section we replace the definition of line shape (4)-(5) by one involving the electromagnetic fields themselves; then both interactions predict the same result. The photon concept is only a tool to solve for the fields. The fields themselves are independent of any method used to determine them and there is no ambiguity if in the specification of a problem only fields are used.

Before we show why the number operators coming from (8) and (10) are different it is instructive to note that there are more alternatives. It is well-known that using the interaction (10) the velocity dipole matrix elements must be converted to length dipole form, using

$$\vec{p}^{mn} = \frac{i}{\hbar} E_{mn} \vec{q}^{mn} \tag{12}$$

to obtain the Einstein A-coefficient in terms of $|\mu|^2$. If one were to use acceleration dipole matrix elements (with acceleration operator $\vec{f}$) we obtain

$$\vec{f}^{mn} = -\frac{1}{\hbar^2} E_{mn}^2 \vec{q}^{mn} \tag{13}$$

and a line shape

$$\frac{(2|\mu|^2/3\hbar c^3)(\omega_E^4/\omega)d\omega}{\pi[(\omega_E - \omega)^2 + (\gamma/2)^2]} \tag{14}$$

would result from

$$H_{\text{EMIT}-\text{RAD}}^{\text{INT}} = -\frac{e}{mc^2}\vec{f} \cdot \vec{Z}(\vec{R}) \tag{15}$$

In (15) $\vec{Z}$ is the Hertz vector for the electromagnetic field. This coupling results from the Kramers-Henneberger transformation [5]. We could call the photons counted in arriving at the distribution (14) Harry's photons!

In writing equation (9) for the minimal coupling form of H^{INT} we used the vector potential $\vec{A}(\vec{r})$ explicitly and not it's mode expansion

$$\vec{A}(\vec{r}) = \sum_{\text{modes}} \sqrt{\frac{2\pi\hbar c}{Vk}} \left(\vec{e} a e^{i\vec{k}\cdot\vec{r}} + \vec{e}^* a^\dagger e^{-i\vec{k}\cdot\vec{r}} \right) \tag{16}$$

On the other hand, in equation (8) for the dipole-length coupling interaction we used the mode expansion

$$\Pi(\vec{r}) = -\frac{1}{4\pi c} \sum_{\text{modes}} \sqrt{\frac{2\pi\hbar ck}{V}} i(\vec{e} a e^{i\vec{k}\cdot\vec{r}} - \vec{e}^* a^\dagger e^{-i\,\vec{k}\cdot\vec{r}}) \tag{17}$$

for the "momentum field" $\vec{\Pi}(\vec{r})$ canonically conjugate to $\vec{A}(\vec{r})$. The canonical commutation relations are

$$[\Pi_i(\vec{r}), \vec{A}_j(\vec{r'})] = -i\hbar\delta_{ij}^\perp(\vec{r} - \vec{r'}) \tag{18}$$

It was a deliberate choice *not* to use an explicit Maxwell field since many quantum opticians would use $-\vec{\mu} \cdot \vec{E}$ which is incorrect. The canonical momentum conjugate to a coordinate need not be the kinetic momentum. This is well known in particle mechanics where $p = m\dot{q}$ only if the Lagrangian has no terms involving $\dot{q}$ other than the kinetic energy $T = \frac{1}{2}m\dot{q}^2$. It is also familiar in electrodynamics where Larmor showed that for a charge e in a magnetic field $\vec{p} = m\vec{q} - (e/c)\vec{A}$. To determine the relationship between $\dot{\vec{A}}$ and Π we use the Hamilton-Heisenberg equation

$$\dot{\vec{A}}(\vec{r}) = \frac{\partial h}{\partial \vec{\Pi}(\vec{r})} \tag{19}$$

where h is the Hamiltonian density in canonical form. If H depends on $\vec{\Pi}$ in $H_{\text{RADIATION}}$ and nowhere else then

$$\dot{\vec{A}}(\vec{r}) = 4\pi c^2 \vec{\Pi}(\vec{r}) \tag{20}$$

i.e.

$$\vec{\Pi}(\vec{r}) = -\frac{1}{4\pi c}\vec{E}^{\perp}(\vec{r}) \tag{21}$$

since the transverse electric field is $\vec{E}^{\perp} = -\frac{1}{c}\dot{\vec{A}}$. However if H has a dependency on $\vec{\Pi}(\vec{r})$ in H^{INT} in addition to $H_{\text{RADIATION}}$ Π is not the transverse electric field. In the case of the uncoupling in equation (9) we find

$$\dot{\vec{A}}(\vec{r}) = 4\pi c^2 \vec{\Pi}(\vec{r}) + 4\pi c \vec{\mu} \cdot \vec{\delta}^{\perp}(\vec{r} - \vec{R}) \tag{22}$$

The last term in (22) is the localized dipole approximation to the transverse polarization $\vec{P}^{\perp}(\vec{r})$. If we use the multipolar Hamiltonian without going to the dipole approximation (22) is replaced by

$$\dot{\vec{A}}(\vec{r}) = 4\pi c^2 \Pi(\vec{r}) + 4\pi c \vec{P}^{\perp}(\vec{r}) \tag{23}$$

which yields

$$-4\pi c\vec{\Pi}(\vec{r}) = \vec{E}^{\perp}(\vec{r}) + 4\pi \vec{P}^{\perp}(\vec{r}) = \vec{D}(\vec{r}) \tag{24}$$

Thus the conjugate field in the multipolar theory is Maxwell's displacement vector not the transverse electric field [6]. We note that for neutral atoms and molecules $\text{div}\vec{D} = 0$ and it is unnecessary to write $\vec{D}^{\perp}$. The multipolar Hamiltonian has an interaction term

$$-\int \vec{P}(\vec{r}) \cdot \vec{D}(\vec{r}) d^3\vec{r} \tag{25}$$

which in electric dipole approximation for an atom at $\vec{R}$ reduces to (8), i.e.,

$$-\vec{\mu} \cdot \vec{D}(\vec{R}) \tag{26}$$

The full Hamiltonian, with no approximations, also has terms in $\text{curl}\vec{A}$. The linear contributions involve the magnetic multipoles and the quadratic terms are required to analyze diamagnetism.

It is now clear than the annihilation and creation operators for minimal and multipolar Hamiltonian are different. Since

$$\vec{E}^{\perp}(\vec{r}) = \vec{D}(\vec{r}) - 4\pi \vec{P}^{\perp}(\vec{r}) \tag{27}$$

the projection of (27) onto transverse plane-waves modes yields

$$a_{\text{MIN}} = a_{\text{MULT}} + i\sqrt{\frac{2\pi}{\hbar ckV}} \int \vec{P}(\vec{r}) \cdot \vec{e}^{\,*}(k)e^{-i\vec{k}\cdot\vec{r}} \tag{28}$$

which, in electric dipole approximation for a single atom at $\vec{R}$, is

$$a_{\text{MIN}} = a_{\text{MULT}} + i\sqrt{\frac{2\pi}{\hbar ckV}} e_i^* \mu_i e^{-i\vec{k}\cdot\vec{R}} \tag{29}$$

Thus

$$
\begin{aligned}
n_{\text{MIN}} &= a_{\text{MIN}}^\dagger a_{\text{MIN}} \\
&= \left(a_{\text{MULT}}^\dagger - i\sqrt{\frac{2\pi}{\hbar ckV}} e_i \mu_i e^{i\vec{k}\cdot\vec{R}}\right)\left(a_{\text{MULT}} + i\sqrt{\frac{2\pi}{\hbar ckV}} e_i^* \mu_i e^{-i\vec{k}\cdot\vec{R}}\right) \\
&= n_{\text{MULT}} + \left(-i\sqrt{\frac{2\pi}{\hbar ckV}} e_i \mu_i e^{i\vec{k}\cdot\vec{R}} a_{\text{MULT}} + \text{h.c.}\right) + \frac{2\pi}{\hbar ckV} e_i \mu_i e_j^* \mu_j
\end{aligned}
\tag{30}
$$

To find the long-term stationary value relationship between n_{MIN} and n_{MULT} it is necessary to determine an approximation for $a_{\text{MULT}}(t)$ to insert into equation (30). This can be done since the solution of equation (6) for $a_{\text{MULT}}(t)$ can be written as a formal integral over the atomic dipole moment

$$a_{\text{MULT}}(t) = a_{\text{MULT}}(0) + \frac{1}{\hbar}\sqrt{\frac{2\pi\hbar ck}{V}} e_i^* e^{-i\omega t} e^{i\vec{k}\cdot\vec{R}} \int_0^t \mu_i(t')e^{i\omega t'}\,dt' \tag{31}$$

If (31) is used in equation (30) which is then evaluated to order $|\mu|^2$, we find the approximate relationship (32).

$$n_{\text{MIN}} = n_{\text{MULT}} + \frac{\vec{\mu}\cdot\vec{e}\vec{\mu}\cdot\vec{e}^{\,*}}{\hbar\frac{2\pi}{V}\left[\frac{1}{\omega_E - \omega - i\gamma/2} + \frac{1}{\omega_E - \omega + i\gamma/2} + \frac{1}{\omega}\right]} \tag{32}$$

Thus

$$
\begin{aligned}
\sum n_{\text{MIN}} &= \sum n_{\text{MULT}} + \frac{(2|\mu|^2/3\hbar c^3)\omega d\omega[2\omega(\omega_E - \omega) + (\omega_E - \omega)^2]}{\pi[(\omega_E - \omega)^2 + (\gamma/2)^2]} \\
&= \left(\frac{\gamma}{2}\right)\frac{\omega_E^2 \omega d\omega}{\pi[(\omega_E - \omega)^2 + (\gamma/2)^2]}
\end{aligned}
\tag{33}
$$

This gives a completely rational explanation of the difference between the results (9) and (11). Photons in the minimal coupling quantization scheme are different from photons in the multipolar formalism. This difference is a function of the source polarization.

The effect of this difference falls away as the inverse third power of the distance from the source; this reflects the non-local nature of the *transverse δ-dyadic*. $\vec{E}^\perp$ and $\vec{D}$ are identical away from the atoms/molecules and the electromagnetic fields in free space obey, of course, the homogeneous Maxwell equations and there are no distinctions for Tom and Dick to worry about.

4. Comments on the Maxwell Field Operators

Although, as we have seen, there are vital distinctions between a and $a^\dagger$ for the minimal and multipolar cases both Hamiltonians lead to the same equations for the electromagnetic field operators. These are Maxwell's equations. The difference between the two Hamiltonians is seen in which the Maxwell field operator equals the canonical momentum field. The Hamilton-Heisenberg equations

$$\dot{\vec{A}}(\vec{r}) = \frac{\partial h}{\partial \vec{\Pi}(\vec{r})}, \quad \dot{\vec{\Pi}}(\vec{r}) = -\frac{\partial h}{\partial \vec{A}(\vec{r})} \tag{34}$$

lead to (35) for $H_{\text{MULTIPOLAR}}$ and to (36) for H_{MINIMAL}.

$$\left(\text{curlcurl} + \frac{1}{c^2}\frac{\partial^2}{\partial t^2}\right)\vec{D}(\vec{r}) = 4\pi\,\text{curlcurl}\vec{P}(\vec{r}) \tag{35}$$

$$\left(\text{curlcurl} + \frac{1}{c^2}\frac{\partial^2}{\partial t^2}\right)\vec{E}^{\perp}(\vec{r}) = -\frac{4\pi}{c^2}\frac{\partial^2\vec{P}^{\perp}(\vec{r})}{\partial t^2} \tag{36}$$

These are equivalent since $\vec{D} = \vec{E}^{\perp} + 4\pi\vec{P}^{\perp}$ which is the connection between the canonical momentum in the two formalisms. For completeness we write down the second order differential equation for three other electromagnetic field operators[2].

$$\left(\text{curlcurl} + \frac{1}{c^2}\frac{\partial^2}{\partial t^2}\right)\vec{A}(\vec{r}) = \frac{4\pi}{c}\frac{\partial\vec{P}^{\perp}(\vec{r})}{\partial t} \tag{37}$$

$$\left(\text{curlcurl} + \frac{1}{c^2}\frac{\partial^2}{\partial t^2}\right)\vec{E}(\vec{r}) = -\frac{4\pi}{c^2}\frac{\partial^2\vec{P}^{\perp}(\vec{r})}{\partial t^2} \tag{38}$$

$$\left(\text{curlcurl} + \frac{1}{c^2}\frac{\partial^2}{\partial t^2}\right)\vec{B}(\vec{r}) = \frac{4\pi}{c}\frac{\partial}{\partial t}\text{curl}\vec{P}(\vec{r}) \tag{39}$$

Solutions for all the wave equations (35) - (39) can be written

$$F_i(\vec{r}, t) = F_i^0(\vec{r}, t) + F_i^s(\vec{r}, t) \tag{40}$$

where $F_i^0(\vec{r}, t)$ in an appropriate solution of the corresponding homogeneous equation and $F_i^s(\vec{r}, t)$ is the contribution from the source. For a field in the Heisenberg picture which identifies with the corresponding Schrödinger operator at $t = 0$, the homogeneous solution is the freely developing vacuum field operator. The source term can be found in using

---

[2]In the present work we are not taking magnetic properties of sources into account, and do not introduce the magnetic auxiliary field $\vec{H}$.

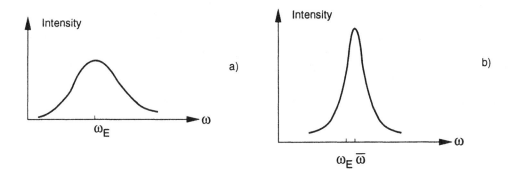

Fig. 1. Lineshapes for (a) emitter and (b) absorber.

Green's function techniques. For example, for a dipole polarization source localized at the origin

$$D_i^s(\vec{r}, t) = (-\nabla^2 \delta_{ij} + \nabla_i \nabla_j) \frac{\mu_j(t - r/c)}{r} \quad (r < ct)$$
$$= 0 \quad (r > ct) \tag{41}$$

a fully retarded causal solution. It should be noted that causal solutions exist for $\vec{D}, \vec{E}$ and $\vec{B}$, however there are precursors in the solutions for $\vec{A}$ and $\vec{E}^\perp$ which arise because the sources for these fields are non-local.

5. A Suggested Direct Operational Meaning of Line Shape

In order to exclude the idea of "photon" from a definition of the line shape we suggest the following gedanken experiment. Set up a series of absorbers each tuned to a different frequency; albeit all the frequencies chosen close to the emitter central frequency ω_E. For the spectral line of the emitter to be resolved it will be necessary for the absorbers' lines to be sharper than the emitter's spectral distribution. (See Figure 1.)

Furthermore, for there to be any appreciable absorption by a particular absorber it is necessary that the radiation shining on it is broad; this is so because the rate of absorption is $B\rho(\omega)$ where B is the Einstein B-coefficient and $\rho(\omega)$ the energy density per unit frequency. It is thus consistent to require the width of the emitter to be larger than the widths of the absorbers. As an approximation we assume that the absorbers are sharp with a spectrum $\delta(\omega - \overline{\omega})$. The prediction is that, for a separation between emitter and absorber $R(>> c/\omega_E)$, the probability that absorber $\overline{\omega}$ is triggered is

$$P(t) \to \frac{|\vec{\mu}(A)|^2}{(\omega_E - \overline{\omega})^2 + (\gamma/2)^2} \times \frac{\text{function of emitter}}{\text{R}^2} \tag{42}$$

The essential result is that the "function of emitter" is independent of $\overline{\omega}$ so that the line shape, as a function of $\overline{\omega}$, is Lorentzian.

It is noteworthy that this prediction is independent of the coupling used between the emitter at the radiation, $-\vec{\mu} \cdot \vec{D}(\vec{R}_E)$ or $e\vec{p} \cdot \vec{A}(\vec{R}_E)/mc + e^2 \vec{A}^2(\vec{R}_E)/2mc^2$. The clear reason for this is that the absorber sees the *electromagnetic field* of the emitter - not individual photons - and, as we have emphasized in Section 4, the Maxwell fields are independent of coupling.

6. Outline of Calculation

To follow the program suggested in Section 5 it is necessary to extend the dynamics to include the absorber and its interaction with the emitter and the radiation. Thus the Hamiltonian for the extended system is

$$H = H_{\text{EMIT}} + H_{\text{RAD}} + H_{\text{EMIT}-\text{RAD}}^{\text{INT}} + H_{\text{ABS}} + H_{\text{ABS}-\text{RAD}}^{\text{INT}} + H_{\text{EMIT}-\text{ABS}}^{\text{INT}} \qquad (43)$$

Whether or not the final interaction term, a direct coupling between emitter and absorber, is present depends upon the formalism adopted [7]. In the multipolar formalism it is absent but for minimal coupling it is the inter-Coulombic interaction energy. This does not affect the final result. It suffices to take a very simple model for the absorber, which is sharp, namely a 2-level system with Hamiltonian

$$H_{\text{ABS}} = \frac{\hbar\bar{\omega}}{2} \begin{bmatrix} 1 & 0 \\ 0 & 1 \end{bmatrix} + \frac{\hbar\bar{\omega}}{2} \begin{bmatrix} 1 & 0 \\ 0 & -1 \end{bmatrix} \qquad (44)$$

The dynamical evolution of the system is analyzed in the Heisenberg picture and the state of the absorber is $\begin{bmatrix} 0 \\ 1 \end{bmatrix}$, i.e. it's ground state. The projector to the excited state is $\frac{1}{2}(I + \sigma_z)$ so that the probability, at time t, that the absorber is excited is

$$P(t) = [0\ 1]\frac{I + \sigma_z(t)}{2} \begin{bmatrix} 0 \\ 1 \end{bmatrix} \qquad (45)$$

It is unnecessary to be precise about $H_{\text{ABS}-\text{RAD}}^{\text{INT}}$ to obtain the main result. Let us assume

$$H_{\text{ABS}-\text{RAD}}^{\text{INT}} = g \begin{bmatrix} 0 & 1 \\ 1 & 0 \end{bmatrix} \times (\text{Maxwell Field}) = g\sigma_x F(\vec{R}_A) \qquad (46)$$

This elementary detector theory (no photons!) predicts that, to order g^2

$$P(t) = \frac{g^2}{\hbar^2} \int_0^t \int_0^t e^{-i\bar{\omega}(t'-t'')} F(\vec{R}_A, t') F(\vec{R}_A, t'') dt' dt'' \qquad (47)$$

In passing we note that there is no intertwining in (47), the full Heisenberg-Maxwell field is operative. Whichever physical electromagnetic field is represented by $\vec{F}(\vec{r}, t)$ it is, equation (40), the sum of the free zero-point field and that field due to the emitter as source. As we have seen in Section 4 the electromagnetic fields are unique — independent of which Hamiltonian is used to calculate them, the minimal coupling or the multipolar coupling. Thus the line shape, as defined here, is the same for both forms of interaction Hamiltonian between the emitter and the radiation. A specific result for the long time value of $P(t)$ will follow if a definite form is assumed for (46). If we use $-\vec{\mu}(A) \cdot \vec{D}(\vec{R}_A)$ for $gF(\vec{R}_A)$, then, from equations (40) and (47),

$$
\begin{aligned}
P(t) = \mu_i(A)\mu_j(A)\frac{1}{\hbar^2} \int_0^t \int_0^t e^{-i\overline{\omega}(t'-t'')} dt' dt'' & \left[D_i^{(o)}(\vec{R}_A,t')D_j^{(o)}(\vec{R}_A,t'') \right] \\
+ D_i^{(o)}(\vec{R}_A,t')D_j^{(s)}(\vec{R}_A,t'') & + D_i^{(s)}(\vec{R}_A,t')D_j^{(o)}(\vec{R}_A,t'') \\
+ \left[D_i^{(s)}(\vec{R}_A,t')D_j^{(s)}(\vec{R}_A,t'') \right] &
\end{aligned}
\tag{48}
$$

The first contribution with two free fields is irrelevant to our purpose as it has no dependence on the emitter! The zero-point fields can excite the absorber virtually over very short times $\approx \omega_E^{-1}$. The three remaining products do contribute to the excitation of the absorber with terms that depend on the state and position of the emitter. From the solution for $\vec{D}^{(s)}(\vec{r},t)$ quoted in equation (41) it is clear that causality is guaranteed: causality in the sense that $P(t) = 0$ for $t < r/c$, where $\vec{R} = \vec{R}_A - \vec{R}_E$. Finally it is possible to obtain an explicit formula for $P(t)$ correct to the order of the square of the dipole moment of the emitter. It is easy to compute the contribution, to this order, from the last, i.e. double source, term in equation (48) It is, for $t > R/c$,

$$
\mu_i(A)\mu_j(A)\mu_k(E)\mu_\ell(E)\frac{1}{\hbar^2} \int_{R/c}^t \int_{R/c}^t e^{-i\overline{\omega}(t'-t'')} dt' dt''
$$

$$
\times (-\nabla^2 \delta_{ik} + \nabla_i \nabla_k)e^{-ik_E(R-ct')} \le -\vec{R}(-\nabla^2 \delta_{j\ell} + \nabla_j \nabla_\ell)\frac{e^{ik_E(R-ct'')}}{R}
\tag{49}
$$

If, as is usual, we impose damping for large t on the oscillating exponentials by adding a small imaginary term to the frequency, this part of the long term value of $P(t)$ is

$$
\frac{\mu_i(A)\mu_j(A)\mu_k(E)\mu_\ell(E)}{\hbar^2[(\omega_E - \overline{\omega})^2 + (\gamma/2)^2]} \left(\frac{\beta_{ik}}{R^3} - \frac{ik_E\beta_{ik}}{R^2} - \frac{k_E^2\alpha_{ik}}{R} \right)
$$

$$
\times \left(\frac{\beta_{j\ell}}{R^3} + \frac{ik_E\beta_{j\ell}}{R^2} - \frac{k_E^2\alpha_{j\ell}}{R} \right)
\tag{50}
$$

where $\beta_{ik} = \delta_{ik} - 3\hat{R}_i\hat{R}_k$ and $\alpha_{ik} = \delta_{ik} - \hat{R}_i\hat{R}_k$. The contribution to $P(t)$ from mixed terms, zero-point and source, is technically more difficult to evaluate. The appropriate form for $D^{(s)}(\vec{r},t)$ is given in our papers [8] and the resulting integral for inserting into (48) is a complicated function of R. However the time dependence of this integral is never resonant with $\overline{\omega}$ and hence to the approximations we are working with there is no contribution to the steady value of $P(t)$. Hence, for large R compared with $\tilde{\lambda} = k_E^{-1}$ the steady state probability is

$$
P(t) \to \frac{\mu_i(A)\mu_j(A)\mu_k(E)\mu_\ell(E)}{\hbar^2[(\omega_E - \overline{\omega})^2 + (\gamma/2)^2]} \frac{\alpha_{ik}\alpha_{j\ell}}{\tilde{\lambda}^4 R^2} ,
\tag{51}
$$

and is Lorentzian as a function of $\overline{\omega}$. For isotropic source and absorber

$$
P(t) \to \frac{2}{9} \frac{|\mu(A)|^2|\mu(E)|^2/\tilde{\lambda}^4}{\hbar^2[(\omega_E - \overline{\omega})^2 + (\gamma/2)^2]} \frac{1}{R^2}
\tag{52}
$$

which was previously discussed following equation (41).

The steady state probability for smaller R, inside the wave-zone region $R \leq \bar{\lambda}$, is interesting since in this region the probability of excitation of the absorber is driven not only by the Maxwell field associated with the excited atom as an actual emitter but by all the associated electrical energy density near it. This latter includes terms that are effective in producing interatomic energies that give rise to the London-Casimir forces. [9]

7. Summary

In non-relativistic quantum electrodynamics, the theory of atoms and molecules interacting with radiation, the ambiguity in predictions that depend on which matter-radiation coupling is used is resolved if the fact that "photon" has a different meaning in the various formalisms is taken into account. If questions are asked that concern the electromagnetic field operators themselves there are no ambiguities. The small differences between minimal and multipolar photons are only apparent near the atomic/molecular sources and, of course, where Maxwell's equations hold in free space the differences vanish. Fingerprints of the sources of the Maxwell field are the distinguishing features between Tom's, Dick's and Harry's quanta.

ACKNOWLEDGMENTS. . This work was completed during a visit to LANL and I thank the Laboratory for the hospitality shown me. Discussions with P.W. Milonni and T. Thirunamachandran are always a pleasure. Finally I am happy to present this paper at the Ed Jaynes celebration at Laramie because his comments on non-relativistic quantum electrodynamics are always of great interest.

REFERENCES

1. A Einstein. Letter to M. Besso, 12th December 1951, published in the Einstein-Besso correspondence 1903-1955; Hermann, Paris, 1972. [Note: free translation by Martin Klein, Les Allen and Peter Knight, Einstein uses 'Lump' which they translate as Tom, Dick or Harry.]
2. E.T. Jaynes, 1976: as cited by Lamb, Schlicher and Scully (see reference 3 following). Quotation appears in D.H. Kobe and A.L. Smirl, Am J. Phys., 46, 628, 1978 from unpublished notes of Jaynes.
3. W.E. Lamb. Phys. Rev. 52, 259, 1952. Semiclassical calculations on the quenching rate have been recently given in: W.E. Lamb, R.R. Schlicher and M.O. Scully, Phys. Rev. A36, 2763, 1987
4. E. Arnous and W. Heitler, Proc. Roy. Soc. A220, 290, 1953; E.A. Power and S. Zienau, Phil. Trans. Roy. Soc., A251, 427, 1959. More recent references in which the line shape problem is addressed include: L. Davidovich. Ph.D. thesis, University of Rochester, N.Y. 1975; L. Davidovich and H.M. Nussenzveig in *Foundations of Radiation Theory and Quantum Electrodynamics* (Plenum, New York, 1980), page 83; V.P. Bykov. Soviet Physics (Uspekhi) 27(B), 631, 1984; P.W. Milonni, R.J. Cook and J.R. Ackerhalt. Phys. Rev. A40, 3764, 1989.
5. H.A. Kramers, Rapport et Discussions, Solvay Congress 1948, Stoops, Bruxelles, 1950; W.C. Henneberger. Phys Rev. Letts, 21, 838, 1968.
6. E.A. Power and T. Thirunamachandran. Phys. Rev. A22, 2894, 1980; Phys. Rev. A26, 1800, 1982 and J. Opt. Soc., B2, 1101, 1985.

7. See, for example, D.P. Craig and T. Thirunamchandran *Molecular Quantum Electrody-namics* (Academic Press, New York & London, 1984), Section 3.6 and Chapter 7.

8. E.A. Power and T. Thirunamachandran, Phys. Rev. <u>A28</u>, 2663, 1983, Phys. Rev. <u>A28</u>, 2671, 1983, and Phys. Rev. <u>A45</u>, 54, 1992.

9. R. Passante and E.A. Power. Phys Rev. <u>A35</u>, 188, 1987.

AN OPERATIONAL APPROACH TO SCHRÖDINGER'S CAT

L. Mandel
Department of Physics and Astronomy
University of Rochester
Rochester, New York 14627

ABSTRACT. It is pointed out that the conclusions drawn from a recent quantum interference experiment with light suggest an operational resolution of the Schrödinger cat paradox.

On this occasion in honor of Prof. E.T. Jaynes, we recall that he devoted some of his research efforts to the interpretation of quantum mechanics, which led him to propose several experimental tests. Although the 'neoclassical' theory he developed was found not to be confirmed by experiment, it nevertheless played a role in encouraging us to think critically about quantum mechanics and to carry out new experiments. This short contribution is concerned with a well-known quantum problem of interpretation.

The quantum paradox known as Schrödinger's cat,[1] in which the cat is cast in a linear superposition of the state of being alive and the state of being dead, has been debated since the beginnings of quantum mechanics. Whereas most physicists are ready to concede the existence of superposition states for a microscopic quantum system like an atom, such states appear to be ruled out for a macroscopic system like a cat by common experience. The question then arises at which level classical concepts take over from quantum mechanical ones.

In a very clear and readable discussion of this and other paradoxes Glauber[2] has pointed out that the noise inevitably associated with the amplification process accompanying the measurement of a microscopic system leaves the cat in a mixed state rather than a pure one. Zurek[3] has argued that the environment surrounding and interacting with the quantum system 'monitors' the latter and thereby causes decoherence of the off-diagonal matrix elements of the density matrix. Superposition states then disappear for the macroscopic system. In the following we adopt an operational point of view, to which one is led by the results of some recent interference experiments with single photons,[4,5] which also suggest a resolution of the cat paradox.

For the sake of consistency, we start by describing the paradox within the domain of quantum optics. A cat is confined within an enclosure containing a device that kills the cat when the device is activated by an electric current, together with a two-level atom. When the atom makes a transition from the upper, excited state $|2\rangle$ to the lower unexcited state $|1\rangle$, it emits a photon. This photon is detected by a nearby photoelectric detector, such as a photomultiplier or avalanche photodiode, with near 100% quantum efficiency. The photocurrent activates the lethal device that kills the cat.

It is clear from the description that there exists a one-to-one correspondence between the excited atomic state $|2\rangle$, the absence of a photon, the absence of a photoelectric current pulse and the live state $|\,\text{🐱}\,\rangle$ of the cat. For the same reason there also exists a correspondence between the lower state $|1\rangle$ of the atom, the emission of the photon, the presence of the photocurrent pulse and the dead state $|\,\text{🐱}\,\rangle$ of the cat (the notation is that of ref. 2).

Now suppose that the atom is in the linear superposition state $|\psi\rangle = \alpha|1\rangle + \beta|2\rangle$, so that its density operator $\hat{\rho}_A$ in the Schrödinger picture is the projector

$$\hat{\rho}_A = (\alpha|1\rangle + \beta|2\rangle)(\alpha^*\langle 1| + \beta^*\langle 2|). \tag{1}$$

In view of the correspondence indicated above, one might then expect the cat to be in the superposition state $\alpha|\,\text{🐱}\,\rangle + \beta|\,\text{🐱}\,\rangle$, with density operator

$$\hat{\rho}_c = (\alpha|\,\text{🐱}\,\rangle + \beta|\,\text{🐱}\,\rangle)\left(\alpha^*\langle\,\text{🐱}\,| + \beta^*\langle\,\text{🐱}\,|\right). \tag{2}$$

We have therefore arrived at a coherent superposition state for a macroscopic object, viz. the cat.

Before proceeding further let us briefly recall the experiment reported in references 4 and 5. Two non-linear crystals NL1 and NL2 act as parametric down-converters (see Fig. 1), such that a pump photon of frequency ω_p incident on crystal j (j=1,2) results in the emission of a pair of simultaneous signal s_j and idler i_j photons, whose frequencies add up to ω_p. Light from the pump laser is divided by the pump beam splitter BS_p into two mutually coherent beams, and directed to NL1 and NL2, as shown. The two crystals are oriented so that the paths of idler 1 and idler 2 are aligned and overlap, while the signal 1 and signal 2 beams are mixed by the movable output beam splitter BS_0 and detected by photodetector D_s. Interference between s_1 and s_2 shows up as a periodic variation of the photodetection probability P_s when BS_0 is translated as shown through a distance of order 1 or 2 optical wavelengths. It is observed that s_1 and s_2 exhibit interference so long as i_1 and i_2 are perfectly aligned. However if i_1 and i_2 do not overlap, or if i_1 is blocked and prevented from reaching NL2, the interference disappears.

Quantum mechanics readily accounts for these observations. Let us oversimplify and treat the signal and idler fields as monochromatic, and suppose that the broken line between NL1 and NL2 represents an attenuator of complex transmissivity $\mathfrak{T}$. Then, by writing down the parametric interaction $\hat{H}_I(t)$ between pump and signal modes, using this to construct the time-evolution operator $\hat{U}(t,0)$ in the interaction picture, letting this act on the initial vacuum state $|vac\rangle_{s_1}|vac\rangle_{i_1}|vac\rangle_{s_2}|vac\rangle_{-i_2}$, and tracing over the idler modes i_1, i_2, we obtain the following density operator $\hat{\rho}_{s_1 s_2}$ for the state of the signal modes[4,5] (up to terms with two photons)

$$\begin{aligned}
\hat{\rho}_{s_1 s_2} = {}&|vac\rangle_{s_1 s_2}\ _{s_1 s_2}\langle vac| \\
&+ |f_1|^2 \langle I_1\rangle |1\rangle_{s_1}|vac\rangle_{s_2}\ _{s_2}\langle vac|_{s_1}\langle 1| + |f_2|^2\langle I_2\rangle|vac\rangle_{s_1} \\
&\qquad \times |1\rangle_{s_2}\ _{s_2}\langle 1|_{s_1}\langle vac| \\
&+ \mathfrak{T}f_1^* f_2\langle V_1(t)V_2^*(t+\tau_0)\rangle e^{-i\theta_0}|1\rangle_{s_1}|vac\rangle_{s_2}\ _{s_2}\langle 1|_{s_1}\langle vac| + h.c.
\end{aligned} \tag{3}$$

Here $V_1(t), V_2(t)$ are the complex amplitudes of the pump waves, treated classically, at NL1 and NL2, respectively, and $\langle I_1\rangle, \langle I_2\rangle$ are the corresponding mean light intensities.

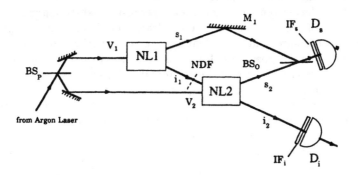

Fig. 1 Outline of the experiment reported in references 4 and 5.

$|f_1|^2, |f_2|^2$ are fractions of the incident pump beams which are down-converted, τ_0 is the propagation time between NL1 and NL2, and θ_0 is the corresponding phase shift. The first three contributions to $\hat{\rho}_{s_1 s_2}$ are diagonal in photon number states. Therefore only the remaining off-diagonal terms can give rise to interference in the detection probability $\langle \hat{E}_s^{(-)} \hat{E}_s^{(+)} \rangle$, where

$$\hat{E}_s^{(+)} = \hat{a}_{s_1} e^{i\theta_1} + \hat{a}_{s_2} e^{i\theta_2} \tag{4}$$

represents the field at the detector D_s. θ_1 and θ_2 are phase shifts associated with the propagation from NL1 to D_s and from NL2 to D_s. We notice that the interference terms are proportional to $|\Im|$ and vanish when $\Im = 0$ and i_1 is prevented from reaching NL2.

Although the formalism correctly predicts what is observed, it is more instructive to try and account for the experimental results in physical terms involving the distinguishability of the path or the source of any detected photon.[4-7] As has recently been shown, there is a precise, quantitative relation between the concepts of mutual coherence and indistinguishability.[7] In the absence of any mutual coherence, the density operator $\hat{\rho}_{s_1 s_2}$ is diagonal in photon number states, and therefore there exists, in principle, the possibility of identifying the source or the state by some auxiliary observation.

When i_1 and i_2 are perfectly aligned, the two sources NL1 and NL2 of each detected photon are indeed intrinsically indistinguishable. However, misalignment or the blocking of i_1 makes it possible to identify the source by an auxiliary measurement with detector D_i (see Fig. 1) that in no way disturbs s_1 and s_2.[4,5] For if D_i registers a photon at the same time as D_s, then the photons must have originated in NL2, whereas if D_i does not register a photon when D_s does, then the photon must have come from NL1. The possibility of identifying the source of each detected photon wipes out the intrinsic indistinguishability and therefore also the mutual coherence of s_1 and s_2,[7] no matter whether the auxiliary measurement is actually carried out or not. There is no need for the introduction of a physical disturbance of the s_1, s_2 quantum system.[6,8] The state of the photon at detector

D_s is then represented by a diagonal density operator in which the off-diagonal matrix elements associated with the interference vanish.

Let us now apply the lessons from the foregoing to the Schrödinger cat experiment. Then we conclude that the existence of a coherent superposition state depends on the intrinsic indistinguishability of the two possible states. Whenever there exists the possibility, in principle, of making a measurement that identifies the state, without physically disturbing the system, the coherent superposition is lost, and the density operator becomes diagonal.

It is now apparent that the state of the photomultiplier current, which typically contains thousands or even millions of electrons, lies essentially in the classical domain, i.e. the photocurrent can be measured, in principle, without disturbing the quantum system. According to the foregoing operational argument the density operator representing the photodetection current is therefore diagonal. As the state of the cat is determined by the state of the photocurrent, the density operator $\hat{\rho}_c$ of the cat is diagonal also, and we may write

$$\hat{\rho}_c = |\alpha|^2 \; | \; \text{🐱} \; \rangle \langle \; \text{🐱} \; | + |\beta|^2 | \; \text{🐱} \; \rangle \langle \; \text{🐱} \; |. \tag{5}$$

Alternatively, we may ignore the detector and look directly into the cat's eye to see whether it is alive or not, which implies the existence of the diagonal form of $\hat{\rho}_c$ by virtue of the distinguishability of the two cat states. There is therefore no coherent superposition and no interference.

Finally, we emphasize, as became apparent from the interference experiments in references 4-6, that it is not necessary for the observations described in the previous paragraph actually to be carried out, in order to destroy the macroscopic superposition state. The mere feasibility of the observations, in principle, is sufficient to achieve the same effect, because the state reflects not only what is known, but also what is knowable in principle. It follows that all classical systems, viz. those which are observable without being disturbed, can not be found in a superposition state.

This operational test for intrinsic indistinguishability appears to be applicable to any macroscopic system, and it will generally lead to the same conclusion, with some well-known exceptions like superconductivity and the states discussed by Yurke and Stoler.[9]

ACKNOWLEDGMENTS. This work was supported by the National Science Foundation and by the U.S. Office of Naval Research.

REFERENCES

1. E. Schrödinger, *Letters on Wave Mechanics* (Philosophical Library, New York, 1967).
2. R.J. Glauber, in *Frontiers in Quantum Optics*, eds. E.R. Pike and S. Sarkar (Adam Hilger, Bristol, 1986) p. 534.
3. W.H. Zurek, *Physics Today* **44**, 36 (1991).
4. X.Y. Zou, X.J. Wang and L. Mandel, *Phys. Rev. Lett.* **67**, 318 (1991).
5. L.J. Wang, X.Y. Zou and L. Mandel, *Phys. Rev.* **A44**, 4614 (1991).
6. A.G. Zajonc, L.J. Wang, X.Y. Zou and L. Mandel, *Nature* (London) **353**, 507 (1991).
7. L. Mandel, *Opt. Lett.* **16**, 1882 (1991).
8. M.O. Scully, B.-G. Englert and H. Walther, *Nature* **351**, 111 (1991).
9. B. Yurke and D. Stoler, *Phys. Rev. Lett.* **57**, 13 (1986).

THE CLASSICAL LIMIT OF AN ATOM

C. R. Stroud, Jr.
The Institute of Optics
University of Rochester
Rochester, New York 14627-0187

ABSTRACT. A series of recent experiments and calculations are described that study the classical limit of an atom. The classical limit of an atom is defined to be the quantum mechanical state that most nearly approximates the ideal of a classical particle traveling in a Kepler orbit. It is found that correspondence between the quantum and classical descriptions is always in the form of ensembles of many realizations so long as the electron is taken to be a point particle. Even in the limit of a wave packet made up of states with all quantum numbers large there are distinctly quantum interference features in the evolution of the quantum ensemble. However, when the classical theory is modified by allowing within the ensemble only the Kepler orbits corresponding to energies and angular momenta allowed by Bohr's old quantum theory, most of the interference features are produced by the ensemble of classical particles.

Ed Jaynes has been characterized variously as a physicist, a statistician, an inventor, a free thinker, a teacher, a mentor, and a rabble rouser. It is hard to dispute his right to any of these mantles he might wish to claim, but it is equally hard to find a finite set of descriptors that do him justice. From my own selfish point of view his most important contribution has been to act as a source of ideas and new ways of looking at some long-standing and fundamental problems. It is not difficult in looking through the titles of the contributions to this symposium to see that there are a lot of us who have been profoundly influenced in our choice of research topics by Ed's ideas.

As a student I was frustrated that in a half an hour discussion Ed would throw out at least a half a dozen exciting new ideas each of which I wanted to follow up immediately. Invariably I got carried away with working on the first one and never got back to the others. Finally, I learned to return to my office after one of these meetings and write out all of the ideas before starting to work on any of them. After a while I developed quite a notebook of these ideas. If I had a difficult time remembering the ideas that he suggested, Ed did not suffer from similar lapses. At more than one party, years after I finished my thesis, Ed has come up and asked when I was going to finish one of those problems he had suggested. In fact, I am not alone in this regard. I have heard him upbraid others of his former students saying that after twenty or thirty years it was really time that they should finish that calculation he had suggested.

The subject of my lecture is an outgrowth of one of those problems that Ed suggested long ago. In the quest to find a theory to supplant the Copenhagen quantum mechanics he suggested that classical mechanics has many desirable features that one would hope to find in the new theory. Of course, classical mechanics itself does not properly describe atomic physics, but, he suggested, the proper theory may lie somewhere between the classical and quantum theory. Neoclassical theory was one attempt to explore this idea. Unfortunately, quantum optical experiments did not confirm the predictions of the new theory. The failures of Neoclassical theory do not prove that Ed's original conjecture was wrong. A theory that is in agreement with experimental observation may still lie somewhere between the full quantum theory and classical mechanics.

One way to explore this idea is to study the regime in which classical and quantum theories should converge - the classical limit of quantum mechanics. In this regime differences between the theories are small and the range of possible alternatives is sharply limited. A great deal of work in this area was carried out in the very earliest days of quantum mechanics. A particular concern at that time was the apparent absence of orbits in the quantum theory of the hydrogen atom. The Hamiltonian is just that of the classical Kepler system, but the elliptical orbits are not evident in the eigenstate wave functions. Schrödinger and Lorentz (Schrödinger, 1926), (Przibram, 1967) studied superpositions of these eigenstate wave functions to see whether they could construct a state in the form of a spatially localized wave packet that traveled along the classical Kepler orbit. In the course of this work they discovered the coherent state of the harmonic oscillator. This state has exactly the property that it remains forever a minimum-uncertainty wave packet whose centroid follows exactly the classical trajectory. They found that in the case of the hydrogen atom there is no such coherent state - any wave packet will inevitably disperse. Following this discovery Heisenberg suggested that the whole idea of orbits had no place in quantum theory and should be given up altogether.

There was relatively little further research on these problems until quite recently when picosecond and femtosecond laser pulses became available as a laboratory tool for exciting wave packet states of atoms and molecules. Due to its finite duration a short laser pulse may have a transform bandwidth of hundreds of wave numbers. When such a pulse is used to excite an atom from its ground state up to the closely spaced Rydberg states, a coherent superposition of many states is produced. This superposition is in the form of a wave packet, localized in the radial coordinate, oscillating periodically from near the nucleus out to a distance of nearly macroscopic size. In the case of an optical pulse tuned so that its central frequency resonantly excites the transition from the ground state to the state with principal quantum number $n = 100$, the radius of the orbit is approximately one half a micron, almost into the regime of size of visible objects. The period of oscillation of the wave packet is just the period of the Kepler orbit of a classical particle with the energy corresponding to the quantum mechanical $n = 100$ state.

The resulting state does not look much like a classical particle oscillating in a Kepler orbit. The period and the radius of the orbit are right, but the localization is only in the radial coordinate, not in the angular coordinates. That this is the case is apparent if we recall that the electric dipole selection rules apply to the optical transition. The angular momentum quantum number changes by one in such a transition. Thus, if the atomic ground state is an S-state, then the states in the wave packet superposition will be P-states with various principal quantum numbers. Each of these states will then have an angular dependence that is the square of a single spherical harmonic; in the case of the $l = m = 1$ state this dependence would be $\sin^2 \theta$. The wave packet is a spherical shell undergoing a breathing motion from the nucleus out to classical orbital turning point and back.

We can connect this wave packet state directly to the classical orbit, but to do this we must generalize our classical theory to treat an ensemble of noninteracting particles. If we imagine an ensemble of classical Kepler orbits with their principal axes distributed according to the $\sin^2 \theta$ distribution, with each particle moving in phase along its orbit so that all are at the nucleus at the same time, and all are at the outer turning point at the same time, then the quantum wave packet motion is accurately reproduced – at least for the first few orbits, but more about that later.

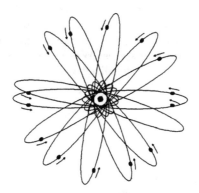

Fig. 1. The ensemble of classical Kepler orbits that make up the radially localized wave packet. The major axes of the ellipses are randomly distributed over all directions in space with a $\sin^2 \theta$ distribution, but each realization has the same orbital phase.

These radially localized wave packets have been studied experimentally by several groups: (ten Wolde, *et al*,1988), (Yeazell, *et al,* 1988, 1990, 1991), (Meacher, *et al,* 1991). The experimental technique makes use of the classical nature of the evolution of the wave packet. The wave packet produces an electric current **J** that is proportional to the velocity of the particles in Kepler orbits. When the particles are near the nucleus they have a large velocity producing a large current, when they are at the outer turning point the current is small. If a second laser pulse is applied to the atom while it is in the wave packet state then the amount of energy absorbed from this probe pulse is proportional to **J·E**, where **E** is the electric field in the laser pulse. Thus a pulse applied when the wave packet is near the nucleus will excite the atom further, ionizing it, while a pulse applied when the wave packet is at the outer turning point will not affect the atom appreciably. A simple measurement of the number of ions versus the delay between the two laser pulses will then map out the wave packet motion.

We may conclude from this something that we already knew, namely, that the quantum mechanical wave function does not generally describe the behavior of a single realization, but rather the average over an ensemble of realizations. By the use of a short laser pulse we are able to reduce the ensemble to those that have a particular phase in their orbital motion. Of course, classical mechanics can go beyond an ensemble average description to describe the evolution of a single realization. To further narrow the gap between the quantum mechanics of our system and the corresponding classical theory we must achieve angular localization as well as radial localization.

The angular dependence is determined by the angular momenta of the states in our superposition. To achieve a strong localization in angle we need many different spherical harmonics in the superposition. The dipole selection rules make the excitation of such a superposition difficult. A single photon carries only one unit of angular momentum. Many photons are required to reach the high angular momentum states from the ground state. But it requires only a few optical photons to ionize the atom, so we must find another source of angular momentum. One way of introducing this angular momentum without ionizing the atom is to use a radio-frequency field. Yeazell, *et al,*(1988) have shown that a radio-frequency field can be used to mix, via multiphoton transitions, the various angular momentum states within a given principal quantum number manifold. A short laser pulse can then be used to excite this mixture of states. When the radio-frequency field is turned off slowly, the atom is left in an angularly localized wave packet oriented in the direction of the rf field.

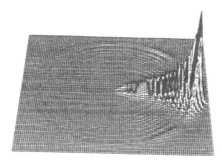

Fig. 2. The angularly localized wave packet. The projection in the equatorial plane of the probability distribution for the electron is localized in a pie-shaped wedge about 20° in width on one side of the nucleus. In the case of a nonrelativistic spinless hydrogen atom the distribution is stationary in space.

The wave packet is localized in the angular direction, but not in the radial direction. This can be interpreted again in terms of an ensemble of classical elliptical orbits. In this case the axes of the ellipses are all aligned, but the phases of the motion along the ellipses are not determined. The result is an elliptical distribution in space that is approximately stationary in time. In the case of a nonrelativistic, spinless hydrogen atom the various angular momentum states, all with the same principal quantum number, in our superposition are degenerate so that the wave packet is indeed stationary. The nucleus is at one focus of the ellipse and the wave packet sticks out in one direction in space.

In the case of a real hydrogen atom or an alkali atom there is a quantum defect splitting the different states in the superposition. In the case of high angular momentum states, l of the order of 30 or so, that Yeazell studied the largest component of the quantum defect is due to special relativity. The angularly localized wave packet will actually precess due to special relativity, just as does the planet Mercury. Of course, in the atomic case, in distinction to the planetary case, the effects of general relativity are negligible.

In order to approach the classical limit of the atom as closely as possible it is desirable to combine the two techniques that we have described and produce full three-dimensional localization. There would seem to be no reason why such localization cannot be achieved in the laboratory, but such an experiment has not been carried out to date. We can, however, carry out a theoretical simulation to see the behavior of such a wave packet.

Gaeta (1990) has carried out such a simulation and produced a three-dimensionally localized wave packet in hydrogen. The wave packet is a superposition of circular orbit states, i.e. states with quantum numbers satisfying $l = m = n - 1$, with the principal quantum number distributed with a gaussian amplitude over the range $n = 320 \pm 5$. The resulting distribution is illustrated in Fig. 3. Immediately after the excitation process the wave packet is localized to approximate a minimum uncertainty wave packet in all three dimensions. In fact, if we calculate the uncertainty products we find

$$\Delta r \Delta p_r = 0.5130 \hbar$$
$$\Delta \theta \Delta L_\theta = 0.5004 \hbar \qquad (1)$$
$$\Delta \phi \Delta L_z = 0.5125 \hbar.$$

The wave packet is illustrated in Figure 3. The dependence on the polar angle for the circular-orbit states is $\sin^l \theta$ so that for $l \approx 320$ the localization in the equatorial plane is extremely good. The plots show the r-ϕ dependence. The orbit is circular with a radius of approximately 10 μm. The wave packet orbits with little change for the first seven or eight orbits, but by the tenth orbit it has spread all the way around the orbit and the head and tail of the packet have begun to interfere. This is no surprise; already by 1927 Schrödinger had discovered this inevitable spreading.

There is more to the evolution than simple spreading however. The interference produces a definite structure in the packet so that by the eighteenth orbit six distinct sub-wave packets can be discerned, all orbiting at the classical velocity along the classical orbit. These disperse, and a bit later at orbit 27 we see that there are now four distinct packets, at orbit 36 three packets, and at orbit 58 two packets. Following this the sequence repeats in reverse forming 3, 4, 5, and 6 wave packets. Finally, on orbit 107 we get an almost perfect revival of the original wave packet.

This complex evolution is easily understood if we consider the energies of the states that make up the superposition. The wave function is given by

$$\Psi(r, t) = \sum_n \psi_n(r) e^{-i\omega_n t}, \qquad (2)$$

where $\psi_n(r)$ is the wave function of the state with principal quantum number n, and ω_n is the energy of that hydrogenic state. It is immediately obvious that the wave packet state is not a periodic function of time because, by Fourier's theorem, the time dependence of the expansion in Eq(2) would have to be in the form of a Fourier series. All of the frequencies

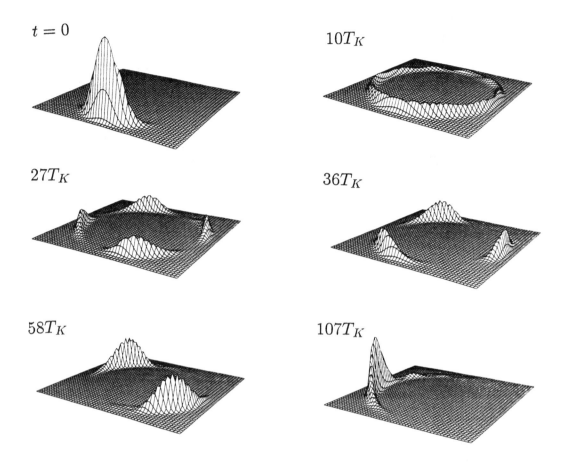

Fig. 3. The evolution of the three-dimensionally localized quantum wave packet. The figures are labeled with the time in Kepler orbital periods. The wave packets are localized in the equatorial plane and travel in a circular orbit at the classical orbital velocity.

ω_n would have to be integer multiples of a fundamental frequency. In the case of a harmonic oscillator the levels are equally spaced so that the wave function is exactly in the form of a Fourier series. Thus the coherent states of a harmonic oscillator can undergo exactly periodic oscillations without spreading.

The energy levels of the hydrogen atom are given by the Rydberg formula, thus

$$\omega_n = -\frac{e^2}{2\hbar a_0 n^2}, \tag{3}$$

where e is the charge of the electron and a_0 is the Bohr radius. Our wave packet is made up of a superposition of a few levels in the vicinity of $n = 320$. We can then express the frequencies ω_n in terms of a Taylor series expansion about the average n that we will denote

$\overline{n}$,

$$\omega_n = \omega_{\overline{n}} + (n - \overline{n})\frac{\partial \omega_n}{\partial n}\bigg|_{n=\overline{n}} + \frac{1}{2}(n - \overline{n})^2 \frac{\partial^2 \omega_n}{\partial n^2}\bigg|_{n=\overline{n}} + \cdots \tag{4}$$

We can alternatively write this expansion in the form

$$\omega_n = \omega_{\overline{n}} + (n - \overline{n})\frac{2\pi}{T_K} + (n - \overline{n})^2 \frac{2\pi}{T_R} \cdots, \tag{5}$$

where

$$T_K \equiv \frac{\pi \overline{n}}{|\omega_{\overline{n}}|} \tag{6}$$

is just equal to the classical Kepler period for a particle with the energy $\hbar\omega_{\overline{n}}$, and

$$T_R \equiv \frac{\overline{n}}{3}T_K \tag{7}$$

is the time for decay and revival of the wave packet. If we substitute this expansion into the wave function we find

$$\Psi(r,t) \approx e^{-i\omega_n t} \sum_m \psi_{\overline{n}+m}(r)e^{-i2\pi mt/T_K} e^{-i2\pi m^2 t/T_R}. \tag{8}$$

In our illustration $\overline{n} = 320$ so that $T_R = 107T_K$. The first exponential inside the sum produces the periodic oscillations at the Kepler period while the second exponential produces the decays and revivals. The occurrence of the fractional revivals with multiple sub-wave packets is due to the m^2 dependence of the second exponential within the sum. This circular-orbit wave packet is the most nearly classical atomic state that we know how to make. For short times it behaves very much like a classical particle with the same energy and angular momentum, but over longer times its behavior is distinctly wave like. It is clear that this quantum state does not describe a particular realization of the electron's motion but rather an average over an ensemble.

If we try to reproduce this complex behavior using an ensemble of classical particles we have no wave-like properties to produce the interference effects. The classical particle can take on a continuous range of values of energy; an ensemble with such a continuous range of orbital periods will spread and never revive. Yeazell (1990, 1991) and (Meacher, 1991) have observed these decays and revivals in the laboratory so any useful quasiclassical theory must include some wave-like properties.

The simplest quasiclassical theory that does indeed include such effects is the old Bohr theory of hydrogen. If we assume that each member of our classical ensemble follows exactly a classical time evolution, but that only certain values of the classical energy and angular momentum are allowed, namely those satisfying the Bohr quantization condition, we find that the resulting classical evolution of the ensemble indeed has decays, revivals and fractional revivals. The classical ensemble is illustrated in Figure 4.

The classical ensemble is distributed over the discrete set of allowed energies with weightings that are identical to those in the quantum ensemble. Within each energy group the initial distribution in space is exactly that of the quantum ensemble, but the evolution after the initial time is strictly classical. The distribution of the classical ensemble in space

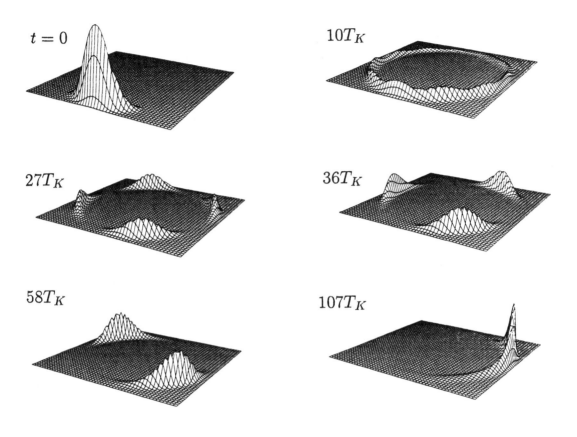

Fig. 4. The evolution of the classical ensemble. By limiting the classical ensemble to the energies and angular momenta allowed in the old quantum theory of Bohr we obtain a time evolution that is almost identical to that of the quantum ensemble.

is remarkably like that of the quantum ensemble with decay, revival and fractional revivals. One distinction can be made however - the phase of the wave packet at the full revival is 180° different in the classical and quantum cases. When the wave packet reassembles in the quantum case it is on the opposite side of the orbit from the location that a classical particle with the same energy would be. One might be tempted to shift the frequencies in the classical ensemble to compensate for this small discrepancy, however a simple linear shift will not work because the one-half fractional revival that was correctly predicted by the original classical ensemble will be shifted to 90° out of phase in the new ensemble.

It would appear that the interference phenomena that persist in the "classical limit" of an atom are more subtle than the simple limitation of classical orbits to those that contain an integer number of de Broglie wave lengths as postulated by Bohr. Further experiments are underway in a number of laboratories exploring this region at the border between quantum and classical physics. At the very least we should end up with a better understanding of the necessary features of some future Neoclassical theory.

ACKNOWLEDGMENTS. The work has benefited enormously from the inspiration of Ed Jaynes, as well as from the hard work of several of my own students. Particularly instrumental have been the contributions of Dr. Jonathan Parker, Dr. John Yeazell, Zagorka Gaeta, Mark Mallalieu, and Jake Bromage. The work was supported by the University Research Initiative through the Army Research Office.

REFERENCES

Gaeta, Z. D., and C. R. Stroud, Jr.:1990, 'Classical and Quantum Mechanical Dynamics of a Quasiclassical State of the Hydrogen Atom,' Phys. Rev. A **42**, 6308.

Meacher, D.R, P. E. Meyler, I. G. Hughes and P. Ewart: 1991, 'Observation of the Collapse and Fractional Revival of a Rydberg Wavepacket in Atomic Rubidium,' J. Phys. B: At. Mol. Opt. Phys. **24**, L63.

Przibram, K.: 1967, *Letters on Wave Mechanics*, Philosophical Library, pp. 55-75.

Schrödinger, E.:1926, 'Der Stetige Übergang von der mikro- zur makromechanik,' Natürwissenschaften **28**, 664.

ten Wolde, A., L. D. Noordam, H. G. Muller, A. Lagaendijk, and H. B. van Linden van den Heuvell:1988, 'Observation of Radially Localized Atomic Electron Wave Packets,' Phys. Rev. Lett. **61**, 2099.

Yeazell, J. A., and C. R. Stroud, Jr.,:1988, 'Observation of Spatially Localized Atomic Electron Wave Packets,' Phys. Rev. Lett. **60**, 1494.

Yeazell, J. A., M. A. Mallaieu, and C. R. Stroud, Jr.:1990, 'Observation of the Collapse and Revival of a Rydberg Electronic Wave Packet,' Phys. Rev. Lett. **17**, 2007.

Yeazell, J. A. , and C. R. Stroud, Jr.:1991, 'Observation of Fractional Revivals in the Evolution of a Rydberg Atomic Wave Packet,' Phys. Rev. A. **43**, 5153.

MUTUAL RADIATION REACTION IN SPONTANEOUS EMISSION

Richard J. Cook
Department of Physics and Geology
Northern Kentucky University
Highland Heights, Kentucky 41076

ABSTRACT. Each transition dipole in a single atom has a radiation reaction field which acts on itself and on the other transition dipoles of the atom. The self interaction on a given transition gives rise to spontaneous emission and radiative decay on that transition. Here we emphasize that the mutual radiative interaction of different transition dipoles can sometimes have important consequences. The mutual radiative interaction can substantially alter radiative decay rates and inhibit the radiative decay of certain expectation values. The hydrogen atom and the harmonic oscillator are cited as examples and a possible experimental test of mutual radiation reaction involving a single electron in a Penning trap is discussed.

1. Introduction

Modern quantum mechanics began with Heisenberg's picture of a bound charge as a "virtual orchestra" of transition dipoles oscillating at different frequencies. It is known from modern quantum theory that each of these dipoles generates a radiation reaction field which, acting on the dipole itself, gives rise to spontaneous emission and radiative relaxation (Milonni, 1976). In this paper we consider the mutual interaction of different transition dipoles within a single "atom". We begin by considering some simple cases of mutual radiation reaction in classical electrodynamics and proceed to the quantum theory of mutual radiation reaction (MRR), citing the hydrogen atom and the charged harmonic oscillator as examples. We point out that the equations describing MRR are not at all new, but can be traced to the earliest treatments of the interaction of atoms with the quantized electromagnetic field. We examine the radiative dynamics of a charged harmonic oscillator and find that mutual radiation reaction between the different transition dipoles of this system is necessary for the classical limit of radiation damping. Finally, we point to a single electron in a Penning trap as a potential testing ground for MRR.

2. Classical Mutual Radiation Reaction

Mutual radiation reaction occurs in classical electrodynamics. It is instructive to consider the classical interaction of radiating dipoles before looking at the quantum-mechanical case where different transition dipoles are represented by noncommuting operators.

Near a classical electron in motion there is a radiation reaction field

$$E_{RR} = \frac{2e\,\dddot{\mathbf{r}}}{3c^3}, \tag{1}$$

where e is the charge of the electron, r its position vector and c is the speed of light (Landau and Lifshitz, 1962). We have in mind that the electron is bound to a center of force and forms a dipole oscillator (dipole moment $\mu = er$) with a well defined frequency ω. Note that the radiation reaction field (1) is independent of position in the near field (at distances from the dipole much less than the radiated wavelength λ).

When the reaction field acts on the charge that generates this field, that charge experiences the reaction force

$$F = eE_{RR} = \frac{2e^2\,\dddot{r}}{3c^3}. \tag{2}$$

Ignoring such difficulties as runaway solutions, the reaction force tends to damp the oscillation of the radiating electron. The classical electron equation of motion reads

$$m\ddot{r} = -\omega^2 r + \frac{2e^2\,\dddot{r}}{3c^3}. \tag{3}$$

Or, in terms of the dipole moment $\mu = er$,

$$\ddot{\mu} + \omega^2\mu = \frac{2e^2\dddot{\mu}}{2mc^3}. \tag{4}$$

For very weak damping, the motion is nearly sinusoidal and the third derivative of the dipole moment may be approximated by $-\omega^2$ times the first derivative. This "slowly varying amplitude approximation" eliminates runaway solutions and yields an equation in which the reaction force is seen to be essentially equivalent to a frictional force, i.e. a damping force opposed to the velocity of the charge. The solution to this equation shows that the dipole oscillates at frequency ω and decays in amplitude at the rate

$$\Gamma = \frac{2e^2\omega^2}{3mc^3}. \tag{5}$$

In this way the <u>self</u> radiation reaction (SRR) of an oscillating charge accounts for radiation damping.

The radiation reaction field of one dipole can act on a different dipole in its neighborhood. We call this interaction "<u>mutual</u> radiation reaction" (MRR) to distinguish it from "<u>self</u> radiation reaction" (SRR). For two dipoles, the reaction fields are

$$E_1 = \frac{2\dddot{\mu}_1}{3c^3}, \quad E_2 = \frac{2\dddot{\mu}_2}{3c^3}, \tag{6}$$

and the total reaction field, $E_1 + E_2$, acts on each of the dipoles. The resulting coupled equations of the motion for the two dipoles oscillators are:

$$\ddot{\mu}_1 + \omega_1^2\mu_1 = \frac{2e^2}{3mc^3}(\dddot{\mu}_1 + \dddot{\mu}_2), \tag{7a}$$

$$\ddot{\mu}_2 + \omega_2^2\mu_2 = \frac{2e^2}{3mc^3}(\dddot{\mu}_1 + \dddot{\mu}_2). \tag{7b}$$

In most systems of interest, the reaction force is quite weak and the dipoles oscillate at approximately the unperturbed frequencies ω_1 and ω_2. Therefore, the radiation reaction field of dipole 1 is sinusoidal at frequency ω_1 and is clearly in resonance with dipole 1. But the reaction field of the second dipole will be in resonance with the first one only if the frequency of dipole 2 is the same as that of dipole 1. If dipole 2 oscillates at a different frequency than dipole 1, then its reaction field E_2 will be out of resonance with dipole 1 and will have little effect on the motion of that dipole. Thus the self reaction field is always in resonance with the dipole that generates it, but the mutual radiative interaction of two dipole oscillators will be in resonance only when the two dipoles have approximately the same frequency. We will see later that the same result occurs in quantum theory, namely that mutual radiation reaction between two different transitions in a single atom is effective only when the two transitions have the same frequency.

MRR is necessary for energy conservation. Recall that the form of the radiation reaction force was first derived from energy conservation, i.e., on the assumption that the time average loss of the mechanical energy of a moving charge is equal to the rate at which energy is radiated as given by the Larmor formula (Jackson, 1962). Similar considerations show that, when two oscillating charges or dipoles are near to one another, mutual radiation reaction is necessary for energy conservation. For instance, when two identical dipoles are oscillating in-phase, the rate of radiation is four times that of a single isolated dipole. Accordingly the mechanical energy of each dipole must decrease at a rate double that of an isolated dipole. The enhanced radiative damping of the dipoles is provided by MRR. If each dipole were damped only by the self interaction, energy would not be conserved. On the other hand, if two identical dipoles oscillate exactly out-of-phase, then the total dipole moment of the system remains zero and there is no dipole radiation. Here energy conservation requires that there be no time-average change in the mechanical energy of the oscillators. Mutual radiation reaction exactly cancels the self radiation reaction, in this case, and there is no radiative damping of either oscillator. Here again, if mutual radiation reaction were not present, energy would not be conserved. Some of the results of the classical theory of MRR carry over into the quantum theory to which we now turn.

3. Quantum Radiation Reaction

The classical dipole oscillator is similar in many respects to the electric dipole expectation value of a two-level atom. In the Heisenberg picture, the dipole operator of a two-level atom oscillates at the transition frequency and the expectation value of the dipole decays as a result of radiation damping. It has been shown by Milonni and others that, in the Heisenberg picture, the quantized electromagnetic field in the neighborhood of the atom contains a source-field term of exactly the same form as the radiation reaction field of classical electrodynamics (Milonni, 1976). As in the classical theory, the reaction field of the dipole acts on the dipole that generates it and influences the atomic dynamics. The self reaction accounts for spontaneous emission and the radiative decay of the upper atomic level as well as decay of the dipole expectation value. The self reaction affects both the diagonal and the off-diagonal elements of the atomic density matrix. The spontaneous decay from the upper to the lower level is described by reaction terms in the equations of motion for the diagonal elements of the density matrix, whereas the reaction terms in the equations for the off-diagonal elements describe the radiative decay of the dipole expectation value. For atoms with more than two levels, a self interaction of the type just described occurs for each dipole-allowed transition.

The present paper is concerned with the mutual radiative interaction of <u>different</u> transition dipoles. We first ask whether different transition dipoles in the same atom experience mutual radiation reaction. In deriving the equations of motion for the atomic density matrix one obtains a number of terms describing the mutual radiative interaction which are not new. As early as 1927 Landau worked out the equations for the atomic density matrix within the framework of Dirac's then new quantum electrodynamics. Since then many other workers have obtained similar results, but have not always attributed the interactions between different transitions to radiation reaction. The density matrix equations read

$$\dot{\rho}_{kj} = -i\omega_{kj}\rho_{kj} + \sum_{n,m}(\beta_{mjnk} + \beta^*_{nkmj})\rho_{nm}$$
$$- \sum_{n,m}(\beta^*_{jnmn}\rho_{km} + \beta_{knmn}\rho_{mj}) \tag{8}$$

where

$$\beta_{ijkl} = \frac{2\mu_{ij}\mu^*_{kl}\omega^3_{kl}\theta(\omega_{kl})}{3\hbar c^3}. \tag{9}$$

Here $\theta(\omega)$ is the unit step function, ω_{kl} is the transition frequency between levels k and l, and μ_{ij} is the dipole transition moment between levels i and j. The terms in (8) representing MRR are those involving β coefficients with different transition dipole moments μ_{ij} and μ_{kl}, whereas the β coefficients involving a single transition moment, i.e. $i = k$ and $j = l$ represent SRR. The β coefficients for SRR can be rewritten in terms of Einstein A-coefficients. In these equations we have included only the terms representing radiation reaction. There are other interactions which give rise to frequency shift terms in the density matrix equations. These are being ignored here for simplicity. In view of Landau's early work, we shall refer to equations (8) as the "Landau equations" and will call the MRR terms in these equations the "Landau terms".

In the absence of radiation reaction, only the first term on the right in (8) is present. Thus the unperturbed frequency of the density matrix element ρ_{ij} is ω_{ij}. Using this fact in a rotating wave approximation, it is easy to see that the density matrix element ρ_{ij} associated with transition $i \to j$ is strongly influenced by the dipole moment of transition $k \to l$ only when the frequencies of these transitions are nearly equal, otherwise, as might have been expected from the classical analogue, the reaction field of the dipole of one transition is not in resonance with the other transition and the interaction is negligible. Accordingly, many of the reaction terms in (8) have only a very small effect in time average and may be ignored. If we keep only those reaction terms associated with self radiation reaction, the density matrix equations reduce to

$$\dot{P}_n = -\left(\sum_{k<n} A_{nk}\right)P_n + \sum_{k>n} A_{kn}P_k \tag{10}$$

$$\dot{\rho}_{nm} = -i\omega_{nm}\rho_{nm} + \frac{1}{2}\left(\sum_{k<n} A_{nk} + \sum_{j<m} A_{mj}\right)\rho_{nm} \tag{11}$$

where

$$A_{nm} = \frac{4|\mu_{nm}|^2\omega^3_{nm}}{3\hbar c^3} \tag{12}$$

are the Einstein A-coefficients. We shall refer to (10) and (11) as the Pauli equations. The first of these equations are the Einstein rate equations for the probabilities to be in the various levels (the diagonal elements of the density matrix), whereas the second set of equations (11) are the equations for the off-diagonal density matrix elements. Notice that the rate equations (10) couple the different probabilities and represent a cascade downward perhaps to the ground state. Each of the off-diagonal elements of the density matrix, on the other hand, decay at a certain rate depending on the rates of spontaneous decay from both of the associated levels without any coupling between different off-diagonal elements. From the way they were derived, it is clear that the Pauli equations describe SRR but not MRR. The question that interests us here, and which has been addressed by a number of authors [(Cohen-Tannoudji, 1977), (Cardimona, Raymer and Stroud, 1982), (Cardimona and Stroud, 1983)], is whether the mutual radiation reaction terms are observable in any quantum system. Are there any quantum systems for which mutual radiation reaction is observable?

To begin answering this question, we look at the density matrix equations for the diagonal elements. From the form of equations (8) and the structure of the β-coefficients (9), it is found that a number of conditions are necessary for two transitions to interact radiatively. First off, there is no interaction unless the two transitions have a common lower level, i.e. the two transitions form a "V" configuration. Secondly, the upper levels of the V-system must be nearly degenerate, so that the two transition frequencies are approximately equal; $E_n \approx E_m$, where n and m are the labels of the two upper states. Thirdly, the two transition dipoles must not be orthogonal; $\mu_{nj} \cdot \mu_{mj}^* \neq 0$, where j is the label of the lower level of the V system. And finally, the upper state density matrix element ρ_{nm} must be nonzero, i.e. there must be coherence between the upper states. If any of these conditions is not satisfied, the mutual radiative reaction between the two transitions is negligible or zero.

As an example of MRR we may consider atomic hydrogen. In the elementary treatment of hydrogen by means of the Schrödinger equation, the energy levels depend only on the principle quantum number n. The lowest transitions which interact via mutual radiation reaction are from the $n = 3$ level to the $n = 2$ level, particularly the transitions from 3d ($m = 0$) and from the 3s ($m = 0$) to the 2p states. Because the 3d and 3s levels are degenerate in the Schrödinger theory, all of these transitions have the same frequency. Notice that the transitions from 3d ($m = 0$) and 3s to, say, 2p ($m = -1$) form a V configuration and one expects mutual radiation reaction to occur here because the dipole transition moments are not orthogonal. Indeed, the decay rate of the 3d ($m = 0$) and 3s levels would depend on the coherence between these levels if the Schrödinger theory were an accurate representation of the hydrogen atom. Unfortunately, this does not appear to be a testable example of MRR because the mutual radiative interaction of these transitions is spoiled by the fine structure splitting when relativistic effects and spin are taken into account. According to the Dirac equation, these effects split the 3s and 3d levels and render MRR ineffective for these transitions in hydrogen.

It seems to be the case in general that, from a given level in an atom, there are at most three transitions to higher lying levels of a given frequency, these being the σ^+, π, and σ^- transitions of that frequency. Since the transition dipole moments for these three transitions are mutually orthogonal, MRR does not usually occur in atomic systems, except perhaps when some accidental degeneracy is present.

As a second example of MRR we consider the charged harmonic oscillator. We look first at the decay of the expectation value of the oscillator energy. The energy levels of the harmonic oscillator form a ladder with rungs separated by $\hbar\omega$. The allowed dipole transitions are between adjacent levels. An harmonic oscillator in a given excited state will decay radiatively by a cascade through each of the lower lying levels to the ground state. This decay cascade is described by the Einstein rate equations

$$\dot{P}_n = (n+1)\Gamma P_{n+1} - n\Gamma P_n \tag{13}$$

for the probabilities P_n to be in levels n, where Γ is the classical energy decay rate (5). From these rate equations the decay of the expectation value of the oscillator energy is readily computed, the result being that the energy decays exponentially at the classical rate. Notice that there are no V configurations of energy levels in the harmonic oscillator and consequently MRR plays no role in the decay of the probabilities or the expectation value of the oscillator energy. This is a statement about the diagonal elements of the density matrix. The off-diagonal elements <u>are</u> influenced by mutual radiation reaction.

Adjacent transitions in the harmonic oscillator ladder have identical frequencies and, according to equations (8), the dipole transition moments on adjacent transition interact via MRR. The equations of motion for the off-diagonal density matrix elements read

$$\begin{aligned}
\dot{\rho}_{nm} = &- i\omega_{nm}\rho_{nm} - \frac{1}{2}\Gamma(n+m)\rho_{nm} \\
&+ \Gamma\sqrt{(n+1)(m+1)}\rho_{n+1,m+1}
\end{aligned} \tag{14}$$

We see that a given off-diagonal element or coherence $\rho_{n,m}$ does not decay independently, as in the Pauli equations (11), but is coupled to the coherence $\rho_{n+1,m+1}$ with each index increased by unity. This coupling term is one of the Landau terms in the density matrix equation of motion. It represents the effect of the reaction field of one transition dipole on the dipole of an adjacent transition. The off-diagonal density matrix elements enter into the expectation value of the oscillator displacement:

$$x = \langle\hat{x}\rangle = \sqrt{\frac{\hbar}{2M\omega}} \sum_{n=0}^{\infty} \sqrt{n+1}(\rho_{n,n+1} + \rho_{n+1,n}) \tag{15}$$

It follows from (14) and (15) that the expectation value of the oscillator displacement obeys the classical equation of motion for a charged oscillator undergoing radiative decay. In other words, the quantum-mechanical expectation value decays in the classical manner. The important point here is that the classical decay of the expectation value depends on mutual radiation reaction. Without the mutual reaction terms in the density matrix equations of motion, the quantum expectation value would not mirror the classical behavior. In this sense we can say that the Landau terms are necessary for the proper classical limit of radiation reaction. It can also be shown that, if the oscillator begins in a coherent state, it remains in a coherent state throughout the radiative decay process (Cook, 1984). This too depends on the Landau terms. Without the MRR terms the coherent state is not preserved as the oscillator decays.

To further illustrate the role played by the Landau terms in the radiative decay of an oscillator, we consider the nature of the decay that would occur if MRR were not present.

In this case, we have the usual Pauli equation (11) for decay of each off-diagonal density matrix element. The density matrix element ρ_{nm} decays exponentially at a rate $(n+m)\Gamma/2$, where Γ is the classical decay rate. For all practical purposes, the density matrix elements which enter the calculation of the displacement expectation value, namely ρ_{nn-1} and ρ_{n-1n}, decay at n times the classical rate. This means that, for an oscillator initially in a high level of excitation, the expectation value of the displacement decays to zero very much more rapidly than its classical counterpart. We can say that MRR greatly inhibits the decay that would occur without it.

The great difference between the decay rates of the oscillator with and without MRR would seem to offer a test of MRR. An almost ideal charged harmonic oscillator that is realizable in the laboratory is a single electron in a Penning trap (Brown and Gabrielse, 1986). The Penning trap consists of cap and ring electrodes which, when kept at different potentials, produces a quadrupole electric field at the center of the electrode system. Superimposed on the quadrupole electric field is a uniform magnetic field directed along the principal axis of the quadrupole (Penning, 1936). For a perfect quadrupole field and exactly uniform magnetic field, an electron at the center of the trap oscillates between the cap electrodes in a simple harmonic manner. The trapping of single electrons for extended periods of time was pioneered by Dehmelt (Wineland, Ekstrom and Dehmelt, 1973). In one of the traps operated at the University of Washington, the frequency of oscillation in the axial mode was 65 MHz. The corresponding classical radiative damping rate for the oscillator $\Gamma = 460$ days. In practice this mode of oscillation is damped by connecting the cap electrodes through a resistor so that the current induced by the oscillating electron charge will dissipate energy in the resistor and thereby damp the motion of the electron. Without such resistive damping, the electron amplitude would presumable decay radiatively in a time of order Γ. It is important to note that realizable Penning traps do not produce a perfect oscillator potential. Any imperfections in the electrode geometry or extraneous external fields will introduce some anharmonicity into the oscillator potential. As a result the stationary states of the oscillator are not equally spaced in energy and adjacent transition frequencies are not exactly equal. This implies that the reaction field of one transition dipole is not in exact resonance with a neighboring transition. Some simple estimates suggest that an exceedingly small anharmonicity will cause the MRR between adjacent oscillator transitions to effectively be turned off. When the detuning between adjacent transitions at level n exceeds $n\Gamma$, the transitions are out of resonance and the effect of MRR is small. The correction of trap imperfections to minimize axial anharmonicity has been studied by VanDyck et al. (VanDyck, Wineland, Ekstrom and Dehmelt, 1976) and by Gabrielse (Gabrielse, 1983).

In the trap we have been considering, excitation of the electron oscillation to level $n = 1000$ is easily achieved. For this degree of excitation, the Pauli decay rate (without MRR) is one thousandth of the classical decay rate. Accordingly, the electron amplitude expectation value would decay in a time of order one half day rather than 460 days. It is to be emphasized that this decay is due to a decay of the off-diagonal elements of the oscillator density matrix, whereas the diagonal elements are unaffected by MRR and decay at a much slower rate. The anharmonicity changes an initial coherent state to an incoherent superposition of the oscillator states. The rapid decay of the displacement expectation value does not mean that the energy of the oscillator is decaying rapidly. The expectation value of the energy decays at the classical rate regardless of the anharmonicity because it is due to

SRR rather than MRR. If the electron is excited from the ground state by a strong external classical field, the oscillator is initially in a coherent state and the expectation value of the position of the electron is quite well defined. That is to say, the probability density for the position of the electron is concentrated near the corresponding classical displacement initially. As the coherences then decay at the Pauli rate (n times the classical rate), the probability density for electron position spreads over the entire energetically allowed region rather than remaining as a coherent state with well defined position. In a perfectly harmonic oscillator, on the other hand, an initial coherent state remains a coherent state throughout the decay process.

There is another motion of an electron in a Penning trap that may serve as a testing ground for mutual radiation reaction. This is the cyclotron motion of the electron in the approximately uniform magnetic field. The quantum-mechanical analysis of the cyclotron motion in a uniform magnetic field shows that the energy levels for this motion, the so-called Landau levels, are equally spaced in energy and are essentially equivalent to a harmonic oscillator but with a high degree of degeneracy. A straightforward analysis shows that again the transition moments for different levels interact via MRR and, if there is a small amount of anharmonicity, say due to inhomogeneities of the magnetic field, adjacent transitions will be out of resonance and MRR will be negligible. As for the harmonic oscillator, the cyclotron motion in a uniform magnetic field admits coherent-state solutions. The radiative decay causes the electron to spiral in toward the center of the cyclotron orbit with the emission of radio-frequency photons. The frequency of the cyclotron motion for existing Penning traps is substantially higher than that for the axial motion and the radiative decay of the cyclotron motion has been observed in a number of experiments. Indeed, the difference between the observed decay rate and the free-space theoretical value led to the conclusion that the radiative decay rate is inhibited by the electrode cavity (Gabrielse and Dehmelt, 1985). For the cyclotron motion, anharmonicity can be introduced by an external inhomogeneous magnetic field which to a certain extent is always present. Once again, only a small anharmonicity is necessary to render MRR ineffective. The lack of MRR causes a decay of the electron cyclotron radius which is much more rapid than expected classically. In some experiments the cyclotron frequency has been of order $\nu = 164$ GHz and the classical damping rate for this motion is 0.5 s, making it accessible to measurement.

ACKNOWLEDGMENTS. The author wishes to thank Dr. P. W. Milonni and Dr. C. R. Stroud for helpful discussions during the course of this work.

REFERENCES

Brown, L. S. and G. Gabrielse: 1986, Rev. Mod. Phys. **58**, 233.

Cardimona, D. A., M. G. Raymer and C. R. Stroud, Jr.: 1982, J. Phys. **B15**, 55.

Cardimona, D. A. and C. R. Stroud, Jr.: 1983, Phys. Rev. **A27**, 2456.

Cohen-Tannoudji, C.: 1977, in <u>Frontiers in Laser Spectroscopy</u>, edited by R. Balian, S. Haroch, and S. Liberman (North-Holland, Amsterdam), p.46.

Cook, R. J.: 1984, Phys. Rev. **A29**, 1583.

Gabrielse, G.: 1983, Phys. Rev. **A27**, 2277.

Gabrielse, G. and H. G. Dehmelt: 1985, Phys. Rev. Lett. **55**, 67.

Jackson, J.D.: 1962, "Classical Electrodynamics", (John Wiley & Sons, Inc., New York), p. 581.

Landau, L.D. and E.M. Lifshitz: 1962 "Classical Theory of Fields", (Addison-Wesley Publishing Co., Mass.), p.231.

Milonni, P.W.: 1976, Physics Reports **25**, 1.

Penning, F. M.: 1936, Physica (Utrecht) **3**, 873.

VanDyck, R. S., Jr., D. J. Wineland, P. A. Ekstrom and H. G. Dehmelt: 1976, Appl. Phys. Lett. **28**, 446.

Wineland,D. J., P. Ekstrom and H. G. Dehmelt: 1973, Phys. Rev. Lett. **31**, 1279.

A MODEL OF NEUTRON STAR DYNAMICS

F.W. Cummings
Department of Physics,
University of California,
Riverside, California 92521

ABSTRACT. The picture of solo pulsars as usually presented is that they are born following a supernova explosion, coincident with the formation of a rapidly spinning neutron star. They emit astonishingly regular pulses of radiation whose periods gradually lengthen over time scales of one to ten million years, taken usually as a pulsar lifetime. The present work advances a substantially different picture than the above. In what follows, a model is presented which makes plausible an alternative view of pulsars which addresses a number of previously puzzling questions. In this new view, a neutron star typically goes through three stages. In the beginning, just after its formation, the magnetic field of the rapidly spinning neutron star tumbles erratically. Only later, on a time scale $<<10^7$ yrs., the tumbling motion gives way to pulsar behavior, an accurate limit cycle behavior whose period is given simply as proportional to the moment of inertia divided by the conserved angular momentum. In a time $<<10^7$ yrs. the pulsar dies, as the magnetic axis aligns itself with the rotation axis, a stable fixed point in the parameter space of the model. There are thus three time scales: the pulsar period, typically of the order of one second, an intermediate time determined principally by mechanical (or viscous) damping, and the third the damping time of the pulsar period, as angular momentum is radiated away over a million years or more. Among the outstanding questions addressed are: 1) What is the source of the immense $(10^{12}G)$ off-axis magnetic field? 2) How can such an astoundingly accurate period, comparable to that of an atomic clock, be achieved by a pulsar period? 3) Why is such a small fraction of pulsars associated with supernova remnants? 4) Why is it that there are no pulsars observed to have periods $> 4s$? 5) Is it reasonable to expect that the erratically tumbling magnetic field of $10^{12}G$ may be the source of the mysterious "bursters", $> 1Mev$ gamma radiation of about 1s duration?

1. Introduction

The usual interpretation of the periodic signals from solo pulsars is that they are the result of a rapidly rotating off-axis magnetic field, where the reference axis is the direction of the conserved angular momentum of the spinning neutron star. The "lighthouse" effect gives rise to a periodic pulse each time the beam of about 15° sweeps over the observer. This interpretation, while supported by a good bit of circumstantial evidence, is nevertheless arrived at principally by elimination of other models of oscillating neutron stars (Lyne

137

and Graham-Smith, 1990; Shapiro and Teukolsky, 1983; Michel, 1991). There does not appear to be a satisfactory underlying model of the lighthouse effect, nor for the remarkable accuracy of the period of the magnetic field precession. While the earth viewed as a clock has an accuracy of about 1 part in 108 per day, the pulsar period is comparable to the best atomic clock, better than the earth's clock by four or five orders of magnitude.

Other worrisome problems exist in the standard interpretation. Foremost perhaps is the problem of the pulsar birthrate. It is in balancing the pulsar production rate with the supernova birthrate that has proved vexing. Supernovas are born at a rate of about two per century, and given the presumed lifetime of pulsars as about 10^6 to 10^7 years, there is expected to be about 10^5 of them just in our galaxy alone. We observe only about 500 pulsars in our galaxy, so that we are seeing only about 1% of the expected number according to the standard picture. Apart from the pulsars found in the Crab and Vela nebula, few clear examples exist of a supernova remnant containing an observable pulsar. The historical Tycho (1572), Kepler (1604), Cassiopeia A (1667), and the recent 1987A are conspicuously lacking pulsars. (Since a typical example of a pulsar has a beam of approximately 15°, it is unlikely that they are not seen because their beams are not sweeping over us.) Where are the missing pulsars?

Equally puzzling is the fact that no pulsar has been observed with a period greater than 4.3 s. The periods cluster strongly between .25 and 2.5 s. According to the received picture that a pulsar is born in a supernova explosion and thereafter gradually winds down as it radiates energy, an upper limit of 4 s is not expected. Where are the long period pulsars?

The simple model of the present paper addresses itself to these questions above. A previous paper (Cummings, Dixon and Kaus, 1992) advanced a similar model with two components instead of the present model's three, whose focus was the mysterious "bursters", gamma ray bursts of about a second duration. We will return to a brief discussion of these bursts later in light of more recent observations from the GRO satellite.

2. The Three Component Model

The model for the dynamical behavior of a collapsed body such as a neutron star (or perhaps also a white dwarf) envisions three differentially rotating bodies. We do not here pursue the interesting question of the makeup of the three bodies, or the mechanism by which they effect a charge separation, although some speculative suggestions will be made in the last section. The innermost spherical body (hereafter #1) is surrounded by a concentric superfluid spherical body (denoted by subscript s), which in turn is encased in the larger concentric spherical shell of the outer body (#2) (See figure 1). The bodies #1 and #2, innermost and outermost, each carry an equal and opposite net charge. (It is convenient but not necessary to picture this charge as residing on the relatively thin outer and inner shells of the inner and outermost bodies, respectively.) The innermost body has an instantaneous angular velocity ω_1, the superfluid body (see e.g., Greenstein, 1974) a (constant) angular velocity ω_s, and the outermost body an instantaneous angular velocity ω_2. The magnetic field of the outer body at the site of the innermost body is presumed to be constant in space, and given as B_2. The magnetization of the innermost body (#1) is taken as M_1. The magnetic field at the site of the innermost body due to the rotation of the outermost body (#2) will be proportional to its angular velocity, where the proportionality constant b will depend on the inner radius and charge of the #2 sphere. Likewise, the magnetization of the innermost body will be proportional to its angular velocity, where its proportionality constant a depends on its radius and net charge. The constants a and b are thus both positive or both negative, so that we have

$$\vec{M}_1 = a\vec{\omega}_1,$$
$$\vec{B}_2 = -b\vec{\omega}_2. \tag{1}$$

There is the further equation of the conservation of total angular momentum, expressed as

$$\bar{I}_1 \cdot \vec{\omega}_1 + \bar{I}_2 \cdot \vec{\omega}_2 + I_s \cdot \vec{\omega}_s = \vec{L} \equiv L\hat{k}. \tag{2}$$

The angular momentum of the superfluid body, made up of a densely packed array of vortices aligned with the total conserved angular momentum will be taken to be constant and proportional to the total angular momentum, so that

$$I_s \cdot \vec{\omega}_s = c\vec{L}, \tag{3}$$

and then eqn.2) can be written as

$$I_1 \cdot \vec{\omega}_1 + I_2 \cdot \vec{\omega}_2 = \vec{L}. \tag{4}$$

Here we have "renormalized" the moments of inertia of #1 and #2 by $(1\text{-}c)^{-1}$, and re-named them. (Very small sudden changes in the period ("glitches", Michel (1991)) may be associated with a decrease in the parameter c as a vortex becomes suddenly "unpinned" (Shapiro and Teukolsky, 1983).)

There will then be three contributions to the time rate of change of the magnetization of the innermost body (#1). The first is the Larmor term, which causes the magnetization vector to precess about the field of #2, and because of the constraints above, eqns. 1) and 4), to precess about the total angular momentum L. The second term, first introduced by Landau in a different context (Landau, 1965), causes the two "magnets" to line up, and again, by the constraints above, eqns.1) and 4), causes them to align themselves opposite to L. The rate of this "magnetic damping" is taken to be $\bar{\lambda}$. The third term is the "mechanical" damping, or viscous damping, and is taken to be proportional to the difference between the angular velocity of body #1 and that of the superfluid, $\omega_1 - \omega_s$, since it vanishes only when the two are equal. The constant (tensor) of proportionality here will be taken to be diagonal, where the "3" direction is along L. Then we write

$$\frac{d}{dt}\vec{M}_1 = -\gamma \cdot (\vec{M}_1 \times \vec{B}_2) - \bar{\lambda}\left(\frac{(\vec{M}_1 \cdot \vec{B}_2)\vec{M}_1}{M_1^2} - \vec{B}_2\right) \bar{\bar{\eta}} \cdot (\vec{\omega}_1 - \vec{\omega}_s). \tag{5}$$

Here the tensor $\bar{\bar{\eta}}$ is diagonal and given by $\{\text{diagonal}(\eta_1, \eta_1, \eta_2)\}$.

What is apparent is that there are two forces which oppose each other. The Landau magnetic damping is attempting to align the two magnets, and to counteralign the corresponding angular velocities, according to eqn. 1); on the other hand, the viscous forces are attempting to align **and** equalize the angular velocities of the two charged bodies and thus counteralign the two magnetizations.

Equation 5) together with the constraint eqns. 1) and 4) constitute a closed system. The elimination of the magnetic field of body #2 in eqn. 5) shows at once that the

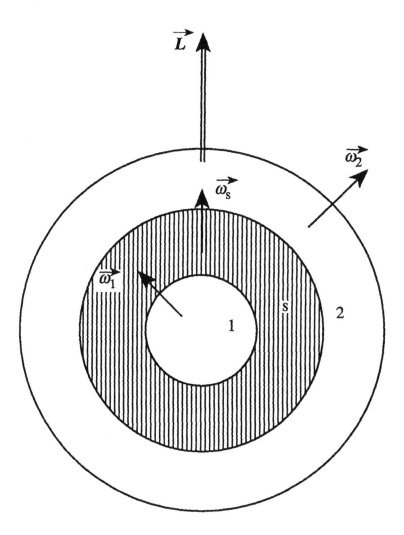

Fig. 1. The three concentric spheres are shown. The middle shell is a superfluid and labelled by "s" with constant angular velocity ω_s. The innermost sphere is labelled #1, and the outermost by #2.

magnetization of #1 precesses about the total angular momentum, and the second term proportional to $\bar{\lambda}$ attempts to damp it to be counteraligned with this same conserved quantity L. Using the fact that the angular velocity of the superfluid is proportional to the total angular velocity, eqn.3), shows that the viscous term attempts, on the other hand, to align the magnetization of #1 with L.

It is remarkable that there are only two parameters which are relevant to the dynamics. Starting with some 11 parameters, $(a,\ b,\ \gamma,\ \bar{\lambda},\ \eta_1, \eta_2, I_1, I_2, I_s,\ c$ and $L)$, after scaling the equation for the magnetization M_1 of eqn. 5), and use of eqns. 1) and 4), this equation is given by

$$\frac{d}{d\tau}\vec{m} = \vec{m} \times \hat{k} + \lambda(\vec{m}\cdot\hat{k})\vec{m}/m^2 - \lambda\hat{k} - \overline{\overline{\epsilon}}\cdot(\vec{m}-\hat{k}). \tag{6}$$

Here we have defined the dimensionless quantities

$$\vec{m} = \vec{M}_1(I_s/acL), \tag{7}$$

$$\tau = (\gamma bL/I_2)t, \tag{8}$$

$$\lambda = \bar{\lambda}(I_s/\gamma acL), \tag{9}$$

$$\overline{\overline{\epsilon}} = \overline{\overline{\eta}}(I_2/ab\gamma L) \equiv \text{diagonal }(\epsilon,\epsilon,\bar{\epsilon}), \tag{10}$$

$$\overline{\overline{\epsilon}}/\lambda = (\overline{\overline{\eta}}/\bar{\lambda})(I_2c/I_sb). \tag{11}$$

A straightforward fixed point stability analysis shows that there are four distinct regions (see Appendix A) of parameter space, in the two-dimensional parameter space labelled by ϵ/λ, the ratio of the transverse viscous damping to the magnetic damping, and $\bar{\epsilon}/\lambda$, the ratio of the longitudinal viscous damping to the magnetic damping respectively. The four regions are shown in figure 2. with the corresponding fixed points and their stability. Regions I and II both have the one stable point corresponding to the three angular velocities ω_1, ω_2, ω_s being all aligned, and the magnetic fields of bodies #1 and #2 counteraligned. This region we will refer to as the "dead" region, that is, where $\epsilon/\lambda > 1$. There is no periodic pulse from this region, and it yields the smallest magnetic field, since the magnetic fields of the two charged bodies are counteraligned along L. In figure 2, the magnitude of the perpendicular component m of the magnetization of body #1 is shown along the "x" axis in the four regions (see Appendix A).

The other regions of parameter space, regions III and IV, are more interesting dynamically. Region III, where $\epsilon/\lambda < 1$ and $\bar{\epsilon}/\lambda < 1$, shows very erratic and unpredictable behavior, characteristic of "chaotic" systems (e.g., Schuster, 1989). As is seen in figure 3., there are occasional very large and directional flipping of the magnetization through the x,y plane, in a time of the order of the inverse rotation frequency, that is, about 1s. These large magnetic fields are unpredictable both in time of occurrence and in angle in the x,y plane. The corresponding figure 4. for the same parameters as figure 3. shows the behavior of the z component vs. the x component of magnetization of body #1. From eqns. 1) and 4) there is a corresponding time behavior of the magnetization of body #2, but not shown in the figures. A forthcoming publication (Dixon, Cummings and Kaus, submitted) discusses some of the more interesting mathematical aspects of the chaotic behavior in this region in more detail. Suffice it to say here that this region exhibits a "generic" chaotic behavior, in the sense that there are no stable points in the space, and only one unstable (saddle) point

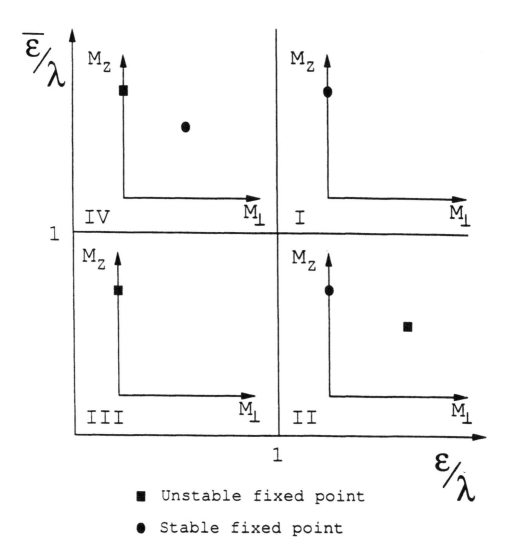

Fig. 2.   The stable (solid circle) and unstable (solid square) fixed points of eqn. 6) are shown
in the two-parameter space of ϵ/λ and $\bar{\epsilon}/\lambda$ discussed in appendix A. The four regions of
parameter space are labelled I, II, III, and IV.

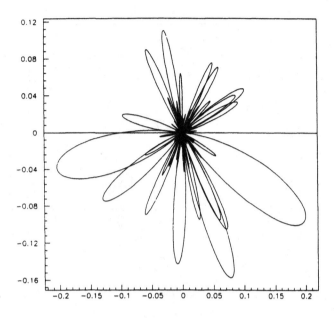

Fig. 3. The dynamical behavior of eqn. 5) or 6) is shown as components of magnetization M_y vs. M_x, that is, in the plane perpendicular to the conserved angular momentum vector L. The parameters are $\epsilon/\lambda = 0.6$ and $\bar{\epsilon}/\lambda = 0.45$.

on the M_z axis; thus the system has no "home" and is condemned to wander forever while in this region of parameter space. The rapidly flipping magnetic field will induce an electric field directed along the equator, and this will give rise to a flow of considerable energy into the magnetoplasma surrounding the star (Cummings, Dixon and Kaus, 1992). We will return to discuss this aspect briefly later, as a possible observable, or as a non-pulsar source of supernova remnant excitation.

Region IV of parameter space corresponds to regular limit cycle behavior, as is seen in figure 2, where the only stable point corresponds to a constant value of both $m_\perp$ and m_z. The period depends only on the conserved (over a time scale T_2, about 10^7 yrs.) total angular momentum, and geometrical factors which are expected to remain constant on this same time scale. In particular, the period T_0 does not depend on any of the numerous parameters introduced in eqn.5) except for L and I_2. This region is clearly to be associated with pulsar behavior.

As is shown in figure 7a., the z components of the angular momentum vectors of the two bodies, #1 and #2, are adding up to the total L, while the x,y components of each are precessing 180° out of phase so as to give zero net x,y component to the total angular momentum; however, because the magnetic fields of the two bodies are aligned and counteraligned respectively to the two angular momentum vectors by eqn. 1), it follows that we **always** have an off-axis precessing total magnetic field in the pulsar region, shown in figure 7b., in accordance with what is **assumed** in the standard picture.

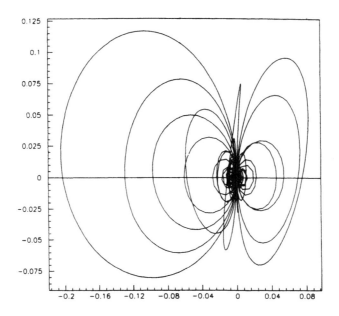

Fig. 4. The values of the parameters are the same as in figure 3. Here the components M_z vs. M_x are shown.

The period T_0 of this pulsar period is given from eqn. 8) by

$$T_0 = 2\pi(\gamma b)I_2/L, \tag{12}$$

An elementary calculation (Appendix B) shows that the quantity (γb) in this expression for the period is unity, in the case that we assume that the charge is distributed on a spherical shell with a thickness much less than the radius of the shell. A fraction $<< 10^{-10}$ of the total number of nucleons present in the star are charged in order to give rise to the immense magnetic fields of a neutron star. This charge need only be spread over a shell of thickness of about one Angstrom if it has the same average mass density as the neutron star (Appendix B), but the present model does not specify the charge distribution, which may be supposed to be much less sharply distributed over the two respective bodies. The pulsar period is thus given by, using $\gamma b = 1$ from Appendix B,

$$T_0 = 2\pi(I_2/L). \tag{13}$$

This period is expected to change on the time scale of 10^6 to 10^7 yrs. as the angular momentum is radiated away, in accordance with the usual picture. This expression for the period T_0 gives us some insight as to why the period is so stable, depending as it does on only the most fundamental constants of the problem, and not, in particular on the two parameters important for the shorter time dynamics in figure 8, (ϵ/λ and $\bar{\epsilon}/\lambda$), or any of the other 10 parameters of the model.

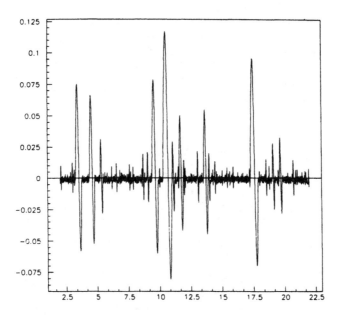

Fig. 5. The Magnetization M_z is shown vs dimensionless time τ for the parameters of figures 3 and 4.

An important aspect of the present model is the implication of a new time scale T_1 in the dynamics of neutron stars, one which we estimate to be of the order of 10^3 to 10^4 yrs. In the standard picture, there are essentially two times, namely, the pulsar period, T_0 of the order of a second, and the time T_2 of loss of total angular momentum, of the order of 1 to 10 million years. The time scale T_1 here arises because of the rate of loss of rotational energy, a quantity independent of the rate of loss of the total angular momentum. These two are usually assumed to be the same rates, since the usual model has only co-rotating components. However, in the present case, as shown in figure 7a., the x,y (transverse) components of the two angular momentum vectors, and thus the corresponding **transverse** rotational energy, will relax to zero in a time T_1 independent of the relaxation time T_2 of the **total** angular momentum. This is because the viscous forces are causing a loss of rotational energy, presumably mostly converting it into heat (not included in the model); thus the system is being propelled toward a state in which all three bodies are co-rotating. This is the inevitable final resting place for the neutron star, and one which we suggest will be reached a good bit sooner than the demise of the total angular momentum. The system is thus doomed to end its life in regions I or II of the parameter space of figure 2. This will lead to a revised estimate of the life of a pulsar as T_1 instead of T_2. If we only see roughly one percent (say) of the pulsars we expect to see based on the rate of supernova explosions and the standard picture, then a lifetime of the pulsar of about ten thousand years is expected as a very crude estimate for T_1 on hand of the present model, very much less, at any rate, than the presently accepted value of millions of years.

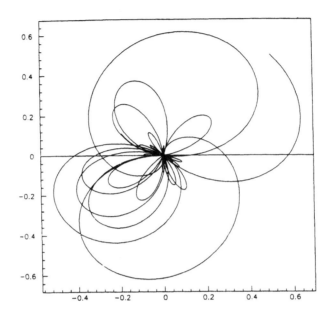

Fig. 6. The x-y components of magnetization are shown for parameter values $\epsilon/\lambda = 0.7$ and $\bar{\epsilon}/\lambda = 0.8$ closer to the pulsar region, where the dynamical behavior becomes a circle.

This leads to an entirely new story of the life history of a neutron star, as shown schematically in figure 8. In the present picture, the neutron star is born most often, befitting its violent beginning, in the chaotic region of parameter space, region III of the space of figure 1. or 8., where the two components giving rise to the two magnetic fields are rotating chaotically (but deterministically!) relative to one another. As they lose rotational energy and make their way in parameter space inevitably to region I or II (the dead regions), they may follow one of two possible paths, shown in figure 8. Path "a" leads through the pulsar region, and during this whole trip through this region the pulsar **period** changes on the time scale of 1 to 10 million years, as observed, while the trajectories "a" or "b" are traversed in the time T_1. A second possible path "b" leads directly from the chaotic region to the dead region. It has not been possible to date to calculate the trajectories of "a" or "b" as the two parameters ϵ/λ and $\bar{\epsilon}/\lambda$ change on a time scale $T_1 \ll T_2$. (One suggestion is that the magnetic damping $\bar{\lambda}$ may be decreasing, perhaps due to the viscous heating; such a decrease gives the positive slopes of figure 8.).

What is clear from the above discussion is that we will expect to see far fewer pulsars than supernovas, and an upper limit on the periods (by observation, 4s) is expected since the pulsar is propelled into the "dead" region long before it has a chance to dissipate its total rotational energy or total angular momentum.

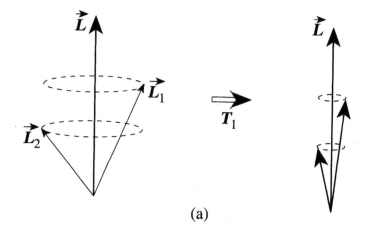

(a)

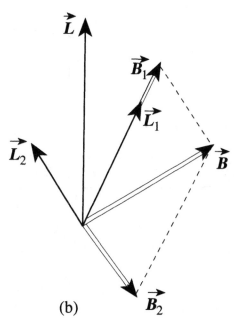

(b)

Fig. 7. a): The decay of the perpendicular components of the two angular momenta L_1 and L_2 is shown schematically to occur in a time T_1 much less than the decay time T_2 of the total angular momentum. There is a corresponding loss of rotational energy. b): The total magnetic moment M is always off-axis relative to the direction of total angular momentum in the pulsar region IV, in view of eqn. 1) of the text.

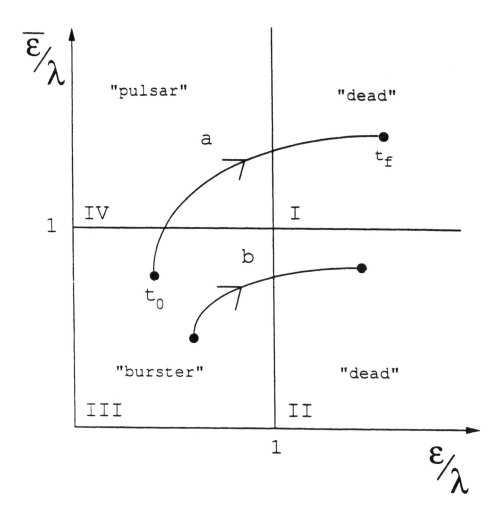

Fig. 8. Two possible trajectories "a" and "b" which occur in a time $T_1 << T_2$, due to the loss of rotational energy without a concomitant loss of total angular momentum. The system must always end up in the "dead" regions I or II. This is the basis in the model for the relatively small number of pulsars observed, as well as for the lack of long period pulsars > 4s.

3. Discussion

As a first point, we notice that the two component model (Cummings, Dixon and Kaus, 1992) is given by ω_s going over to ω_2 and I_s/c over to I_2.

The gamma ray "bursters", more than twenty years after their discovery, still have no generally accepted explanation (Hurley, 1990). It is tempting to suggest (Cummings, Dixon and Kaus, 1992) that the rapid tumbling of the immense magnetic field as depicted in figures 3, 4, and 5 could supply enough energy to the magnetoplasma surrounding the neutron star to allow a small fraction of that energy to be converted to high energy gammas by such processes as inverse Compton scattering. A simple estimate given in the CDK (1992) reference above shows that approximately 10^{41} ergs/sec will be available to the magnetoplasma, and that if as small a fraction of this energy as 10^{-4} is converted to gammas, then the observed flux on earth of 10^{-5} ergs/cm^{2}s will place the burster easily in our own galaxy. However, the Gamma Ray Observatory (GRO) which started to relay data to earth in about March, 1992, shows a very surprising isotropic distribution of the bursters, as well as a cutoff of intensity, which is in marked distinction to the distribution of pulsars, which is predominately in the galactic plane. Thus CDK can claim to have a nice example of a falsifiable model of bursters, à la Karl Popper; even better, one which in fact has already been falsified, according to these recent GRO data.

However, a prediction of observability of the rapidly reversing magnetic field of young neutron stars remains. For example, it seems possible that this energy could be supplying the luminosity of the supernova remnant in a number of cases. It is even possible that the Crab, so often believed to have begun pulsing in 1054 AD. could in fact have started pulsing a mere thirty years ago; a pulsar is no longer needed to supply the remnant with its glow. A similar suggestion may be relevant for SN1987A.

It is interesting that the dominant opinion before launch of the GRO was that the bursters were emanating from neutron stars; the fast rise time of the order of millisecs. argued for a compact body; there is evidence for very strong magnetic fields; and the single well localized burster, seen in 1979 (Hurley, 1990), came from a well known SN remnant in the Magellanic cloud, some 170,000 l.y. away. This all makes the GRO isotropic observations even more perplexing.

There are many questions which remain concerning this model (as for all others). Foremost perhaps is the difficult question of the origin and maintenance of the charge separation proposed here. It would seem that a charge separation is required, for how else arises the huge magnetic field? As we have shown in appendix B., only a very small fraction of the total number of nucleons in the star need be charged to give such huge fields. However, if we spread this charge of one sign uniformly over a sphere of radius the order of 10^6 cm. we find that the repulsive electrostatic force overwhelms the gravitational forces by a factor of about 10^{17}. This means that we must look to electrostatic means of confining the charge to the sphere. A suggestion is that the charge stays confined to the sphere because the underlying medium becomes strongly polarized. In the case of the neutron star, the neutrons at a given density are in equilibrium with a much smaller number of protons and electrons at any time. A given net charge distribution on the surface, say of negative sign, will polarize these underlying protons and electrons, forming a kind of three layered sandwich with the protons in the middle layer. Equilibrium is obtained at that density and radius which supplies sufficient numbers of protons and electrons to confine the charge layer. Yet another variation of this polarization theme (and even more speculative) has the inner body radius at an "Oppenheimer-Volkov" phase boundary with an inner core of hadrons (quarks) (Olive, 1991) which are polarized sufficiently to confine the charged layer. Perhaps one mechanism works for the outer body, and the O-V scenario for the inner. On the other hand it may be more realistic to picture the charge to be distributed throughout the inner sphere rather than concentrated on the surface. Future calculations will hopefully show whether these brief speculative remarks are realistic.

Appendix A

It is convenient, because of the axial symmetry of eqn.6) to introduce the perpendicular and z components of m, $m_\perp$ and m_z. The fixed points of eqn. 5) are determined by setting

$$m_x = m_\perp \cos\Theta, \quad m_y = m_\perp \sin\Theta. \tag{A.1}$$

Then eqn.6) of the text takes the form of the three equations

$$\frac{d}{d\tau} m_\perp \equiv \dot{m}_\perp = \frac{\lambda m_\perp m_z}{m_\perp^2 + m_z^2} - \epsilon m_\perp, \tag{A.2}$$

$$m_\perp \dot{\Theta} = m_\perp, \tag{A.3}$$

$$\dot{m}_z = \frac{\lambda m_z^2}{m_\perp^2 + m_z^2} - \bar{\epsilon} m_z - (\lambda - \bar{\epsilon}). \tag{A.4}$$

To find the fixed points of eqn. 6), let $\dot{m}_\perp = 0$ and $\dot{m}_z = 0$ in eqns. A.2) and A.4). This results in the two solutions

$$a) \quad (m_\perp)_0 = 0, \quad \text{and } (m_z)_0 = 1; \tag{A.5}$$

and

$$b) \quad (m_z)_0 = \frac{\lambda - \bar{\epsilon}}{\epsilon - \bar{\epsilon}}, \quad \text{and } (m_\perp)_0^2 = (\lambda/\epsilon)(m_z)_0 - (m_z)_0^2$$

$$= \left[\frac{(\lambda - \bar{\epsilon})(\epsilon - \lambda)}{(\epsilon - \bar{\epsilon})^2}\right]\frac{\bar{\epsilon}}{\epsilon} > 0. \tag{A.6}$$

The relevant parameters are easily seen to be ϵ/λ and $\bar{\epsilon}/\lambda$, and the various fixed points in the two dimensional parameter space are shown in figure 2). Here the parameter space is divided into four regions, labelled I, II, III, and IV, moving clockwise. The stable fixed points are denoted by the filled circles and the unstable fixed points by the filled squares. The two parameters are ratios of the transverse mechanical damping to the magnetic damping, ϵ/λ, and the second parameter is the ratio of the parallel (to the spin axis) mechanical damping to the magnetic damping, $\bar{\epsilon}/\lambda$. A summary of the main results follows.

Region I, where $\epsilon/\lambda > 1$ and $\bar{\epsilon}/\lambda > 1$ has only one fixed point, and that one is the stable point $(m_\perp)_0 = 0$ and $(m_z)_0 = 1$. This is a region of phase space for which the star's magnetic axis is aligned with the spin axis, or the direction of the conserved angular momentum. This is a region in which, as in region II, the angular velocities of the two bodies are equal, and both aligned with the m_z direction. The magnetic field in both regions I and II is thus a constant. Region II, for which $\epsilon/\lambda > 1$ and $\bar{\epsilon}/\lambda < 1$ also has only the stable fixed point $(m_\perp)_0 = 0$ and $(m_z)_0 = 1$, but also has an unstable fixed point at

$(m_\perp)_0 > 0$. From the point of view of the present model, a region I or region II star will be invisible, or "dead".

Region IV for which $\epsilon/\lambda < 1$ and $\bar{\epsilon}/\lambda > 1$ shows two fixed points, one of which is stable and the other unstable. The stable fixed point has $(m_\perp)_0 > 0$ and $(m_z)_0 > 0$. This stable dynamical behavior, by reference to eqn. A.6) or eqn. 5) is thus a circle in the m_x, m_y plane, and corresponds to precession of the magnetization vector about the z direction at a constant angle, $\tan(\alpha) = (m_\perp)_0/(m_z)_0$, and with the constant period of 2π. The dynamical behavior of the system in this region of parameter space is thus to be associated with "pulsar" behavior.

Appendix B

An estimate of the number of charge carriers required to give a constant internal magnetic field of 10^{12} Gauss when rotating on a thin spherical shell of radius R at constant angular speed of ω may be found as follows. The current density on the sphere (current/length) to give an equivalent constant internal magnetization is given by the well known expression

$$\vec{j} = c(\vec{M} \times \vec{n}) = cM \cdot \sin\vartheta \cdot \vec{e}_\varphi. \tag{B.1}$$

Integrating this over a length from north to south pole gives the total current as

$$I = \int_0^\pi (cM \sin\vartheta)R d\vartheta = 2cM \cdot R. \tag{B.2}$$

This total current may be equated with the total charge times the rotational frequency $\omega/2\pi$ to give

$$I = Q\frac{\omega}{2\pi} = 2cMR. \tag{B.3}$$

The magnetic field inside the sphere is related to the magnetization by $B = (8\pi/3)M$, allowing us to write the total number of charges $N = Q/e$ as

$$N = \frac{3cBR}{2e\omega} = (3e/2mc)\left(\frac{B}{\omega}\right)\left(\frac{R}{(e^2/mc^2)}\right) \equiv \gamma_0\left(\frac{B}{\omega}\right)\frac{R}{r_0}. \tag{B.4}$$

Here we have defined the classical electron radius $r_0 \approx 10^{-13}$ cm. and the convenient "magnetron" $\gamma_o = 3 \cdot 10^8/(\text{Gauss sec})$. Using $B = 10^{12}$ Gauss, $\omega = 1$ sec^{-1}, and $R = 10^6$ cm. (Michel, 1991) gives a value for the number of required charged particles as

$$N \simeq 10^{36}. \tag{B.5}$$

The total number of neutrons in a neutron star may be estimated as 1.4 times the sun's mass divided by the neutron mass. This gives a value of about 10^{55} neutrons or more. Clearly then, the ratio (number of charges needed to reside on a thin shell to give the requisite magnetic field) to (the total number of neutrons in the neutron star) is an extremely small number, certainly less than 10^{-10}. Using these same numbers and assuming that the thin shell has about the same density as the rest of the neutron star allows us to easily estimate the thickness of the shell as about an Angstrom, where we may assume that the ratio of the shell radius to the neutron star radius is approximately 10^{-1}.

The crucial time scale T_0 of the process introduced in eqn.8) of the text may be seen to be the inverse angular speed of the outer sphere, that is $(L/I_2)^{-1}$, where L is the conserved angular momentum and I_2 is the moment of inertia of the outer body. From eqn. 8) we have that

$$T_0^{-1} \equiv \frac{b\gamma L}{I_2}. \qquad (B.6)$$

Reference to eqns. 1) and 5), and realization that the rotational motion about the z axis arises solely from the first term on the right of eqn.5), that is, from the term $\sim \gamma$, shows that $\gamma = b^{-1} \approx a^{-1}$. Further, using the value obtained in the previous paragraph for the magnetic field $\mathbf{B}$ in the first term in eqn. 5) of the text gives us b and an estimate of a as well, that is

$$\gamma = \gamma_0 \frac{(R/r_0)}{N} = b^{-1} \simeq 10^{-12}/(\text{Gauss sec}). \qquad (B.7)$$

We notice that this value for γ when used in eqn. 5) in conjunction with eqn. 1) gives a consistent result, namely that a field of 10^{12} Gauss will give a precession frequency of the order of $1 \ \text{sec}^{-1}$.

ACKNOWLEDGMENTS. I owe a great debt to E.T. Jaynes for the encouragement and support that he has given over the many years. Although this particular document may not be **directly** inspired by him (and of course, he is never liable for my mistakes), he has been my principal physics inspiration, teacher and model over the past thirty five years. I would also like to remember friends Mike Duggan and Larry Davis who, sadly, could not attend this conference.

REFERENCES

Cummings, F.W., D.D. Dixon and P.E. Kaus,: 1992, Astrophy. Jour. **386**, 221.

Dixon, D., F.W. Cummings and P.E. Kaus (submitted, 5/92).

Greenstein, G.: "Superfluidity in Neutron Stars" in *Physics of Dense Matter* (C.J. Hansen, ed., 1974, IAU)

Hurley, K.A.,: 1990, Sky & Telescope **80**, 143.

Landau, L.,: 1965, *Collected Papers of L.D. Landau*, (Pergamon Press, Oxford).

Lyne, A.G. and F. Graham-Smith,: 1990, *Pulsar Astronomy* (Springer-Verlag, Berlin).

Michel, F.C.,: 1991 *Theory of Neutron Star Magnetospheres*, (Univ. of Chicago Press, Chicago and London).

Olive, K.A.,: 1991, Science **251**, 1194.

Schuster, H.G.,: 1989, *Deterministic Chaos*, (VCH, Weinheim, 2nd ed.).

Shapiro, S.L. and S.A. Teukolsky,: 1983, *Black Holes, White Dwarfs and Neutron Stars; the physics of compact objects* (John Wiley & Sons, New York).

THE KINEMATIC ORIGIN OF COMPLEX WAVE FUNCTIONS

David Hestenes
Department of Physics and Astronomy
Arizona State University
Tempe, Arizona 85282

ABSTRACT. A reformulation of the Dirac theory reveals that $i\hbar$ has a geometric meaning relating it to electron spin. This provides the basis for a coherent physical interpretation of the Dirac and Schödinger theories wherein the complex phase factor $\exp(-i\varphi/\hbar)$ in the wave function describes electron zitterbewegung, a localized, circular motion generating the electron spin and magnetic moment. Zitterbewegung interactions also generate resonances which may explain quantization, diffraction, and the Pauli principle.

> *You know, it would be sufficient to really understand the electron.*
> —Albert Einstein [1]

1. Introduction

Edwin T. Jaynes is one of the great thinkers of twentieth century science [2]. More than anyone else he has deepened and clarified the role of statistical inference in science and engineering. To my mind, his greatest accomplishment has been to recognize that in the evolution of statistical mechanics the principles of physics had gotten confused with principles of statistical inference, and then to show how the two can be cleanly separated to produce a simpler yet more powerful theoretical system.

I share with Ed Jaynes the belief that quantum mechanics suffers from an analogous muddle of probability with physics, which is at the root of the perennial controversy over physical interpretation.

Though a Jaynesian revolution of the "quantum muddle" remains elusive, I will report here on a promising possibility that has been overlooked.

One of the puzzling features of quantum mechanics is the fact that "probability amplitudes" are complex numbers whereas probabilities are real. This has inspired a belief that quantum mechanics somehow involves a generalization of the "classical probability concept." On the contrary, I contend that complex phase factors have a physical origin that has noting to do with the probability concept *per se*.

The words of Einstein heading this article serve to remind us that the main ideas, as well as the greatest successes, of quantum mechanics have come from studying the electron. To the electron, therefore, we should look for the physical clues needed to resolve the quantum muddle. Now, the Dirac electron theory and its extension to *quantum electrodynamics* is universally recognized as the most well substantiated domain of physics. Strangely, however, it is rarely involved in the discussions of the foundations of quantum mechanics. This is a grievous error, for the *Dirac theory entails an irreducible relation between spin and complex numbers* with undeniable implications for the interpretation of quantum mechanics. Analysis of this relation strongly suggests that the *complex phase factor in the complex function describes a kinematic feature of electron wave motion* and therefore has a physical, rather than statistical, origin.

The argument in support of these contentions has been elaborated elsewhere [3], so we can be satisfied here with an outline of the main ideas. The argument and its implications are developed in three separate stages. Stage I is a reformulation of the Dirac theory which makes its geometric structure explicit, and reveals a connection between spin and the complex imaginary. Stage II assigns a coherent physical interpretation to the Dirac theory which is consistent with its geometric structure. Stage III speculates on the possibility of a deeper theory of electrons to which the Dirac theory is only an approximation.

In the text that follows, major assertions are set off in boxes and supported by just enough discussion to make them intelligible or, at least, stimulating to the reader familiar with the Dirac theory.

2. STAGE I. Explicating the geometric structure of the Dirac theory

The crucial first step is to recognize that the set of Dirac matrices γ_μ can be interpreted as an orthonormal frame of vectors in spacetime, rather than mysterious matrix components of a single vector, as is ordinarily done. This is possible because products of the γ_μ have well-defined geometric meanings. Thus, the familiar symmetrized product

$$\tfrac{1}{2}\left(\gamma_\mu\gamma_\nu + \gamma_\nu\gamma_\mu\right) = \gamma_\mu \cdot \gamma_\nu = g_{\mu\nu} \tag{1}$$

is nothing other than the *inner product* of vectors defining the metric tensor $g_{\mu\nu}$. On the other hand, the antisymmetrized product

$$\tfrac{1}{2}\left(\gamma_\mu\gamma_\nu - \gamma_\nu\gamma_\mu\right) = \gamma_\mu \wedge \gamma_\nu \tag{2}$$

can be identified with the *outer product* of Grassmann algebra, with the $\gamma_\mu \wedge \gamma_\nu$ composing a basis for the space of *bivectors* (i.e., skew-symmetric tensors of rank two). These two products are thus parts of a single *geometric product*

$$\gamma_\mu\gamma_\nu = \gamma_\mu \cdot \gamma_\nu + \gamma_\mu \wedge \gamma_\nu. \tag{3}$$

Two other geometric entities are generated by this product: the trivectors (or *pseudovectors*)

$$\gamma_\mu \wedge \gamma_\nu \wedge \gamma_\alpha = \epsilon_{\mu\nu\alpha\beta}\gamma^\beta\gamma_5 \tag{4}$$

and the *pseudoscalar*

$$\gamma_5 = \gamma_0\gamma_1\gamma_2\gamma_3 = \gamma_0 \wedge \gamma_1 \wedge \gamma_2 \wedge \gamma_3. \tag{5}$$

All of these geometric entities are perfectly defined without regarding the γ_μ as matrices. Every element of the Dirac algebra can be expressed as a linear combination of the 16 basis elements (with $\mu, \nu = 0, 1, 2, 3$)

$$I, \quad \gamma_\mu, \quad \gamma_\mu \wedge \gamma_\nu, \quad \gamma_\mu\gamma_5, \quad \gamma_5 \tag{6}$$

where I is the identity matrix if the γ_μ are matrices, but $I = 1$ is the unit scalar if the γ_μ are regarded as vectors. If the coefficients are real, the matrix algebra generated by these basis elements is called the *real Dirac algebra*, but the algebra is called the *spacetime algebra* if the γ_μ are vectors. Thus, it can be concluded that:

> The real Dirac algebra is a matrix representation of a generic geometric algebra describing properties of spacetime with no special relation to the quantum mechanics of spin.

The spacetime algebra was first employed in the formulation of electrodynamics and general relativity in [4].

Naturally, it takes some practice to become fully conversant with the spacetime algebra and its geometric significance. In the meantime, readers may continue to regard the γ_μ as matrices when considering algebraic manipulations.

The next step in explicating the geometric structure of the Dirac theory is to reformulate it in terms of the spacetime algebra. One easy way to do that is to choose a fixed unit spinor u satisfying the eigenvalue equations

$$\gamma_0 u = u, \tag{7}$$

$$\gamma_2 \gamma_1 u = i u, \tag{8}$$

where i is the usual unit imaginary of the complex field. Equation (8) is especially significant because it relates i to the bivector $\gamma_2 \gamma_1$, and thus reveals that i has an implicit geometric meaning in the Dirac theory. This meaning can be made explicit by eliminating i from the theory in favor of the geometric quantity $\gamma_2 \gamma_1$. It is easy to prove that each Dirac spinor Ψ can be written in the form

$$\Psi = \psi u, \tag{9}$$

where ψ is a unique *even* element of the *real* Dirac algebra. In other words, ψ can be written in the form

$$\psi = \alpha_1 + \alpha^{\mu\nu} \gamma_\mu \wedge \gamma_\nu + \alpha_2 \gamma_5, \tag{10}$$

where the eight alphas are real coefficients. Since there is an isomorphism between Ψ and ψ, it is fair to refer to ψ as a spinor or, more specifically, as an *operator representation* of a Dirac spinor.

In terms of ψ the Dirac equation takes the form

$$\gamma^\mu \left(\partial_\mu \psi \gamma_2 \gamma_1 \hbar - \frac{e}{c} A_\mu \psi \right) = mc\psi\gamma_0 \tag{11}$$

Multiplication by u on the right shows immediately that this is equivalent to the usual Dirac equation

$$\gamma^\mu \left(i\hbar\partial_\mu - \frac{e}{c} A_\mu \right) \Psi = mc\Psi \tag{12}$$

The significant difference between these isomorphic equations is that (11) involves the real Dirac algebra exclusively. Note how the imaginary factor $i\hbar$ in (12) has been replaced by the bivector $\gamma_2 \gamma_1 \hbar$ in (11).

The geometric structure of the Dirac wave function is expressed by the following assertion.

> The Dirac wave function has an invariant operator representation
>
> $$\psi = \left(\rho e^{\beta\gamma_5}\right)^{\frac{1}{2}} R, \tag{13}$$
>
> where R is a unimodular spinor describing the kinematics of electron motion.

Here unimodularity means that R can be written in the form

$$R = e^{\frac{1}{2}B}, \tag{14}$$

where B is a bivector. The relation to kinematics comes from the fact that R determines a Lorentz transformation of the frame $\{\gamma_\mu\}$ into a "frame of observables" $\{e_\mu\}$ given by

$$e_\mu = R\gamma_\mu \widetilde{R}. \tag{15}$$

The timelike unit vector e_0 is the direction of the Dirac current, while e_3 is the direction of the "spin," or polarization, vector. This can be proved by establishing the following relations to standard expressions in the Dirac theory. The components J_μ of the *Dirac current*

$$J = \psi\gamma_0\widetilde{\psi} = \rho e_0 \tag{16a}$$

are given by

$$J_\mu = J \cdot \gamma_\mu = \widetilde{\Psi}\gamma_\mu\Psi. \tag{16b}$$

The components of the *spin vector*

$$s = \tfrac{1}{2}\hbar e_3 \tag{17a}$$

are given by

$$\rho s_\mu = \tfrac{1}{2}\hbar\gamma_\mu \cdot (\psi\gamma_3\widetilde{\psi}) = \tfrac{1}{2}\hbar\widetilde{\Psi}\gamma_5\gamma_\mu\Psi. \tag{17b}$$

The γ_μ (or the $\gamma_0\gamma_\mu$) have been interpreted as velocity operators in the Dirac theory. As (16b) shows, this is tantamount to identifying the direction of Dirac current as the local electron velocity, so the γ_μ are operators only in the trivial sense of picking-out vector components by the inner product with basis vectors.

Spin angular momentum is actually a bivector quantity, though it can be proved from angular momentum conservation in the Dirac theory that the bivector spin S is related to the spin vector s in (17a) by

$$S = \gamma_5 s e_0 = \tfrac{1}{2}\hbar\gamma_5 e_3 e_0. \tag{18a}$$

Moreover, $\gamma_5 e_3 e_0 = e_2 e_1$, so

$$S = \tfrac{1}{2}\hbar e_2 e_1 = \tfrac{1}{2}R(\gamma_2\gamma_1\hbar)\widetilde{R}. \tag{18b}$$

This relates the bivector $\gamma_2\gamma_1\hbar$ in the Dirac equation (11) to the electron spin. To express it as a relation in the standard Dirac language, multiply (9) by (18b) and use (13) and (8) to get

$$S\Psi = \tfrac{1}{2}i\hbar\Psi \qquad (19)$$

This proves unequivocally that the "imaginary" factor $i\hbar$ in the Dirac equation is a representation of the electron spin angular momentum S by its eigenvalue, and the electron wave function is always an "eigenstate" of the spin.

This *fact* that the ubiquitous factor $i\hbar$ is a representation of electron spin necessarily applies to the Schrödinger equation as well, in as much as it is an approximation to the Dirac equation. It enriches Dirac's conclusion that *the most fundamental aspect of quantum mechanics is the role of the unit imaginary i* [5].

3. STAGE II. The zbw interpretation

The identification of the bivector S as spin rests on the prior identification of

$$\underline{p}_\mu = i\hbar\partial_\mu - \frac{e}{c}A_\mu \qquad (20)$$

as *energy-momentum operator*, which is surely one of the fundamental postulates of quantum mechanics. But what shall we make of the "spin" factor $i\hbar$ in (20)? An answer is suggested by examining the electron kinematics.

In terms of $\underline{p}_\mu$ the energy-momentum tensor of the Dirac theory is given by

$$T_{\mu\nu} = \tilde{\Psi}\gamma_\mu\underline{p}_\nu\Psi \qquad (21)$$

To see what this implies about electron kinematics, note that the derivatives of the kinematics factor R in (13) can be written in the form

$$\partial_\mu R = \tfrac{1}{2}\Omega_\mu R, \qquad (22)$$

where Ω_μ is a bivector representing the rotational velocity of the frame $\{e_\nu\}$ under a displacement in the direction γ_μ. The local electron momentum p_μ is defined by the energy-momentum flux in the direction of the electron velocity, that is,

$$\rho p_\mu = (e_0 \cdot \gamma^\nu)T_{\mu\nu}, \qquad (23)$$

where ρ is the probability density defined by (13). Evaluated with (21) and (22), (23) yields

$$p_\mu = S \cdot \Omega_\mu - \frac{e}{c}A_\mu \qquad (24)$$

The term $S \cdot \Omega_\mu$ has precisely the form of a rotational kinetic energy. Thus, the intrinsic energy (mass) of the electron is associated with rotational motion in the spacelike plane of the spin S.

To give this fact a literal physical interpretation wherein the spin S is generated by electron motion, the velocity of the electron must be redefined to maintain a component in the spin plane. Accordingly, the most natural definition of electron velocity is the *null vector*

$$u = e_0 - e_2. \tag{25}$$

This leads to *a self-consistent interpretation of the Dirac theory with the following features:*

(a) The electron is modeled as a structureless point particle travelling at the speed of light along a helical lightlike trajectory in spacetime.

(b) The helical trajectory has a diameter on the order of a Compton wavelength, and a circular frequency on the order of twice the de Broglie frequency $mc^2/\hbar \approx 10^{21}\mathrm{s}^{-1}$.

(c) The helical motion generates electron spin and may be attributed to magnetic self-interaction.

(d) Each solution of the Dirac equation determines an infinite family of such helices and a probability distribution for the electron to be found on any given helix.

(e) The center of curvature for each helix lies on a streamline of the Dirac current.

All these assertions are consistent with the mathematical form of the Dirac theory, and they supply the mathematical structure of the theory with the most complete and coherent physical interpretation available. They constitute a generalization of the *zitterbewegung interpretation* originally proposed by Schrödinger.

The zitterbewegung (zbw) is reflected in the structure of the kinematic factor R of the Dirac wave function (13) by writing it in the form

$$R = R_0 e^{-\gamma_2\gamma_1\varphi/\hbar}. \tag{26}$$

The helical motion of the electron can be visualized as a particle moving in a circle lying in the spacelike plane of the spin S while the center of the circle is translated along a streamline of the Dirac current. The *phase angle* φ in (26) represents the angular displacement on the circle (in units of angular momentum), lying in $\gamma_2\gamma_1$-plane, while R_0 represents a Lorentz transformation which rotates this plane into the spin plane. Inserting (26) into (22) and using (18b), one finds

$$\Omega_\mu = 2(\partial_\mu R)\widetilde{R} = 2(\partial_\mu R)\widetilde{R}_0 + (\partial_\mu\varphi)S^{-1}. \tag{27}$$

The term involving R_0 describes relativistic effects as the electron is accelerated by external forces, and it is always smaller than the last term. By keeping only the dominant term, then, insertion of (27) into (24) yields

$$p_\mu = \partial_\mu\varphi - \frac{e}{c}A_\mu, \tag{28}$$

which shows explicitly that the electron energy-momentum can be attributed to the *circular zitterbewegung*. In the same approximation, we can set $R_0 = 1$ and the Dirac wave function reduces to

$$\psi_0 = \rho^{\frac{1}{2}} e^{-\gamma_2 \gamma_1 \varphi / \hbar} \tag{29}$$

This is exactly the form of the Schrödinger wave function. Thus, we conclude that

> The complex phase factor $\exp(-i\varphi/\hbar)$ in both the Dirac and Schrödinger wave functions describes kinematics of electron motion, specifically the circular zitterbewegung.

Schrödinger theory describes the dominant component of the zitterbewegung. That's why it is such a successful approximation to the Dirac theory.

Of course, the challenge to this interpretation is to devise experimental tests which show that the helical zitterbewegung is a real physical phenomenon. One such test is a prediction of deviations from the Mott-scattering cross section when the impact parameter is on the order of a Compton wavelength [6].

4. STAGE III. Zitterbewegung Interactions

If the circular zitterbewegung is a real physical phenomenon and not merely a picturesque metaphor, then we should expect its presence to be manifest in the electron's electromagnetic field. Indeed, to explain the electron's static magnetic dipole field as a consequence of the zitterbewegung, we must regard the electron as a point charge for which the average motion over a zitterbewegung period is an effective current loop. But the same assumption implies that the zitterbewegung must generate high-frequency fluctuations about that average; call this the *zitterbewegung field* of the electron. The frequency of these fluctuations ($\sim 10^{21} \text{s}^{-1}$) is too high to observe directly, but it has been suggested that zitterbewegung fields are responsible for some of the most peculiar features of quantum mechanics [7].

The possibility of zbw interactions is not contemplated in the Dirac theory, though to some degree it may be inherent in the theory and its extention to quantum electrodynamics. In any case, it appears that a complete mathematical treatment of zbw interactions requires new physical assumptions which have not yet been formulated and analyzed, so the best that can be done at this time is a qualitative analysis of possible consequences.

We begin with the picture of each electron as the seat of a bound electromagnetic field fluctuating with the electron's zbw frequency, and we set aside the question of what theoretical assumptions are necessary to make this possible. Every electron, of course, is perturbed by the zbw fields of other electrons, with doppler-shifted frequencies due to their motions. Conceivably, this random background of electromagnetic fluctuations can play the role of the vacuum field in quantum electrodynamics, and a stochastic term in the electron equations of motion is needed to account for its effect. In that case, the Casimir effect could be explained by adapting well-known arguments.

A more significant and prominent feature of zbw interactions is the likelihood of resonances. Three kinds of resonances are of special interest:

(1) *Electron diffraction* is usually "explained" by invoking an interference metaphor. However, an explanation in terms of "quantized momentum exchange" is equally consistent with the formalism of quantum mechanics as well as a strict particle interpretation. Moreover, the zbw field provides a physical mechanism for this exchange. When an electron is incident on a crystal, it is preceded by its zbw field which is reflected from the crystal back to the electron. The conditions for *resonant momentum exchange* are met when the reflected field is resonant with the electron's zbw, which must occur at the Bragg angles if the explanation is correct.

(2) *Quantized energy states* appear when the frequency of an electron's orbital motion is resonant with a harmonic of its zbw frequency, as is implied by a zbw interpretation of solutions to Schrödinger's equations. As in diffraction, the underlying causal mechanism may be a resonance of the electron's zbw with the reflection of its own zbw field off the atomic nucleus.

(3) *The Pauli principle* may be explained as a zbw resonance between two electrons, mediated by their zbw fields. Such a resonance can occur only when the electrons are in the same state of translational motion, as the Pauli principle requires.

Other quantum phenomenon, such as barrier penetration and the Lamb shift, can also be explained qualitatively by the zbw. The problem remains to make these explanations fully quantitative.

REFERENCES

[1] This quotation was recollected and explained by Valentine Bargmann. In H. Woolf (ed.), *Some Strangeness in Proportion*, a Centennial symposium to celebrate the achievements of Albert Einstein. Addison-Wesley, Reading, MA (1980).

[2] D. Hestenes, "E.T. Jaynes: Papers on Probability, Statistics and Statistical Physics," *Found. Phys.*, **14** 187–191 (1984).

[3] D. Hestenes, "The Zitterbewegung Interpretation of Quantum Mechanics," *Found. Phys.*, **20**, 1213–1232 (1990). Further references are contained therein.

[4] D. Hestenes, *Space-Time Algebra*, Gordon and Breach. New York (1966, reprinted with corrections 1992).

[5] C.N. Yang, "Square root of minus one, complex phases and Erwin Schrödinger." In C.W. Kilminster (ed.), *Schrödinger*, Cambridge (1987), p. 53–64.

[6] M.H. MacGregor, "KeV Channeling Effects in the Mott Scattering of Electrons and Positrons," *Found. Phys. Letters*, **5**, 15–24 (1992).

[7] D. Hestenes, "Quantum Mechanics from Self-Interaction," *Found. Phys.* **15**, 63–87 (1985).

ON RADAR TARGET IDENTIFICATION

C. Ray Smith
Advanced Sensors Directorate
Research, Development, and Engineering Center
U.S. Army Missile Command
Redstone Arsenal, Alabama 35898-5293

1. Introduction

Humans have long been interested in devices which extend their senses, the telescope being an early example of such a device. We have since progressed to systems which operate in domains beyond the normal senses, as in the use of x-rays in medical imaging or seismic waves to probe the earth. Nevertheless, most existing remote sensing systems operate with considerable human input and interpretation; naturally, there is great interest in developing remote sensing systems which can with greater autonomy identify or recognize salient aspects of their environments. Potential applications of such systems abound:

1. Air-traffic control (identification of incoming aircraft).
2. Air defense (recognizing friendly or hostile aircraft).
3. Quality assurance (assembly line inspection).
4. Medical screening (identifying anomalies in radiographic or tomographic images).
5. Security and surveillance systems.
6. Robotics (navigation, carrying out tasks).
7. Exploration geophysics (interpreting seismic data).
8. Analytical chemistry (neutron activation analysis, nuclear magnetic resonance, other kinds of spectroscopy).
9. Detection of unexploded subsurface munitions.

All of these systems have in common the detection and processing of signals. In some cases (e.g., radar), the system must also emit a signal which interacts with the environment and is subsequently detected in modified form.

Though our understanding of human perception and information processing is incomplete, the very fact that recognition and identification are biologically possible demonstrates that the problem is not fundamentally intractable. And while understanding biological processes may well provide insight into designing such systems, certainly, efforts in this area should not be solely devoted to mimicking living systems. Other strategies—utilizing, for example, appropriate mathematical algorithms—may prove more suited to the technological materials at hand.

The identification of objects by active radar (such objects are termed, by convention, *targets*) is particularly interesting, because of the unique and valuable attributes of radar: It can operate in almost any weather conditions, it can provide nearly a limitless variety of waveforms, and its principles are thoroughly described and well understood. Further, radar systems are already in place at airports, on aircraft, boats and ships, and on satellites. Although identification by radar would be of immense value, such capabilities have lagged in the overall development of this technology.

In this paper, we introduce the reader to the radar target identification problem (active radar being understood), outline a theory of radar target identification (henceforth, RTID) and discuss aspects of the radar target identification problem which are making progress slow and arduous. In the next paragraph, we provide a brief review of our research program.

A small research program on RTID was begun in the Radar Technology Area of the Advanced Sensors Directorate, U.S. Army Missile Command, in 1988. The research has been conducted jointly by the author and collaborators affiliated with various universities. The initial studies of RTID were intended to be a theoretical analysis of the feasibility of target identification with active radar, wherein we would evaluate the relationships between waveforms, the information content of the returns, and noise. To our astonishment, we found that such studies could not be the "initial" studies, because the necessary theoretical framework for conducting the studies had not been assembled. Lacking the necessary analytical tools to attack the RTID problem, we assigned priority to the development of those tools. The theory of RTID outlined in Secs. 2 and 3 below is due to E. T. Jaynes; the complete development is available in [Jaynes and Smith, 1991].

2. General Formulation of the Problem

Suppose a target is known to be present and we are to infer from measured radar data the identity of the target (the level of identification need not be specified here; it could refer to a broad classification such as aircraft type or to the assignment of an individual aircraft "tail number"). Each target that can be present is assigned a hypothesis or proposition, where, by definition, a *proposition* is an unambiguous statement that is either true or false. To express target identification as an inference problem, we associate a proposition T_k with an index k labelling a target:

$$T_k = \text{"Target k is present."} \tag{1}$$

Hence, *target identification* is concerned with inferring which of the hypotheses T_k is true. In this discussion, we consider m different target classes and let k refer to a target class. In order to concentrate on the identification problem, we assume here that detection has indeed occurred and that exactly one target is present; thus, the propositions T_k are exhaustive (at least one is true) and mutually exclusive (only one can be true)—both of these assumptions must be eventually relaxed. Finally, it should be noted that our usage of the term identification is consistent with the more general definition in which *identification* refers to the process of selecting a system response function from a set of such functions, which is, in turn, isomorphic to the model selection problem [Bretthorst, 1988].

In practice, it is desired that target identification result in a decision or declaration that a specific target is present, but we do not consider the decision-theoretic aspect of this subject.

Consider the hypothesis T_k, and suppose we have radar data D that are related to some properties of the target (D will be defined explicitly in the next section). The identification problem is to assess the extent to which D supports the hypothesis T_k. Thus, we must calculate the probability for T_k based upon all information in our possession; this information consists of D and our prior information pertinent to T_k. This probability will be denoted by $p(T_k|DE)$, which is to be read as the probability for T_k given D and E, the prior information. In practice, it is usually $p(D|T_kE)$ that one can assign; this is normally the means by which the physics of the problem is introduced, relating the radar signal and noise with the hypothesis of interest. Once $p(D|T_kE)$ is in hand, we need a formula that will allow us to calculate the probability of interest, $p(T_k|DE)$. This formula follows from the rules of probability theory, as we now show.

Let A and B represent any two propositions (that are not self-referential, completely ridiculous or the like). Then, the rules of probability theory are

$$p(AB|E) = p(A|E)p(B|AE) \tag{2}$$

$$p(A|E) + p(\overline{A}|E) = 1 \tag{3}$$

$$p(A + B|E) = p(A|E) + p(B|E) - p(AB|E), \tag{4}$$

where the last rule is implied by the first two and our notation follows Jaynes [Jaynes, 1992], with AB the logical product and $A + B$ the logical sum of A and B and $\overline{A}$ the negation of A. The logical product is commutative, so that $AB = BA$ and $p(AB|E) = p(BA|E)$. Using this property in Eq.(2), we obtain, for $p(B|E) \neq 0$,

$$p(A|BE) = \frac{p(A|E)p(B|AE)}{p(B|E)}, \tag{5}$$

the familiar *Bayes' theorem*. Finally, we use Bayes' theorem to relate the two probabilities which prompted our review of probability theory:

$$p(T_k|DE) = \frac{p(T_k|E)p(D|T_kE)}{p(D|E)}. \tag{6}$$

The terminology is standard: (1) $p(T_k|DE)$ is called the posterior probability for T_k, given the data D and the prior information E. (2) $p(T_k|E)$ is called the prior for T_k. (3) When D represents potential outcomes of an experiment (that is, D represents a measurable quantity, not just a formal proposition), $p(D|T_kE)$ is called the sampling distribution for D, given that T_k is true. When T_k is unknown, but D represents actual data, $p(D|T_kE)$ is called the likelihood function. (4) Although $p(D|E)$ is the prior for D, its role is primarily as a normalization factor. We will show next how one usually calculates $p(D|E)$, though in many applications its value is not needed.

Introduce a set of exhaustive and mutually exclusive propositions $A_1, ... A_n$, judiciously selected such that we know $p(A_i|E)$ and $p(D|A_iE)$—usually, one would select $\{T_k\}$ as this set. One can show easily that

$$p(D|E) = \sum_{i=1}^{n} p(DA_i|E) \tag{7a}$$

$$= \sum_{i=1}^{n} p(A_i|E)p(D|A_iE). \tag{7b}$$

This result is often called the chain rule, and the introduction of the set of exhaustive propositions is sometimes referred to as "extending the conversation." It follows from

Eq.(7b) that a sum over k of the numerator in Eq.(6) will lead to the normalization factor (the denominator) in that equation.

The expressions in Eq.(7) are general (no specific meaning is attached to the hypotheses) and are extremely useful in applications involving nuisance parameters—A nuisance parameter is a quantity on which a proposition, hypothesis or model involved in a probability is dependent, and further is unknown and of no interest (e.g., the angular orientation of the target). For example, suppose we have the probability distribution $\{p(DA_i|E)\}$, but only D is of interest, not the hypotheses A_i; then, Eq.(7) can be used to eliminate the A_i. This procedure is called marginalization, and $p(D|E)$ is called a marginal distribution. Marginalization provides us with a powerful tool for dealing with nuisance parameters. Some of the nuisance parameters we encounter are continuous, so we will also need the generalizations of Eqs.(7a) and (7b) for a probability dependent upon a continuous parameter, which we denote by α. Let $p(A\alpha|E)$ represent the probability density for any proposition A and a parameter α. Applying the sum and product rules for probability densities, we obtain

$$p(A|E) = \int d\alpha\, p(A\alpha|E) \tag{8a}$$

$$= \int d\alpha\, p(\alpha|E)\, p(A|\alpha E), \tag{8b}$$

where the proposition A may depend upon a continuous or discrete parameter or upon both—our notation will not distinguish between probability densities and probabilities for discrete propositions. We have presented marginalization in some detail, because nuisance parameters are an unavoidable component of signal analysis for RTID—cf. discussions in Secs. 3 and 4.

3. Theory of Radar Target Identification

In Sec. 2, we provided the conceptual background needed to develop a theory of RTID. In the first part of this section, we outline the derivation of that theory; then, we apply the theory to an example devised to be simple and instructive. We assume the reader is moderately familiar with a number of concepts related to radar, but we attempt to supply enough background on the physics involved in radar target identification so the general reader can readily follow our analysis.

THE RETURN SIGNAL

The transmitted signal will be denoted by $f(t)$; this signal represents some field, voltage, or current quantity in the radar transmitter. In this discussion, it is assumed that we have a replica of $f(t)$ available for any subsequent signal analysis. Note that the basic theory we outline makes no assumptions about the inherent properties of the waveform (e.g., the bandwidth).

The waveform $f(t)$ contains no information relevant to the identity of a target. Rather, as will emerge in the following analysis, the transformation of this waveform induced by the target provides the means for acquiring information about the target. If, as we assume here, a target has no moving parts that interact with the radar signal, the linear response of the kth target to the radar waveform can be described by the following equation in which convolution is denoted by an asterisk ($*$):

$$g_k(t) = a_R(t) * a_P(t) * r_k(t) * a_P(t) * a_T(t) * f(t). \tag{9}$$

Here, $g_k(t)$ is the *target signature* (return signal) of target k, $r_k(t)$ is the *target impulse response function* of target k, $a_T(t)$ is the impulse response of the transmitter and transmit antenna, $a_R(t)$ is the impulse response of the receiver and receive antenna, R is the range from the radar to the target, c is the speed of light, and $a_P(t)$ accounts for propagation of the signal between the radar and target and is taken to be $a_P(t) = R^{-1}\delta(t - R/c)$. Parameters and errors can be introduced through any of the six factors in Eq.(9). We do not wish to get into the issue of errors (except insofar as described by system noise); however, we do want to address the problem of dealing with the time delay described by $a_P(t)$ and the aspect angles in $r_k(t)$—this will be done at the end of this section.

Using the commutative property of convolutions, we obtain from Eq.(9)

$$g_k(t) = \left\{ \frac{a_T(t) * a_R(t) * f(t - 2R/c)}{R^2} \right\} * r_k(t), \tag{10}$$

where we have chosen to account for the total time delay in $f(t)$. The signal $g_k(t)$ represents some field, voltage, or current quantity in the radar receiver. The quantities in the curly brackets in Eq.(9) are assumed known, so it is convenient to combine them and define them as a single quantity. In this paper, this quantity is referred to as the *equivalent transmitted signal*; denoted by $\tilde{f}(t)$, it is defined as follows:

$$\tilde{f}(t) = \frac{a_T(t) * a_R(t) * f(t - 2R/c)}{R^2} \tag{11}$$

and

$$g_k(t) = \tilde{f}(t) * r_k(t). \tag{12}$$

The equivalent transmitted signal $\tilde{f}(t)$ includes the effect of the receive antenna and receiver, the transmit antenna and transmitter, and the range to the target. The target impulse function $r_k(t)$ isolates the scattering characteristics of the target; however, it is also a function of the transmit and receive polarization.

Because the target impulse response function plays a central role in our approach to RTID and has itself become the object of study, we want to add a few remarks about this object. For some purposes, we must focus on the target impulse response function as a problem in electromagnetic theory, independently of any radar system. In such cases, it is desirable to start with the full vector formulation of the scattering problem. Let $\boldsymbol{E_i}(t)$ represent the electric field of the radar signal incident on the target and $\boldsymbol{E_s}(t)$ the electric field scattered by the target, the former in the far-field region of the radar, the latter in the far-field region of the target; further, let $\boldsymbol{\Theta}$ denote the *aspect angles* (e.g., yaw, pitch and roll) specifying the orientation of the target relative to a set of space-fixed axes. Then, in the linear regime [Cho, 1990]

$$\boldsymbol{E_s}(t, \boldsymbol{\Theta}) = \int \mathbf{r_k}(t - t', \boldsymbol{\Theta}) \cdot \boldsymbol{E_i}(t') \, dt',$$

where $r_k(t, t', \Theta)$ is the target impulse response *dyadic* and all spatial dependence, except that on Θ, is suppressed. The use of the target impulse dyadic provides a description of polarization effects [Cho, 1990]. In the analysis that follows, we need only $g_k(t)$.

The expression in Eq.(12) describes symbolically, and mathematically, the physical process by which the signal can acquire target-specific information—such relations carry with them certain background and assumptions that we presume are familiar to the reader (e.g., causality). The target impulse response function and so the target signature will depend upon the orientation of the target relative to the incident and reflected waves. As noted above, the target orientation will be described by aspect angles; for high range-resolution radar waveforms, the target impulse function is a rapidly varying function of the aspect angles. The enormous complications caused by the dependence of the target reflection function on the aspect angles will be discussed in Sec. 4.

The signature $g_k(t)$ involves several parameters which originate from $r_k(t)$ and $\tilde{f}(t)$, as noted above. Most of our information about the target is contained in these parameters. Examples of these parameters are: target range (initialization time), amplitudes and locations of target scattering centers, aspect angle and Doppler-modulated frequencies. These parameters are present in the return waveform only implicitly; if we need their numerical values, we must obtain them through parameter estimation.

Probabilistic Analysis

The reception and recording of signals are invariably accompanied by noise and by measurement errors. Noise arises in the antenna and receiver system, in the background of the target and from the target itself, while errors arise, for example, in the measurement (association of numbers with signals), model definition, quantization and sampling processes. One will strive to reduce the errors (which are, after all, somewhat under our control) so that they are negligible compared to the noise (which is not entirely under our control in most cases), and we assume in this discussion that this has been achieved, so that all noise is system noise. When the signal is $g_k(t)$, the radar return is

$$d(t) = g_k(t) + n(t), \tag{13}$$

where $n(t)$ denotes the noise and is taken to be independent of k, the target index. The received signal $d(t)$ can be measured continuously over a time interval or discretely at sampling times $t_1, ..., t_N$, not necessarily equally spaced. Sampling provides the necessary format for digital signal analysis and is followed in this presentation. We use D to represent the proposition describing the entire data record $d(t_1), ..., d(t_N)$, where

$$d(t_s) = g_k(t_s) + n_s, \tag{14}$$

with $n_s = n(t_s)$ and $s = 1, ..., N$.

Our information concerning the noise is usually very limited. If we know something about the physical origin of the noise, then we will try to make good use of that information. A useful and simple procedure for learning something quantitative about the noise is to record the signal in the receiver when no target is present—Eq.(13) reduces to $d(t) = n(t)$. Suppose we have done this, but instead of recording the noise as a time series, we measure its average amplitude (e.g., a voltage) and its average power (amplitude squared), finding the average noise voltage is zero and the average noise power (into a one-ohm resistor) is σ^2. On the basis of this meager information, we must characterize the noise for all time so that we can take into account the range of values $n(t)$ might exhibit. The result will be the probability density for $n(t)$. We prefer to approach the problem from an information-theoretic point of view, where we assign this probability density in a way that expresses what we know about $n(t)$ in the most conservative manner possible. Such an analysis is beyond the scope of this paper, so we refer to well-known treatments of this topic for the results we need [Jaynes, 1987, 1992].

Given the average noise voltage is zero and the average noise power is σ^2, we assign the following probability density for any noise component $n(t_s)$:

$$p[n(t_s)|E] = (2\pi\sigma^2)^{-1/2} \exp[-n^2(t_s)/2\sigma^2], \tag{15}$$

which is the familiar Gaussian density. The rationale for this probability assignment is based on the principle of maximum entropy (assign the most conservative probability consistent with our information about the noise) and is supported by other considerations (e.g., the central-limit theorem). Because our information about the noise tells us nothing about how the noise varies in time, the probability density for the noise vector $(n_1, n_2 ..., n_N)$ is composed of products of Eq.(15):

$$p(n_1, n_2, ..., n_N|E) = \prod_{s=1}^{N} (2\pi\sigma^2)^{-1/2} \exp(-n_s^2/2\sigma^2). \tag{16}$$

According to this result, the most conservative noise prior (based only on σ^2) is uncorrelated in time and is said to be white Gaussian. This is not a statement about the physics of the noise, only how we allow for noise variations in the analysis of this problem. When we have more information about the noise, we can expect our predictions to be sharper. For cogent arguments as to why it is useful to assign white Gaussian noise even when we know that physics would suggest a sharper characterization of the noise, see Jaynes [Jaynes, 1987]. In the RTID problem based on high range-resolution radar, this discussion is inadequate, especially with respect to aspect angles.

The preceding discussion defines clearly the problem we have set for ourselves: Extract from our prior information and from measurements, relevant information for inferring which target is present. We have provided the concepts and notation that are needed to translate our problem into symbolic form in Sec. 2, so we can now proceed to develop a quantitative theory. From our discussion on target identification in Sec. 2, we see that our problem is to calculate $p(T_k|DE)$, the probability for T_k given D and E. As seen from Eq.(6), the mathematical relation between the data D and the propositions T_k that we need is found by first calculating the sampling distributions $p(D|T_kE)$.

To determine the explicit expression for the sampling probability $p(D|T_kE)$, we first rearrange Eq.(14) to read $n_s = d(t_s) - g_k(t_s)$. Then, with $g_k(t)$ assumed given, we use this expression to carry out a change of variables from $n_1, \ldots, n_N$ to $d(t_1), \ldots, d(t_N)$ in Eq.(16). The result is

$$p(D|T_kE) = (2\pi\sigma^2)^{-N/2} \exp[-(d \cdot d - 2d \cdot g_k + g_k \cdot g_k)/2\sigma^2], \tag{17}$$

where we have adopted the notation

$$u \cdot v = \sum_{s=1}^{N} u(t_s)\, v(t_s). \tag{18}$$

Another way to arrive at Eq.(17) is to note that the probability for the data D given that the target response function is g_k, is just the probability for $n(t_s)$ to provide the difference between $d(t_s)$ and $g_k(t_s), s = 1, \ldots, N$. Combining Eqs.(6) and (17), we obtain the probability for target k to have produced the data:

$$p(T_k|DE) = \frac{(2\pi\sigma^2)^{-N/2} p(T_k|E)}{\sum_{k=1}^{m} p(T_k|E) p(D|T_k E)} \exp[-(d \cdot d - 2d \cdot g_k + g_k \cdot g_k)/2\sigma^2] \tag{19a}$$

$$= K p(T_k|E) \exp(-g_k \cdot g_k/2\sigma^2) \exp(d \cdot g_k/\sigma^2), \tag{19b}$$

where K contains all factors not dependent upon k and $g_k \cdot g_k$ is independent of the data set. We have in Eq.(19) the formal solution of our RTID problem. So far, it has not been necessary to say much about the waveform $\tilde{f}(t)$, so the preceding discussion is general. In fact, our current efforts are concerned with a specific waveform which we describe briefly next. The trend in radar has been to higher frequencies (e.g., X and K bands) and to greater bandwidths, or, as commonly expressed, to high range-resolution radar (HRR). Besides certain desirable attributes from the perspective of radar engineers, HRR is of interest to us because it can provide a vast amount of target-specific information. The reason that RTID based on HRR is not a *fait accompli* stems primarily from the complicated dependence of the HRR returns upon the orientation of the target. Many of us are attracted by the challenge to deal with, and even exploit, this sensitive angular dependence. In Sec. 4, we discuss some potential consequences of this complicated angular dependence for RTID.

A SIMPLE EXAMPLE

In order to illustrate how one applies the preceding theory, we have devised a simple example. The discussion will introduce some concepts and issues that arise in the full-blown problem.

One of a specified set of very simple aerial targets is present; our objective is to identify the target. We have already stated that radar will be used to acquire information relevant to target identity; this leaves a great deal of latitude in the frequency spectrum of the radar waveforms that can be used, and this feature should be exploited. Waveform design should take into account those attributes of the targets that are target specific. Ideally, one would measure the target impulse response function $r_k(t)$ in Eq.(11) for each k; then, one should select waveforms $f(t)$ so that $g_k(t)$ is as target-specific as possible (of course, additional constraints on the waveforms will enter these considerations). Suppose we have carried out detailed studies on $r_k(t)$ for all potential targets and have learned the following: (1) The radar cross section (RCS) of each target is dominated by a scatterer at the nose and another at the tail. (2) Each target is characterized by its radar length (the distance between its two scatterers as measured by the radar pulse). To circumvent several complicating factors we assume further that the target advances toward the radar with a low, constant velocity at a known range (in particular, we are assuming that the orientation of each target is known and that Doppler modulation of the signals can be ignored or handled easily). To minimize the number of parameters we have to deal with, we assume the transmitter and receiver of the radar are collocated (this is the so-called monostatic configuration); for the same reason, we constructed the problem to be one-dimensional with all targets approaching the radar nose-on.

Given the attributes of the targets which make them identifiable, we would want to choose a radar waveform which enhances the radar's performance. This is a subject unto itself which we cannot get into here; however, Jaynes [Jaynes and Smith, 1991] has derived an integral equation which such waveforms must satisfy. Here, we can proceed by using well known properties of radar signals. By choosing a short pulse for $f(t)$ in Eq.(10) and assuming a correspondingly wide bandwidth for the transmitter and receiver (that is, for HRR), we can acquire information about the length of a target.

For HRR waveforms, the target impulse response function for the kth target can be approximated by

$$r_k(t - t') = \sum_\ell A_{k\ell}\delta(t - t' - t_{k\ell}), \tag{20}$$

which we discuss further in Sec. 4. For generality, we have given this expression for a target consisting of any number of scattering centers (labeled by ℓ). The amplitudes $A_{k\ell}$ are related to the radar cross section of the ℓth scattering center. The time delay $t_{k\ell}$ is proportional to the range (from the radar) of the ℓth scattering center; the time delay associated with the first scattering center is used here to define the range R between the target and radar, so that $t_{k1} = 2R/c$.

Now, we return to the specifics of the problem we are considering. We assume $A_{kc} = A'$, with A' known, so that Eq.(20) reduces to

$$r_k(t - t') = A'[\delta(t - t' - t_{k1}) + \delta(t - t' - t_{k2})]. \tag{21}$$

Using Eqs.(12) and (21) and the notation $v(t) = \tilde{f}(t - t_{k1})$, we obtain the signal returned by the target to the radar system:

$$g_k(t) = A[v(t) + v(t - 2L_k/c)], \tag{22}$$

where we noted that $t_{k2} - t_{k1} = 2L_k/c$, with L_k the radar length of the target. Because we have assumed that the range as well as A' in Eq.(21) are known, we can treat A as known. We have deliberately cast the expression in Eq.(22) in a form that reveals the presence of the parameter L_k. So, as desired, the selected radar waveform does indeed lead to returns containing target-specific information. Next, we show how we extract that information toward our stated objective of target identification.

The final phase of the analysis is easy: Assuming a uniform prior [i.e., that $p(T_k|E)$ is independent of k], we combine Eqs.(6) and (22) to obtain

$$p(T_k|DE) = K' \exp\left(A\frac{d \cdot v_k}{\sigma^2} - \frac{g_k \cdot g_k}{2\sigma^2}\right), \tag{23}$$

where the notation is similar to that in Eq.(18),

$$d \cdot v_k = \sum_{s=1}^{N} d(t_s)v(t_s - 2L_k/c), \tag{24}$$

and K' contains all of the contributions that are independent of k—the data and the copy are properly aligned in Eq.(24), because the range of the target is known; this problem is addressed more carefully in the next subsection. Finally, we assume the energy in the sampled signal $\{g_k(t_s)\}$ is the same for all targets (this is consistent with preceding assumptions), so that Eq.(23) reduces to

$$p(T_k|DE) = Q \exp\left(A\frac{d \cdot v_k}{\sigma^2} \right), \qquad (25)$$

where Q is independent of k, but is a function of A. It is important to note that the probability for the kth target depends upon the data only through the cross-correlation $d \cdot v_k$. Based on intuition or on maximum likelihood, this has been assumed for a long time; here, we have established the necessary rationale for this assumption as well as mathematical expressions placing this quantity in its proper role. For example, the posterior probability for T_k depends upon the exponential of the cross-correlation. Further, we have all of the formalism needed for dealing with nuisance parameters. We return to this topic when we discuss aspect angles and initialization time.

Although many details were omitted in the discussion of the theory and many simplifying assumptions were made in the example, the general structure of the RTID problem has nevertheless been exhibited. Next, we outline how a simple algorithm based on Eq.(25) might proceed (the algorithm for the full-blown problem will be considerably more complicated and sophisticated). We assume that we know what targets can be present, that the necessary target signatures are in the library and that the noise level σ is known (though, if necessary, we could estimate this parameter). After obtaining data D, we would insert each hypothesis T_k into Eq.(25) [this is achieved by substituting each target signature $v_k(t)$ in the library into Eq.(25)]. Next, the normalization factor is calculated to arrive at the posterior probabilities $p(T_k|DE)$; in turn, these probabilities are ordered. If the largest exceeds a preselected threshold, a declaration is made; if no probability exceeds the threshold, more pulses must be transmitted (preferably, the additional pulses will be tailored to enhance the discrimination between the major contenders for the target identity). If unknown targets can be present and the "unknown target" declaration must be a possible outcome of the algorithm, the target signature library must include a generic signature that will fit the data as well as any target signature in the library. Larry Bretthorst has provided us with an excellent start on this problem, but further study remains.

NUISANCE PARAMETERS

The algorithm just outlined is inadequate on several grounds. First, the evaluation of the cross-correlation $d \cdot v_k$ in Eq.(25) requires the data $d(t)$ and the signatures $v_k(t)$ to be aligned in time (i.e., synchronized); but if we do not know the target range extremely accurately, there will be uncertainty in the time at which we begin receiving real data rather than just noise. This confronts us with an important problem in signal processing, which we want to examine in sufficient detail to exhibit its structure and outline its proper treatment. Specifically, we consider the data-alignment problem and the aspect-angle problem [Bretthorst, et al., 1991].

To keep the discussion reasonably general, we return to Eq.(19b). In analogy with our preceding algorithm, the evaluation of Eq.(19b) proceeds as follows: Data $d(t_1), \ldots, d(t_N)$

are obtained; then, sampled values of a stored signature $g_k(t)$ for target k are combined, via a scalar product, with the data—$g_k \cdot g_k$ in Eq.(19b) can be evaluated in advance and is ignored here (this may not be viable when aspect angles appear as nuisance parameters). The scalar product $d \cdot g_k$ assumes that the sample times in the data and the signature are identical [cf. Eq.(24)]. To achieve the proper alignment of the data and a signature, one must be rigidly translated relative to the other. For example, writing $g_k(t' + t_0)$ for a translated signature, we can synchronize the data and signature by selecting t_0 such that $t' + t_0 = t$, where the initialization time t_0 is related to the range [cf. the time delay in Eq.(10)] and t' refers to the time appearing in the stored signature. But this entire exercise has transpired because we do not know t_0; further, it is assumed here that the value of t_0 that achieves alignment is of no interest, though it allows us to estimate the range. Hence, assuming t_0 is a nuisance parameter, we remove it from the posterior probability for T_k by marginalization [Bretthorst, 1988; Jaynes, 1992], as we demonstrate next.

If the signature is $g_k(t + t_0)$, then instead of Eq.(20) the probability for the data D is

$$p(D|T_k t_0 E) \propto \exp[d \cdot g_k(t_0)/\sigma^2] \tag{26}$$

and the posterior probability for D and t_0 [cf.Eq.(6)] becomes

$$p(T_k t_0 | DE) = \frac{p(T_k t_0 | E) p(D | T_k t_0 E)}{p(D|E)}. \tag{27}$$

Using Eq.(8), we can arrive at the desired posterior probability for T_k by marginalization as follows:

$$p(T_k | DE) = \int p(T_k t_0 | DE)\, dt_0 = \int p(T_k | D t_0 E) p(t_0 | DE)\, dt_0. \tag{28}$$

This problem was considered by Woodward [Woodward, 1964, Sec. 4.6]. The assignment of the prior $p(t_0|DE)$ depends upon our information about the range of the target—a Gaussian is used most frequently [Jaynes,1987]. Further discussion of this topic is beyond the scope of this article.

Finally, we come to the complicated aspect-angle problem. As indicated earlier, the target impulse response function depends upon the aspect angles symbolized by Θ, and therefore so does the signature in Eq.(12). Writing $g_k(t, \Theta)$ for the signature, we can use the results in the preceding paragraph, *mutatis mutandis*, to obtain [Jaynes and Smith, 1991]

$$p(T_k | DE) = \int p(T_k | D\Theta E) p(\Theta|DE)\, d\Theta. \tag{29}$$

Though this problem is formally identical to the problem with the initialization time, there is one significant difference: It is not necessary to store $g_k(t + t_0)$ for a range of values of t_0, while $g_k(t_k, \Theta)$ must be available for all aspect angles. Actually, for storing $g_k(t, \Theta)$, Θ must be discretized. Letting ξ, η, ζ represent yaw, pitch and roll (or heading, attitude and bank), we will have to measure and store sampled values of

$$g_k(t, \xi_\ell, \eta_m, \zeta_n), \tag{30}$$

where ℓ, m, n are integers. Thus, the integral in Eq.(29) must be replaced by a sum as in Eq.(7). The discussion of aspect angles is continued in the next section. Our purpose in the above discussion was to indicate how one deals with nuisance parameters, an unavoidable feature of target identification with radar.

4. The Signature-Library Problem

In the last section, we avoided any assumptions about the radar waveform, and the use in our example of the scattering-center model, and hence HRR, should not be construed as a premise of the theory. As pointed out in Sec. 3, we are currently concentrating on HRR for RTID, so in the remainder of this article we want to identify the challenges posed by HRR and offer some thoughts on how the challenges may be met.

Target identification by means of HRR requires a complete set of signatures of all potential targets to be available in the target library. But this could have some adverse consequences as we discuss briefly next—the subsequent remarks are restricted to airborne targets. At this time, it is held by many that about 10^5 signatures per target will be necessary to allow for all possible orientations of a target. Further, each signature corresponds to a data string containing at least 10^4 bits (this estimate is based upon calculated signatures for actual aircraft). Hence, the set of signatures for each target could require 10^9 bits of storage (all estimates may be high or low by an order of magnitude, or perhaps more). Thus, for a target library applicable to 200 targets, it would appear that about 10^{11} bits of data must be stored for subsequent use by the target-identification algorithm. Currently available computer storage capabilities can meet this level of storage requirement. But storage problems pale in comparison with two other problems: (1) The acquisition of these signature data by measurement is not realistically a possibility (because, among other things, of cost). (2) The retrieval from memory of, say, 10^7 signatures and their subsequent processing by the identification algorithm in quasi-real time would appear to be rather formidable. Thus, we cannot, and given the magnitude of the problem, ought not appeal to technology alone to deal with the computational burden imposed by the direct approach just outlined. Further inquiry into the physics of the problem is cleary indicated; then, we would expect some sophisticated computer programming will be necessary to produce an acceptable algorithm.

Clearly, we need to think about the problem of target identification based on HRR target signatures at a deeper level, starting with the physical origin of signatures. Sophisticated electromagnetic computer codes have been developed for calculating HRR target signatures. At this time, the calculated signatures do not replicate measured signatures in every detail, but it is expected that measured and calculated signatures will be interchangeable in the near future (in which case most signatures will be calculated rather than measured, solving the data acquisition problem). The calculations rely on the equation for the exterior surface of the target. Noting that the surface can be represented by orders of magnitude fewer data than 10^9 bits, we would argue that the 10^9 bits of signature data for each target must be highly redundant. The redundancy can be reduced by formulating the problem at the most fundamental, tractable level possible (most likely at the level of a boundary value problem supplemented by measurements).

Analytical representations of either HRR signatures or target impulse response functions are highly desirable for the efficient implementation of the radar identification theory presented in the last section. We have seen such a representation in action in Eqs.(20) –

(25), but we have no guarantee such neat and simple results will always obtain. At any rate, an analytical representation of either $g_k(t)$ or $r_k(t)$ can simultaneously ameliorate some of the storage, retrieval and data-processing problems discussed above. Both theoretical and empirical studies indicate that analytical representations suitable for HRR can be based upon the concept of *scattering center*, and the example in the last section already used a scattering-center model for the target impulse response function. In the high-frequency region, the assumed domain of HRR, there are six types of physical processes, falling into one of three broad classes, that are responsible for returning the radar pulse [Ruck, et al., 1970; Bechtel,1976]:

Local mechanisms

  1. Specular reflection

  2. Diffraction by discontinuities

Multiple mechanisms

  3. Localized multibounce

  4. Separated multibounce

  5. Cavity and duct

Extensive processes

  6. Surface waves

Specular reflection is usually the dominant process; however, for some types of targets and for most targets at certain orientations, the other processes become the source of the return. The physical interaction of electromagnetic waves with targets does not cleanly break up into these cases; rather, the mathematical description of the physics plus various approximations allow us to identify these processes. Except for cavities and ducts, the other processes transform the signal in a discrete fashion that can be represented by scattering centers.

The Scattering-Center Representation

As we have argued above, target identification by means of HRR requires target signatures to be available for signal analysis, analytical representations for these signatures would be ideal, and the scattering-center concept can provide the basis for such representations. It is important, therefore, that we understand the origin and meaning of the scattering-center concept at a level deeper than the phenomenological or intuitive one. One would expect the appearance of scattering centers, in some limit, in the formal description of radar target response based on electromagnetic theory. This is the case, though further work is desirable [Ruck, et al., 1970]. We should note from the outset that there is no consensus on the meaning of the scattering-center concept or on the form of the corresponding mathematical model. The best we can do is explain what we understand by the term "scattering-center model."

We say that target k consists of m scatterers if its impulse response function can be expressed in the form

$$r_k(t) = \sum_{\ell=1}^{m} r_{k\ell}(t - t_{k\ell}), \tag{31}$$

where $r_{k\ell}(t - t_{k\ell})$ vanishes for $t - t_{k\ell}$ outside some range. Further, we say that the *scattering-*

center model obtains if the target impulse response fuction can be represented as follows:

$$r_k(t) = \sum_{\ell=1}^{m} A_{k\ell}\delta(t - t_{k\ell}).$$

(32)

We have referred to the *scattering-center representation* several times; by this term we mean Eq.(32)—cf. Eqs.(20) and (21). The amplitudes $A_{k\ell}$ and the delay times $t_{k\ell}$ apply to the ℓth scattering center, and it is these quantities that we would determine if describing a target impulse response function in the scattering-center representation; one should note, however, that no assertion is made that a scattering center is a physically significant point on a target, or, for that matter, even lies on the target surface. The physics giving rise to a scattering center can be nonlocal; yet, the scattering-center representation may adequately describe the situation. By taking m large enough, we can approximate any target impulse response with Eq.(32); but we would not consider this a desirable analytical representation. Further, by choosing $r_{k\ell}(t - t_{k\ell})$ in Eq.(31) to be a suitable set of functions, we can find other representations of $r_k(t)$—this is the approach we expect to take in describing returns from cavities, ducts and the like. It should be mentioned that a generalized scattering-center representation, involving derivatives and integrals of the delta function, can provide a more accurate representation, and is even physically more appropriate, than Eq.(32) [Altes, 1976]; however, our experience so far indicates that Eq.(32) adequately represents HRR signatures for the purposes of target identification—this excludes the portions of the return due to cavities and ducts. We are currently studying whether the scattering-center representation is sufficiently accurate for the purposes of target identification and to what extent it can lead to a reduction in the lengths of the data vectors representing target signatures.

5. Concluding Remarks

We have given a brief account of some of our research in RTID, presented a theory of radar target identification, and outlined the status of the project. We have argued that HRR offers promise for RTID and have identified a number of problems that must be solved, or circumvented, before HRR can reach its potential. Ironically, the very complexity of HRR returns, the characteristic allowing target-specific information to be contained in the returns, is currently the greatest obstacle to RTID based on HRR.

ACKNOWLEDGMENTS. It is a pleasure to acknowledge the following persons for offering suggestions that have led to the improvement of this paper: Dr. G. L. Bretthorst, Prof. D. G. Dudley, Prof. J. D. Echard, Prof. P. M. Goggans, Dr. J. M. Loomis, Prof. W. M. Nunn, Jr., and Mr. M. A. Smith. Finally, the start in the right direction provided by Ed Jaynes is gratefully acknowledged.

REFERENCES

Altes, R. A.: 1976, 'Sonar for generalized target description and its similarity to animal echolocation systems,' *J. Acoust. Soc. Am.* **59**, pp. 97 – 105.
Bechtel, M. E.: 1976, 'Short-pulse target characteristics,' in H. E. G. Jeske (ed.), *Atmospheric Effects on Radar Target Identification and Imaging*, D. Reidel Publishing Company, Dordrecht.
Bretthorst, G. L.: 1988, *Bayesian Spectrum Analysis and Parameter Analysis*, Springer-Verlag, Berlin.

Bretthorst, G. L., C. R. Smith and L. S. Riggs: 1991, 'Radar target discrimination using probability theory,' in Bruce Noel (ed.), *Ultra-Wideband Radar: Proceedings of the First Los Alamos Symposium*, CRC Press, Boca Raton, pp. 417 – 434.

Cho, S. K.: 1990, *Electromagnetic Scattering*, Springer-Verlag, New York.

Jaynes, E. T.: 1987, 'Bayesian spectrum and chirp analysis,' in C. R. Smith and G. J. Erickson (eds.), *Maximum-Entropy and Bayesian Spectral Analysis and Estimation Problems*, D. Reidel Publishing Company, Dordrecht.

Jaynes, E. T.: 1992, *Probability Theory – The Logic of Science*. Preliminary copies available from the author.

Jaynes, E. T. and C. R. Smith: 1991, 'Theory of radar target discrimination,' U. S. Army Missile Command Tech. Rept. RD-AS-91-6, February, 1991.

Ruck, G. T., D. E. Barrick, W. D. Stuart and C. K. Krichbaum: 1970, 'Theory,' Chapter 2 in G. T. Ruck (ed.), *Radar Cross Section Handbook*, Vol. 1, Plenum Press, New York.

Woodward, P. M.: 1964, *Probability and Information Theory, with Applications to Radar*, Pergamon Press, Oxford.

ON THE DIFFERENCE IN MEANS

G. Larry Bretthorst
Washington University
Department of Chemistry
St. Louis, Missouri 63130-4899

ABSTRACT. Given two sets of data that are repeated measurements of the same physical quantity, one "control" and one "trial," there are three problems of interest to the experimenter: (1) determine if something changed, (2) if something changed, what? and (3) estimate the magnitude of the change. These three problems are addressed using probability theory as extended logic. In the first section, the probability that the data sets differ is computed independent of what changed, i.e., independent of whether or not the means or standard deviations changed. In the second section, two probability distributions are computed: first, the probability that the means changed is computed independent of whether or not the standard deviations changed. Then second, the probability that the standard deviations changed is computed independent of whether or not the means changed. In the third section, the problem of estimating the magnitude of the changes is addressed. Here the probability density functions for both the difference in means and the ratio of standard deviations is computed. The probability for the ratio of standard deviations is computed independent of whether or not the means are the same, just as the probability for the difference in means is computed independent of whether or not the standard deviations are the same. This last calculation generalizes the solution of both the two-sample problem (different means and same but unknown standard deviations) and the Behrens-Fisher problem (different means and different unknown standard deviations). The calculations are illustrated with a numerical example in the fourth section.

This paper addresses one of the most fundamental problems that can occur in experimental science, that of analyzing two independent measurements of the same physical quantity under slightly different experimental conditions when the measurements are assumed uncorrelated. This problem has a long history going back to at least 1929, when Behrens [1] proposed a solution to the problem of estimating the difference in means when the standard deviations are assumed unequal and unknown. Fisher rederived the same result using fiducial probabilities in 1937 [2,3], and last Jeffreys arrived at the same distribution in 1939 using Bayesian probability theory [4]. However, the Behrens test became quite controversial because it called into question some of the basic tenets of orthodox sampling theory. For this reason the Behrens test is essentially not used today. Instead a series of *ad hoc* tests are in use. For a good discussion of the Behrens-Fisher controversy see Lee, [5]. For a review of the Behrens-Fisher problem see Refs. [6,7,8] and for a description of how this problem is addressed in orthodox sampling theory see Refs. [9,10,11].

The problem is more complex than just determining the difference in means or the ratio of standard deviations. For example, the standard deviations might be the same, or they might be different; one simply may not know which condition applies and indeed, it is certainly possible, that the data may not strongly favor either hypothesis (as they do not in the numerical example) and so neither the two-sample test nor the Behrens-Fisher test is truly applicable. Additionally, estimating the difference in means is a parameter estimation problem, and it implicitly assumes that the means are different. Before one attempts to estimate the difference in means, it would be reasonable to ask if the means are different? The same statements apply to estimating the ratio of the standard deviations.

An early attempt to solve part of this problem was made by Dayal and Dickey [12,13], when they computed the probability for four basic hypotheses. These hypotheses, the means and standard deviations are the same, the means are the same and the standard deviations differ, the means differ and the standard deviations are the same, and the means and standard deviations differ, are of fundamental importance in this problem; but they do not directly tell one if the data sets changed, and if they changed how? To address the questions "Did the data sets change?" one needs to compute the probability that the data sets differ independent of what changed. To address the other question, "What changed?" one needs to compute two probabilities: the probability that the means changed, and the probability that the standard deviations changed. A complete list of all of the hypotheses addressed in this paper are given in Table 1. Before proceeding to discuss these hypotheses, the problem being addressed is described in more detail.

In two-sample and Behrens-Fisher like problems there are two data sets, D_1 and D_2, one "control," and one "trial." Data set D_1 has N_1 data items labeled d_{1i}, and similarly for data set D_2. These data sets are repeated measurements of the same physical quantity. This quantity is designated as A in data set D_1, and B in data set D_2. The parameters, A and B, will be referred to as the means of the data sets, although one should keep firmly in mind that this is a colloquial use. What one really means by A and B are two hypotheses. These hypotheses are of the form "the constant signal in data set D_1 had value A" and similarly for B. Each data set is contaminated by additive noise of standard deviations σ_1 and σ_2. Similarly σ_1 and σ_2 will be referred to as the standard deviations of the noise, but what is really meant is again two hypotheses of the form "the noise in data set D_1 has standard deviation σ_1" and similarly for D_2. With these definitions the hypotheses and the data are related by

$$d_{1i} = A + \text{noise of standard deviation } \sigma_1, \qquad (1)$$

$$d_{2i} = B + \text{noise of standard deviation } \sigma_2. \qquad (2)$$

Table 1 lists the hypotheses addressed in this paper and assigns an abbreviation, to each of them. In this paper, simple hypothesis will be labeled with a single letter. For example, from Table 1, the hypothesis v means the standard deviations are the same. Two hypotheses next to each other should be read with an 'and' between them. For example, sv stands for the means are the same and the variances are the same. Last, the negation of a hypothesis is represented with a bar over the hypothesis. For example, $\overline{v}$ means the variances are not the same. The simple hypotheses will generally be addressed first. There is one exception to this, $\overline{s} + \overline{v}$, the means or the standard deviations are not the same. This hypothesis is the one that answers the question "did something change?" and it is the first hypothesis addressed in this paper. The hypothesis δ means the difference

Table 1: The Hypotheses Addressed

Hypotheses	Abbreviation
The means are the same	s
The means are not the same	$\bar{s}$
The standard deviations are the same	v
The standard deviations are not the same	$\bar{v}$
The means and the standard deviations are the same	sv
The means are the same, and the standard deviations differ	$s\bar{v}$
The means are not the same, and the standard deviations are the same	$\bar{s}v$
The means and standard deviations are not the same	$\bar{s}\,\bar{v}$
The means or the standard deviations are not the same	$\bar{s}+\bar{v}$
The difference, $A - B$, is equal to δ	δ
The ratio σ_1/σ_2 is equal to r	r

between the constant in the first data set, and the constant in the second data is equal to δ. And similarly the hypothesis r means that the ratio of the standard deviation in the first data set divided by the standard deviation in the second data set is equal to r. The four compound hypotheses, sv, $s\bar{v}$, $\bar{s}v$, and $\bar{s}\,\bar{v}$ are the four hypotheses addressed by Dayal and Dickey [12,13].

1. Did Something Change?

The "trial" data set can only differ from the "control" data set in two important ways: either the means or the standard deviations can change; there are no other possibilities. This hypothesis is abbreviated $\bar{s}+\bar{v}$. The probability that represents this state of knowledge is denoted by $P(\bar{s}+\bar{v}|D_1 D_2 I)$. In words, this is the probability that the means or standard deviations are not the same given the two data sets, D_1 and D_2, and the prior information I. The prior information I is all of the assumptions that have gone into making this a well posed problem. At present I includes the separation of the data into a signal plus additive noise, and that A and B are constants.

To compute the probability that the means or the standard deviations differ, note that

$$P(\bar{s}+\bar{v}|D_1 D_2 I) = P(\overline{sv}|D_1 D_2 I) = 1 - P(sv|D_1 D_2 I). \tag{3}$$

It is sufficient to compute the probability that the means and the standard deviations are the same and from that one can compute the probability that the means or the standard deviations differed. This is the first hypothesis studied by Dayal and Dickey. $P(sv|D_1 D_2 I)$ does not depend on any parameters, it is a marginal probability density function. The hypothesis sv assumes the means and the standard deviations are the same, so two parameters (a constant A, and a standard deviation σ_1) have been removed by marginalization:

$$P(sv|D_1 D_2 I) = \int dA d\sigma_1 P(svA\sigma_1|D_1 D_2 I) \tag{4}$$

where $B \to A$ and $\sigma_2 \to \sigma_1$. The right hand side of this equation may be factored using Bayes' Theorem to obtain

$$P(sv|D_1 D_2 I) \propto \int dA d\sigma_1 P(svA\sigma_1|I)P(D_1 D_2|svA\sigma_1 I). \tag{5}$$

Assuming logical independence of the parameters and the data, this may be further simplified to obtain

$$P(sv|D_1 D_2 I) \propto \int dA d\sigma_1 P(sv|I) P(A|I) P(\sigma_1|I) P(D_1|svA\sigma_1 I) P(D_2|svA\sigma_1 I) \quad (6)$$

where $P(sv|I)$ is the prior probability that the means and the standard deviations are the same, $P(A|I)$ is the prior probability for the amplitude, $P(\sigma_1|I)$ is the prior probability for the standard deviation, and $P(D_1|svA\sigma_1 I)$ and $P(D_2|svA\sigma_1 I)$ are the direct probabilities or likelihoods of the two data sets.

In this calculation uninformative prior probabilities will be used. However, care will be taken to ensure that all prior probabilities are fully normalized. For the amplitude, A, a bounded uniform prior will be used:

$$P(A|I) = \begin{cases} \dfrac{1}{R_a} & \text{If } L \leq A \leq H \\ 0 & \text{otherwise} \end{cases} \quad (7)$$

where $R_a \equiv H - L$, and H and L are the limits on the constant A and are assumed known. A bounded Jeffreys prior will be used for the standard deviation of the noise:

$$P(\sigma_1|I) = \begin{cases} \dfrac{1}{\sigma_1 \log(R_\sigma)} & \text{If } \sigma_L \leq \sigma_1 \leq \sigma_H \\ 0 & \text{otherwise} \end{cases} \quad (8)$$

where R_σ is the ratio σ_H/σ_L, and σ_H and σ_L are the limits on the standard deviation σ_1 and are also assumed known.

As noted earlier there are four fundamental hypotheses: sv, $s\overline{v}$, $\overline{s}v$, and $\overline{s}\,\overline{v}$, and probability theory requires us to assign a prior probability to each of them. Here it is $P(sv|I)$ that must be assigned. No assumption will be made about which of these four possibilities are present, but all four of these possibilities are assumed to occur and so a probability of 1/4 is assigned to each.

The only quantities that remain to be assigned are the two likelihood functions. The prior probability for the noise will be taken to be Gaussian. If the limits on the A integral extend from minus infinity to plus infinity, and the limits on the σ_1 integral extend from zero to infinity, then both integrals can be evaluated in closed form. However, with finite limits either of the two indicated integrals may evaluated, but the other must be evaluated numerically. Evaluating the integral over the amplitude, and neglecting a factor of $(2\pi)^{-N/2}$, which is common to all the models, we obtain:

$$P(sv|D_1 D_2 I) \propto \frac{\sqrt{\pi/2N}}{4 R_a \log(R_\sigma)} \int_{\sigma_L}^{\sigma_H} d\sigma_1 \sigma_1^{-N} \exp\left\{-\frac{z}{2\sigma_1^2}\right\} \left[Q\left(\frac{1}{2}, \frac{U_L^2}{\sigma_1^2}\right) \pm Q\left(\frac{1}{2}, \frac{U_H^2}{\sigma_1^2}\right)\right] \quad (9)$$

where $\overline{d}$ and $\overline{d^2}$ are the mean and mean-square of the pooled data, $N = N_1 + N_2$,

$$U_H = \sqrt{\frac{N}{2}}(H - \overline{d}), \qquad U_L = \sqrt{\frac{N}{2}}(L - \overline{d}), \qquad z = N[\overline{d^2} - (\overline{d})^2], \quad (10)$$

and $Q(r, x)$ is the complimentary Gamma function of index r and argument x. The sign is chosen to be minus if both U_H and U_L are of the same sign, and plus if U_H and U_L are of different sign.

Equation (6) can be used to tell one if the data sets are the same, and by using Eq. (3) one can determine the probability that the means or the standard deviations changed, and thus, answers the first question of interest, "did something change?" But as soon as one knows that something changed, ones interest in the problem changes and one wants to know "What changed?" and it is this problem that is addressed in the next section.

2. Given That Something Changed, What Changed?

Given that something changed, there are only two possibilities: either the means or the standard deviations changed. To determine if the means changed one computes $P(s|D_1 D_2 I)$. Similarly, to determine if the standard deviations changed, one computes $P(v|D_1 D_2 I)$. Using the sum rule, these probabilities may be written

$$P(s|D_1 D_2 I) = P(sv|D_1 D_2 I) + P(s\overline{v}|D_1 D_2 I) \tag{11}$$

and

$$P(v|D_1 D_2 I) = P(sv|D_1 D_2 I) + P(\overline{s}v|D_1 D_2 I). \tag{12}$$

where $P(s|D_1 D_2 I)$ is computed independent of whether or not the standard deviations are the same; while $P(v|D_1 D_2 I)$ is independent of whether or not the means are the same. Note that three of the four hypotheses studied by Dayal and Dickey (sv, $s\overline{v}$, and $\overline{s}v$) have appeared. The fourth, $\overline{s}\,\overline{v}$, appears whenever either $P(\overline{s}|D_1 D_2 I)$ or $P(\overline{v}|D_1 D_2 I)$ is computed.

DID THE MEANS CHANGE?

The probability that the means are the same, $P(s|D_1 D_2 I)$, is a marginal probability. It is computed from two terms, $P(sv|D_1 D_2 I)$, and $P(s\overline{v}|D_1 D_2 I)$. $P(sv|D_1 D_2 I)$ was computed in the last section and $P(s\overline{v}|D_1 D_2 I)$ is addressed in this subsection. Notice that $P(s\overline{v}|D_1 D_2 I)$ assumes the constants are the same in both data sets, but the standard deviations are different. Neither the constant nor the standard deviations appear in $P(s\overline{v}|D_1 D_2 I)$. Consequently, $P(s\overline{v}|D_1 D_2 I)$, is a marginal probability density, where the constant and the two standard deviations were removed as nuisances:

$$
\begin{aligned}
P(s\overline{v}|D_1 D_2 I) &= \int dA d\sigma_1 d\sigma_2 P(s\overline{v} A \sigma_1 \sigma_2 | D_1 D_2 I) \\
&\propto \int dA d\sigma_1 d\sigma_2 P(s\overline{v} A \sigma_1 \sigma_2 | I) P(D_1 D_2 | s\overline{v} A \sigma_1 \sigma_2 I) \\
&= \int dA d\sigma_1 d\sigma_2 P(s\overline{v}|I) P(A|I) P(\sigma_1|I) P(\sigma_2|I) \\
&\quad \times P(D_1 | s\overline{v} A \sigma_1 I) P(D_2 | s\overline{v} A \sigma_2 I).
\end{aligned}
\tag{13}
$$

The prior, $P(\sigma_2|I)$, must be assigned; and for logical consistence it must be assigned the same prior range as σ_1:

$$P(\sigma_2|I) = \begin{cases} \dfrac{1}{\sigma_2 \log(R_\sigma)} & \text{If } \sigma_L \leq \sigma_2 \leq \sigma_H \\ 0 & \text{otherwise} \end{cases}. \tag{14}$$

To evaluate the two integrals over σ_1 and σ_2, one substitutes Eqs. (7,8,14), a Gaussian noise prior for the likelihoods, and $P(s\overline{v}|I) = 1/4$. Then evaluating the integral over σ_1 and σ_2, one obtains

$$P(s\overline{v}|D_1 D_2 I) \propto \frac{\Gamma(N_1/2)\Gamma(N_2/2)}{16 R_a [\log(R_\sigma)]^2} \int_L^H U_1^{-\frac{N_1}{2}} dA U_2^{-\frac{N_2}{2}}$$
$$\times \left[Q\left(\frac{N_1}{2}, \frac{U_1}{\sigma_H^2}\right) - Q\left(\frac{N_1}{2}, \frac{U_1}{\sigma_L^2}\right) \right] \left[Q\left(\frac{N_2}{2}, \frac{U_2}{\sigma_H^2}\right) - Q\left(\frac{N_2}{2}, \frac{U_2}{\sigma_L^2}\right) \right]$$

(15)

where

$$U_1 = \frac{N_1}{2}(\overline{d_1^2} - 2A\overline{d_1} + A^2), \qquad U_2 = \frac{N_2}{2}(\overline{d_2^2} - 2A\overline{d_2} + A^2)$$

(16)

and $\overline{d_1}$, $\overline{d_1^2}$, $\overline{d_2}$, $\overline{d_2^2}$ are the means and mean-squares of D_1 and D_2 respectively.

DID THE STANDARD DEVIATIONS CHANGE?

To obtain the probability that the standard deviations changed, $P(v|D_1 D_2 I)$, two terms must be computed. The first of these term, $P(sv|D_1 D_2 I)$, has already been computed, Eq. (6), and the second term, $P(\overline{s}v|D_1 D_2 I)$, like the first, is a marginal probability given by

$$P(\overline{s}v|D_1 D_2 I) = \int dAdBd\sigma_1 P(\overline{s}vAB\sigma_1|D_1 D_2 I)$$
$$\propto \int dAdBd\sigma_1 P(\overline{s}vAB\sigma_1|I)P(D_1 D_2|\overline{s}vAB\sigma_1 I)$$
$$= \int dAdBd\sigma_1 P(\overline{s}v|I)P(A|I)P(B|I)P(\sigma_1|I)$$
$$\times P(D_1|\overline{s}vA\sigma_1 I)P(D_2|\overline{s}vB\sigma_1 I).$$

(17)

The prior probability, $P(B|I)$, must be assigned; and for logical consistency it must be assigned the same prior ranges as A:

$$P(B|I) = \begin{cases} \dfrac{1}{R_a} & \text{If } L \leq B \leq H \\ 0 & \text{otherwise} \end{cases}.$$

(18)

To evaluate the integrals over A and B, one substitutes $1/4$ for $P(\overline{s}v|I)$, Eqs. (7,8,18), and a Gaussian noise prior is used to assign the two likelihoods. Then evaluating the integrals over A and B one obtains,

$$P(\overline{s}v|D_1 D_2 I) \propto \frac{\pi}{8 R_a^2 \log(R_\sigma)\sqrt{N_1 N_2}} \int_{\sigma_L}^{\sigma_H} d\sigma_1 \sigma_1^{-N+1} \exp\left\{ -\frac{z_1 + z_2}{2\sigma_1^2} \right\}$$
$$\times \left[Q\left(\frac{1}{2}, \frac{U_{1L}^2}{\sigma_1^2}\right) \pm Q\left(\frac{1}{2}, \frac{U_{1H}^2}{\sigma_1^2}\right) \right] \left[Q\left(\frac{1}{2}, \frac{U_{2L}^2}{\sigma_1^2}\right) \pm Q\left(\frac{1}{2}, \frac{U_{2H}^2}{\sigma_1^2}\right) \right]$$

(19)

where

$$z_1 = N_1[\overline{d_1^2} - (\overline{d_1})^2], \qquad z_2 = N_2[\overline{d_2^2} - (\overline{d_2})^2],$$

(20)

$$U_{1H} = \sqrt{\frac{N_1}{2}}(H - \overline{d_1}), \qquad U_{1L} = \sqrt{\frac{N_1}{2}}(L - \overline{d_1}), \tag{21}$$

$$U_{2H} = \sqrt{\frac{N_2}{2}}(H - \overline{d_2}), \qquad U_{2L} = \sqrt{\frac{N_2}{2}}(L - \overline{d_2}), \tag{22}$$

and the sign is chosen to be minus when U_{1H} and U_{1L} are of the same sign and plus when they are of different sign, and similarly for U_{2H} and U_{2L}.

Using the calculations presented so far one can determine if something has changed, and then determine what has changed; either the means or the standard deviations. But after determining what changed, again one's interest in the problem changes. Now one wants to estimate the magnitude of the changes. To estimate the magnitude of the changes one needs to compute $P(\delta|D_1 D_2 I)$ and $P(r|D_1 D_2 I)$, and these calculations are the subject of the next section.

3. Estimating The Magnitude Of The Changes

Estimating the difference in means and the ratio of the standard deviations is where most research on this problem has been concentrated. These calculations are based on two assumptions: that one of the parameters differs, and that the other parameter is either the same or not. Then using these assumptions, one estimates the difference in means or the ratio of standard deviations. For the two-sample problem, one assumes that the means are different and that the standard deviations are the same. The difference in means is then estimated. The Behrens-Fisher problem comes about when the standard deviations are assumed unknown and different. In real problems, when a lot of data are available, one does not need probability theory or statistics. The evidence in the data is usually so overwhelming that one can draw correct conclusions without any formal statistical procedures. It is only in the case where the evidence in the data is meager, that one needs any formal statistical theory. But it is precisely in these cases, that the assumptions being made by both the two-sample calculations and the Behrens-Fisher calculations are most questionable. In the next few subsections, the problem of estimating the difference in the means and the ratio of standard deviations is addressed independent of whether or not the other parameter is the same. For the difference in means, this results is a weighted average of the two-sample calculation and the Behrens-Fisher calculation, where the weights are just the probability that the standard deviations are the same or not. A similar result holds for the ratio of the standard deviations.

ESTIMATING THE DIFFERENCE IN MEANS

To estimate the difference in means, one must first introduce this difference into the problem. Defining δ and β to be the difference and sum of the constants A and B, one has

$$\delta = A - B, \qquad \beta = A + B. \tag{23}$$

The two constants, A and B, are then given by

$$A = \frac{\delta + \beta}{2}, \qquad B = \frac{\beta - \delta}{2}. \tag{24}$$

The model equations, Eqs. (1–2), then become

$$d_{1i} = \frac{\delta + \beta}{2} + \text{noise of standard deviation } \sigma_1, \tag{25}$$

and

$$d_{2i} = \frac{\beta - \delta}{2} + \text{noise of standard deviation } \sigma_2. \tag{26}$$

The probability for the difference, δ, is then given by

$$
\begin{aligned}
P(\delta|D_1 D_2 I) &= P(\delta v|D_1 D_2 I) + P(\delta \overline{v}|D_1 D_2 I) \\
&= P(v|D_1 D_2 I) P(\delta|v D_1 D_2 I) + P(\overline{v}|D_1 D_2 I) P(\delta|\overline{v} D_1 D_2 I).
\end{aligned}
\tag{27}
$$

This is a weighted average of the probability for the difference in means given that the standard deviations are the same (the two-sample problem) and the probability for the difference in means given that the standard deviations are different (the Behrens-Fisher problem). The weights are just the probabilities that the standard deviations are the same or different. Two of these four probabilities, $P(v|D_1 D_2 I)$ and $P(\overline{v}|D_1 D_2 I) = 1 - P(v|D_1 D_2 I)$, have already been computed, Eqs. (12). The other two probabilities, $P(\delta|v D_1 D_2 I)$ and $P(\delta|\overline{v} D_1 D_2 I)$, must now be addressed.

The Two-Sample Problem

$P(\delta|v D_1 D_2 I)$, is essentially the two-sample problem. This probability is a marginal probability where the standard deviation and β have been removed as nuisance parameters:

$$
\begin{aligned}
P(\delta|v D_1 D_2 I) &= \int d\beta d\sigma_1 P(\delta \beta \sigma_1 | v D_1 D_2 I) \\
&\propto \int d\beta d\sigma_1 P(\delta \beta \sigma_1 | v I) P(D_1 D_2 | v \delta \beta \sigma_1 I) \\
&= \int d\beta d\sigma_1 P(\delta|I) P(\beta|I) P(\sigma_1|I) P(D_1|v \delta \beta \sigma_1 I) P(D_2|v \delta \beta \sigma_1 I)
\end{aligned}
\tag{28}
$$

where $P(\delta|I)$ and $P(\beta|I)$ are assigned bounded uniform priors:

$$
P(\delta|I) = \begin{cases} \dfrac{1}{2R_a} & \text{If } L - H \leq \delta \leq H - L \\ 0 & \text{otherwise} \end{cases},
\tag{29}
$$

and

$$
P(\beta|I) = \begin{cases} \dfrac{1}{2R_a} & \text{If } 2L \leq \beta \leq 2H \\ 0 & \text{otherwise} \end{cases}.
\tag{30}
$$

To evaluate the integral over σ_1 one substitutes Eqs. (8,29,30), and a Gaussian noise prior is used to assign the two likelihoods. Then evaluating the indicated integral, one obtains

$$
P(\delta|v D_1 D_2 I) \propto \frac{\Gamma(N/2)}{8 R_a^2 \log(R_\sigma)} \int_{2L}^{2H} d\beta V^{-\frac{N}{2}} \left[Q\left(\frac{N}{2}, \frac{V}{\sigma_H^2}\right) - Q\left(\frac{N}{2}, \frac{V}{\sigma_L^2}\right) \right]
\tag{31}
$$

as the probability for the difference in means given that the standard deviations are the same, where

$$V = \frac{N}{2}\left[\overline{d^2} - 2\delta b - \beta\overline{d} + \frac{\beta^2}{4} + \frac{\delta^2}{4} + \frac{\delta\beta\Delta}{2}\right], \tag{32}$$

$$\Delta = \frac{N_1 - N_2}{N}, \quad \text{and} \quad b = \frac{N_1\overline{d_1} - N_2\overline{d_2}}{2N}. \tag{33}$$

THE BEHRENS-FISHER PROBLEM

The Behrens-Fisher problem is essentially given by $P(\delta|\overline{v}D_1D_2I)$, the probability for the difference in means given that the standard deviations are not the same. This probability is a marginal probability where both the standard deviations and the sum of the means, β, have been removed as nuisance parameters:

$$
\begin{aligned}
P(\delta|\overline{v}D_1D_2I) &= \int d\beta d\sigma_1 d\sigma_2 P(\delta\beta\sigma_1\sigma_2|\overline{v}D_1D_2I) \\
&\propto \int d\beta d\sigma_1 d\sigma_2 P(\delta\beta\sigma_1\sigma_2|\overline{v}I)P(D_1D_2|\overline{v}\delta\beta\sigma_1\sigma_2I) \\
&= \int d\beta d\sigma_1 d\sigma_2 P(\delta|I)P(\beta|I)P(\sigma_1|I)P(\sigma_2|I) \\
&\quad \times P(D_1|\overline{v}\delta\beta\sigma_1 I)P(D_2|\overline{v}\delta\beta\sigma_2 I)
\end{aligned}
\tag{34}
$$

where all of the terms appearing in this probability density function have been previously assigned.

To evaluate the integrals over σ_1 and σ_2, one substitutes Eqs. (8,14,29,30) and a Gaussian noise prior is used in the two likelihoods. Evaluating the integrals, one obtains

$$
\begin{aligned}
P(\delta|\overline{v}D_1D_2I) &\propto \frac{\Gamma(N_1/2)\Gamma(N_2/2)}{16R_a^2[\log(R_\sigma)]^2}\int_{2L}^{2H} d\beta W_1^{-\frac{N_1}{2}}W_2^{-\frac{N_2}{2}} \\
&\quad \times \left[Q\left(\frac{N_1}{2},\frac{W_1}{\sigma_H^2}\right) - Q\left(\frac{N_1}{2},\frac{W_1}{\sigma_L^2}\right)\right]\left[Q\left(\frac{N_2}{2},\frac{W_2}{\sigma_H^2}\right) - Q\left(\frac{N_2}{2},\frac{W_2}{\sigma_L^2}\right)\right]
\end{aligned}
\tag{35}
$$

where

$$W_1 = \frac{N_1}{2}\left[\overline{d_1^2} - \overline{d_1}(\delta + \beta) + \frac{(\delta + \beta)^2}{4}\right], \tag{36}$$

and

$$W_2 = \frac{N_2}{2}\left[\overline{d_2^2} - \overline{d_2}(\beta - \delta) + \frac{(\beta - \delta)^2}{4}\right]. \tag{37}$$

With the completion of this calculation the probability for the difference in means, Eq. (27), is now complete. Before turning our attention to a numerical example there is one final calculation, the probability for the ratio of the standard deviations independent of whether or not the means are the same, that must be completed.

ESTIMATING THE RATIO OF THE STANDARD DEVIATIONS

To estimate the ratio of the standard deviations, this ratio must be introduced into the problem. Defining r and σ to be

$$r = \frac{\sigma_1}{\sigma_2}, \qquad \sigma = \sigma_2 \tag{38}$$

and substituting these into the model, Eqs. (1,2), one obtains

$$d_{1i} = A + \text{noise of standard deviation } r\sigma, \tag{39}$$

and

$$d_{2i} = B + \text{noise of standard deviation } \sigma. \tag{40}$$

The probability for the ratio of the standard deviations, $P(r|D_1 D_2 I)$, is then given by

$$
\begin{aligned}
P(r|D_1 D_2 I) &= P(rs|D_1 D_2 I) + P(r\bar{s}|D_1 D_2 I) \\
&= P(s|D_1 D_2 I)P(r|sD_1 D_2 I) + P(\bar{s}|D_1 D_2 I)P(r|\bar{s}D_1 D_2 I).
\end{aligned} \tag{41}
$$

This is a weighted average of the probability for the ratio of the standard deviations given the means are the same plus the probability for the ratio of the standard deviations given that the means are different. The weights are just the probabilities that the means are the same or not. Two of the four probabilities, $P(s|D_1 D_2 I)$ and $P(\bar{s}|D_1 D_2 I) = 1 - P(s|D_1 D_2 I)$, have already been computed, Eqs. (11). The other two probabilities, $P(r|sD_1 D_2 I)$ and $P(r|\bar{s}D_1 D_2 I)$, must now be addressed.

ESTIMATING THE RATIO, GIVEN THE MEANS ARE THE SAME

The first term to be addressed is $P(r|sD_1 D_2 I)$. This probability is a marginal probability where both σ and A have been removed as nuisance parameters:

$$
\begin{aligned}
P(r|sD_1 D_2 I) &= \int dA d\sigma\, P(rA\sigma|sD_1 D_2 I) \\
&\propto \int dA d\sigma\, P(rA\sigma|sI)P(D_1 D_2|srA\sigma I) \\
&= \int dA d\sigma\, P(r|I)P(A|I)P(\sigma|I)P(D_1|srA\sigma I)P(D_2|srA\sigma I)
\end{aligned} \tag{42}
$$

where the prior probability for the ratio of the standard deviations is taken to be a bounded Jeffreys prior:

$$
P(r|I) = \begin{cases} \dfrac{1}{2r \log(R_\sigma)} & \text{If } \sigma_L/\sigma_H \leq r \leq \sigma_H/\sigma_L \\[2mm] 0 & \text{otherwise} \end{cases}. \tag{43}
$$

To evaluate the integral over A, one substitutes Eq. (8,14,43), and a Gaussian noise prior probability is used to assign the two likelihoods. Evaluating the integral, one obtains

$$
P(r|sD_1 D_2 I) = \frac{\sqrt{\pi/8w}\; r^{-N_1-1}}{R_a [\log(R_\sigma)]^2} \int_{\sigma_L}^{\sigma_H} d\sigma \exp\left\{-\frac{x}{2\sigma^2}\right\} \left[Q\left(\frac{1}{2}, \frac{X_L^2}{\sigma^2}\right) \pm Q\left(\frac{1}{2}, \frac{X_H^2}{\sigma^2}\right) \right] \tag{44}
$$

where

$$X_H = \sqrt{\frac{w}{2}}[A_H - v/w], \qquad X_L = \sqrt{\frac{w}{2}}[A_L - v/w], \tag{45}$$

$$u = \frac{N_1\overline{d_1^2}}{r^2} + N_2\overline{d_2^2}, \qquad v = \frac{N_1\overline{d_1}}{r^2} + N_2\overline{d_2}, \tag{46}$$

$$w = \frac{N_1}{r^2} + N_2, \qquad x = u - \frac{v^2}{w}, \tag{47}$$

and the sign is again chosen to be a minus if X_L and X_H are of the same sign, and plus if they are of different sign.

ESTIMATING THE RATIO, GIVEN THE MEANS ARE DIFFERENT

The second term that must be computed is $P(r|\overline{s}D_1D_2I)$, the probability for the ratio of standard deviations given that the means are not the same. This is a marginal probability where σ, A, and B have been removed as nuisance parameters:

$$
\begin{aligned}
P(r|\overline{s}D_1D_2I) &= \int dA dB d\sigma \, P(rAB\sigma|\overline{s}D_1D_2I) \\
&\propto \int dA dB d\sigma \, P(rAB\sigma|\overline{s}I)P(D_1D_2|\overline{s}rAB\sigma I) \\
&= \int dA dB d\sigma \, P(r|I)P(A|I)P(B|I)P(\sigma|I)P(D_1|\overline{s}A\sigma I)P(D_2|\overline{s}B\sigma)
\end{aligned}
\tag{48}
$$

where all of the terms appearing in this probability density function have been previously assigned.

To evaluate the integral over A and B one substitutes Eq. (7,8,18,43), and a Gaussian noise prior is used in assigning the two likelihoods. Evaluating the indicated integrals, one obtains

$$
P(r|\overline{s}D_1D_2I) \propto \frac{\pi}{4R_a^2[\log(R_\sigma)]^2\sqrt{N_1N_2}} \int_{\sigma_L}^{\sigma_H} d\sigma \, r^{-N_1}\sigma^{-N+1} \exp\left\{-\frac{z_1}{2r^2\sigma^2} - \frac{z_2}{2\sigma^2}\right\}
$$
$$
\times \left[Q\left(\frac{1}{2}, \frac{U_{1L}^2}{r^2\sigma^2}\right) \pm Q\left(\frac{1}{2}, \frac{U_{1H}^2}{r^2\sigma^2}\right)\right] \left[Q\left(\frac{1}{2}, \frac{U_{2L}^2}{\sigma^2}\right) \pm Q\left(\frac{1}{2}, \frac{U_{2H}^2}{\sigma^2}\right)\right]
\tag{49}
$$

where the minus sign is chosen when U_{1H} and U_{1L} have the same sign and the plus sign is chosen when the sign of U_{1H} and U_{1L} differ and similarly for U_{2H} and U_{2L}.

4. Numerical Example

With the completion of this calculation, the probability for all of the hypotheses appearing in Table 1 have been computed. It is now time to apply these calculations in an example. The example used is taken from Jaynes [14,15]. This article is a series of examples contrasting orthodox and Bayesian methods. In one example, Jaynes applied both the two-sample and the Behrens-Fisher calculations to the same sets of data. This example will be extended here using the procedures developed in this paper. The data in this example are the mean lifetimes and standard deviations for a certain component from two different manufactures. Manufacturer A supplies 9 units for test, which turn out to have a (mean $\pm$ standard deviation) lifetime of (42 ± 7.48) hours. Manufacturer B supplies 4 units, which yield (50 ± 6.48) hours. The calculations presented here will first determine if the data sets differ; then if they differed, how; and last, given that they differed, the magnitude of the difference will be estimated.

In the analysis performed by Jaynes the nuisance parameters were removed using improper prior probabilities. This could be done, because the problem was treated as a parameter estimation problem, and the infinites introduced cancel when the distributions were normalized. However, in the calculation presented here, improper priors cannot be used; because the infinites do not cancel. The prior range on the amplitudes will be taken as $34 \leq A, B \leq 58$ and for the standard deviations $3 \leq \sigma_1, \sigma_2 \leq 10$. Because the data and the type of components are not stated, assigning these prior ranges is more difficult than normal. Consequently, at the end of this example, the calculations will be repeated using wider ranges to see what effect this has on the conclusions.

The data consists of the means and the standard deviations of each data set as well as the number of data values. But how the standard deviations were computed was not given. Here it will be assumed that $(N - 1)$ was in the standard deviation calculation. With this assumption the mean-square data values may be computed, and thus, all of the calculations presented in this paper may be performed. These calculations have been implemented as a general fortran program that will analyze any two-sample or Behrens-Fisher like data sets. This program is available from the author. The output from this program is given in Table 2 for the data presented in this example.

NUMERICAL EXAMPLE – DID SOMETHING CHANGE?

The first question of interest is whether or not the data sets are the same. Jaynes, essentially takes it as a given that the data sets differ. He indicates "I think our common sense tells us immediately, without any calculation, that this constitutes fairly substantial (but not overwhelming) evidence in favor of B." To arrive at this conclusion, one must first conclude that the data sets are different, and second that they differ in the means, independent of whether or not the standard deviations are the same. In the analysis done by Jaynes, the calculations are first done using the assumption that the standard deviations are different, from which he demonstrated that there is a 0.92 probability that $B > A$. And he went on to demonstrate that essentially the same conclusions would be drawn when the standard deviations are assumed equal.

In this paper, the probability that the data sets are the same, has been explicitly computed, Eq. (3). Similarly the probability that the means differed is given by $P(\overline{s}|D_1 D_2 I) = 1 - P(s|D_1 D_2 I)$, and $P(s|D_1 D_2 I)$ is given by Eq. (11) and these probabilities are given in Table 2. Consulting Table 2, the probability the data sets are different is 0.83, thus supporting Jaynes' conclusion that the data sets differed. And confirms Jaynes' intuition that it is good evidence, but not overwhelming.

NUMERICAL EXAMPLE – WHAT CHANGED?

Now that one knows that the data sets are not the same, or at the very least are probably not the same, one would like to know what changed. Did the means change or did the standard deviations change? Consulting Table 2, the probability that the means are different is 0.72. So Jaynes' conclusion that the means are different is being supported.

Table 2: Numerical Example

```
Enter The Amplitude Lower Bound:   34
Enter The Amplitude Upper Bound:   58

Enter The Variance Lower Bound:   3
Enter The Variance Upper Bound:   10
```

No.	Standard Deviation	Average	Data Set
4	6.4800	50.000	Jaynes.1
9	7.4800	42.000	Jaynes.2
13	7.9099	44.462	Combined

---------------Model----------------	Probability
Same Constant, Same Variance	0.1689119
Different const, Same Variance	0.4153378
Same Constant, Different Variances	0.1077223
Different Const, Different Variances	0.3080280

```
The probability the constants are the same is:  0.27663
The probability the constants are different is:  0.72337
The odds ratio is 2.61 to 1 in favor of different constants

The probability the variances are the same is:  0.58425
The probability the variances are different is:  0.41575
The odds ratio is 1.41 to 1 in favor of the same variances

The probability the data sets are the same is:  0.16891
The probability the data sets are different is:  0.83109
The odds ratio is 4.92 to 1 in favor of different
    means or variances
```

Table 2 summarizes the output from the fortran implementation of this calculation. This program produces three different types of output: (1) the probability for the four hypotheses examined by Dayal and Dickey; (2) the probability that the data sets differ, the probability that the means are different and the probability that the variances are different; and finally (3) the probability for the difference in means and the ratio of the standard deviations is computed – see Figs. 1 and 2.

In the calculations performed by Jaynes both a two-sample like calculation and a Behrens-Fisher like calculation were performed to indicate the relative evidence in favor of the hypothesis that the means were different. But to perform a two-sample calculation one must assume that the standard deviations are the same. Similarly to perform a Behrens-Fisher calculation, one must assume that the standard deviations are different. The probability that the standard deviations are the same is also given in Table 2. Consulting Table 2, one finds the probability that the standard deviations are the same is 0.58; neither confirming nor denying the hypothesis. So neither a two-sample calculation nor a Behrens-Fisher calculation is justified for this data; the data do not support either hypothesis. Given how close these probabilities are to 50/50 one might expect that a weighted average of the results from a two-sample calculation and a Behrens-Fisher calculation would be a better indicator of the difference in means, and this is exactly what probability theory tells one to do.

NUMERICAL EXAMPLE – ESTIMATING THE CHANGES

The probability that the means are different is 0.72, indicating good but not overwhelming evidence in favor of different means. In this subsection it will be assumed that the means are different and the problem to be addressed is one of estimating the magnitude of this difference. The probability for the difference in means is given by Eq. (27). To compute this probability both a two-sample calculation and a Behrens-Fisher calculation must be performed. So it is easy to have the program report the two-sample calculation (dotted line, Fig. 1), the Behrens-Fisher calculation (dashed line, Fig. 1), and weighted average derived in this paper (solid line, Fig. 1).

The two-sample calculation assumes the standard deviations are the same. There is only a 0.58 probability of this, so the two-sample model must not fit the data much differently than a Behrens-Fisher model. Here the two-sample model estimates the standard deviation of the noise to be higher than the Behrens-Fisher calculation, because the pooled standard deviation is larger than that of either data set separately. Consequently, this distribution is more spread out, less certain, of the difference (dotted line, Fig. 1). The Behrens-Fisher calculation has a standard deviation for each data set, and can reduce the overall estimate of the noise; so the Behrens-Fisher distribution is more sharply peaked (dashed line, Fig. 1). But probability theory tells one to use a weighted average of these two distributions. Here both models fit the data about equally well. Under these conditions probability theory will prefer the simpler model. The probabilities that the standard deviations are the same is 0.58 and 0.42 that they are different. So the weighted average follows the two-sample calculation a little more closely than the Behrens-Fisher calculation (solid line, Fig. 1).

NUMERICAL EXAMPLE – THE RATIO OF STANDARD DEVIATIONS

When estimating the difference in means, there was not much evidence in favor of different standard deviations. Consequently one would not expect either the two-sample calculation or the Behrens-Fisher calculation to be very different and this is exactly what Fig. 1 shows. But there is fairly strong evidence in favor of the means being different. If the problem is to estimate the ratio of the standard deviations, one would expect the two calculations to be substantially different. That is to say, the probability for the ratio of

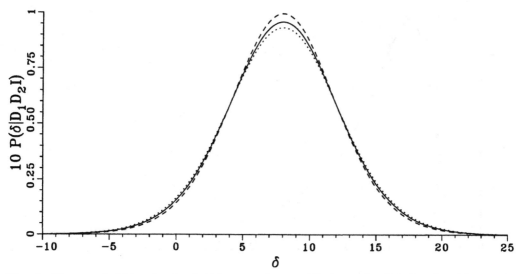

Fig. 1. Three probability density function are shown: (1) the probability for the difference in means given the standard deviations are the same (dotted line), (2) the probability for the difference in means given that the the standard deviations are different (dashed line), and (3) the probability for the difference in means independent of whether or not the standard deviation are the same (solid line).

the standard deviations given the same means should be substantially different from the probability for the ratio of the standard deviations given that the means are different.

These two distributions, as well as the weighted average are shown in Fig. 2. The probability for the ratio of the standard deviations given the means are the same is shown as the dotted line. This model does not fit the data well (the pooled data have a larger standard deviation than either data set separately). Consequently, the uncertainty in this probability distribution is large compared to the other models and the distribution is more spread out. The probability for the ratio of standard deviations given different means is shown as the dashed line. This model fits the data better, and results in a more strongly peaked probability distribution. But probability theory tells one to take a weighted average of these two distributions, solid line. The weights are just the probabilities that the means are the same or different. Here those probabilities are 0.28 and 0.72 respectively. So the weighted average follows $P(r|\bar{s}D_1D_2I)$ more closely than $P(r|sD_1D_2I)$.

Numerical Example – The Effect Of The Prior Ranges

It has been noted several times that the prior ranges do not cancel. This occurs when models are being considered that contain either different types of parameters or different numbers of parameters of the same type. Here the models all contain the same types of parameters, constants and standard deviations, but they contain differing numbers of these parameters. Consequently the prior ranges are important and will affect the conclusions.

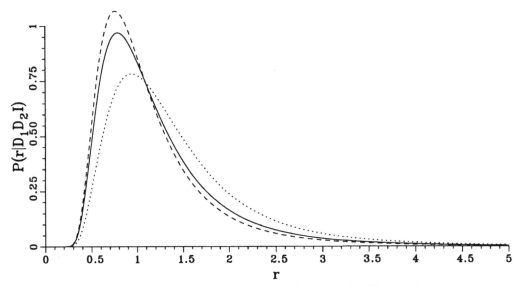

Fig. 2. Three probability density functions are shown: (1) the probability for the ratio of standard deviations given that the means are the same (dotted line), (2) the probability for the ratio of standard deviations given that the means are different (dashed line), (3) the probability for the ratio of standard deviations independent whether or not the means are the same (solid line).

In the example just given the prior range on the constants, A and B, was $42 - 8 \leq A, B \leq 50 + 8$, where 8 was a rough estimate of the standard deviation. This assumes, in effect, that all of the data values are roughly within one standard deviation of the mean. This would be an unreasonable assumption if the number of data items were large, but with only 4 and 9 data items, who knows? The question investigated here is just how strongly do the conclusions depend on the prior ranges?

Suppose that $8 \rightarrow 16$, so that the prior ranges become $26 \leq A, B \leq 66$. What effect would this have on the conclusions? Rerunning the program using these ranges (for the time being holding the prior ranges on the standard deviations constant) one finds the probability that the data sets are different is 0.765. This compared to 0.83, previously. So changing the uncertainty in the amplitudes by a factor of 2, has lowered this probability by roughly 8%; a small change. Second, the probability that the standard deviations changed is now 0.411, compared to 0.415 previously; only a 1% change. And third, the probability that the means are different is 0.61, compared to 0.72 previously; an 18% change. In all cases the results have changed slightly, but not nearly by a factor of 2 and *none of the conclusions were changed.*

Now, suppose that the prior range on the standard deviations were changed by a factor of 2: $(3 \leq \sigma_1, \sigma_2 \leq 10) \rightarrow (3/\sqrt{2} \leq \sigma_1, \sigma_2 \leq 10 * \sqrt{2})$; what would the effect of this be? Here the prior range on the constant will be returned to their original value. Rerunning the program, one obtains a probability of 0.81 that the data sets are different. This compares to 0.83 obtained previously, about a 2% change. For the probability the means are different, one obtains 0.695 compared to 0.72 previously, about a 3.6% change. And last the probability that the standard deviations are different was 0.38 compared to 0.41, about an 8% change. Again none of the major conclusions are changed.

Apparently, the effect of these prior ranges is small. Here the size of the effect is of the order of the square root of the logarithm of change or smaller. The magnitude of the effect depends on just how strongly a proposition is being supported by the data: when the evidence in the data is large, changing the prior ranges by factors of thousands has essentially no effect on the conclusions. The only reason one could see an effect here was that the evidence in the data was small and then the prior information is important.

5. Summary And Conclusions

The calculations presented in this paper are generalizations of the Behrens-Fisher and two-sample problems and the traditional F and student t-distributions. They allow the experimenter to investigate the problems of interest in a way never before possible. First, they allow one to determine if the data sets differ. Second, they allow one to determine how they differ; either in the means or in the standard deviations. And third, one can estimated the magnitude of these changes without additional assumptions. The difference in mean can be estimated independent of whether or not the standard deviations are the same or not; while the ratio of the standard deviations can be estimated independent of whether or not the means are the same. Last, these probability density functions allow one to picture the results in a way never before possible: not by stating the results as a single number, but graphically; so that one can see the evidence with ones own eyes.

ACKNOWLEDGMENTS. The encouragement of Professor J. J. H. Ackerman is greatly appreciated as are extensive conversations with Professor E. T. Jaynes. This work was partially supported by a gift from the Monsanto Company and NIH grant GM30331, J. J. H. Ackerman principal investigator.

REFERENCES

[1] Behrens, W. V., "Ein Beitrag zur Fehlerberechnung bei weinige Beobachtungen," *Landwirtschaftliche Jahrbücher*, **68**, pp. 807-837 (1929).

[2] Fisher, R. A,. "The comparison of samples with possibly unequal variances," *Ann. of Eugenics*, **9**, pp. 174-180 (1937).

[3] Fisher, R. A., *Statistical Methods and Scientific Inference*, Hafner Publishing Co., New York (1956).

[4] Jeffreys, H., *Theory of Probability*, Oxford University Press, London (1939).

[5] Lee, Peter M., *Bayesian Statistics: An Introduction*, Oxford University Press, New York (1989).

[6] Patil, V. H., "The Behrens-Fisher problem and its Bayesian solution," *Journal of Indian Statistical Assoc.*, **2** pp. 21-31 (1964).

[7] Robinson, G. K., "Properties of Student's t and of the Behrens-Fisher solution to the two means problem." *Ann. Statist.*, **4** (1976).

[8] Robinson, G. K., "Properties of Student's t and of the Behrens-Fisher solution to the two means problem." *Ann. Statist.*, **10** (1982).

[9] Roberts, Norman A., *Mathematical Methods in Reliability Engineering*, McGraw-Hill Book Co. Inc., New York pp.86-88 (1964).

[10] Smith, H. F., "The Problem of Comparing the Results of Two Experiments With Unequal Errors," *J. Sci. Ind. Research (India)*, **9**, pp. 211-212 (1936).

[11] Scatterthwaite, F. E., "An Approximate Distribution Of Estimates Of Variance Components," *Biometrics Bull.*, **2**, pp. 110-114 (1946).

[12] Dayal, Hari H., "Bayesian statistical inference in Behrens-Fisher Problems," Ph.D. dissertation, State University of New York at Buffalo (1972).

[13] Dayal, Hari H., and James M. Dickey "Bayes Factors for Behrens-Fisher Problems" *The Indian Journal of Statistics*, **38** pp. 315-328 (1976).

[14] Jaynes, E. T. "Confidence Intervals vs Bayesian Intervals," in *Foundations of Probability Theory, Statistical Inference, and Statistical Theories of Science*, **2**, pp. 175-257 (1976).

[15] Jaynes, E. T., *Papers on Probability, Statistics and Statistical Physics*, a reprint collection, D. Reidel, Dordrecht the Netherlands (1983).

BAYESIAN ANALYSIS, MODEL SELECTION AND PREDICTION

Arnold Zellner and Chung-ki Min
H.G.B. Alexander Research Foundation
Graduate School of Business, University of Chicago
Chicago, Illinois 60637

1. Introduction

In this paper, we shall consider (1) the derivation of prior distributions for use in Bayesian analysis, (2) Bayesian and non-Bayesian model selection procedures and (3) an application of Bayesian model selection procedures in the context of an applied forecasting problem. With respect to derivations of prior distributions, Jaynes (1980, p.618) has stated:

"It was from studying these flashes of intuition in Jeffreys that I became convinced that there must exist a general formal theory of determination of priors by logical analysis of the prior information–and that to develop it is today the top priority research problem of Bayesian theory."

As is well known, many including Jeffreys (1967), Lindley (1956), Savage (1961), Hartigan (1964), Novick and Hall (1965), Jaynes (1968), Box and Tiao (1973), Bernardo (1979), and Zellner (1971, 1977, 1990) have worked to solve the "top priority research problem of Bayesian theory," to use Jaynes' words. Herein we shall consider further the maximal data information prior (MDIP) distribution approach put forward and discussed in Zellner (1971, 1977, 1990). A MDIP is a solution to a well-defined optimization problem and its use provides maximal data information relative to the information in a prior density. Also, as shown below, a MDIP maximizes the expectation of the logarithm of the ratio of the likelihood function, $\ell(\theta|y)$, to the prior density, $\pi(\theta)$, i.e. $\ln[\ell(\theta|y)/\pi(\theta)]$. Several explicit MDIPs will be presented including those for parameters of time series processes–see Phillips (1991) for a lively discussion of this topic.

With respect to model selection procedures, posterior odds and several non-Bayesian model selection criteria will be compared and discussed along with corrections for "data-mining"–see e.g. Jeffreys (1967, p. 254). After Jeffreys' seminal work many have employed posterior odds to evaluate alternative hypotheses and models. Herein, we shall show how posterior odds can be used in variable selection problems–see Geisel (1975), Zellner (1971, 1984), Zellner and Siow (1980), Leamer (1978), and the studies in Poirier (1991) for earlier discussions and applications of posterior odds in model selection and Geisel (1975), Diebold (1990), Min and Zellner (1990), and Palm and Zellner (1991) for their use in model and forecast combination. Finally, some of the techniques mentioned above will be applied in analyzing U.S. annual data to determine how to formulate a forecasting model for the growth rate of real gross national product. Posterior odds and forecasting results for 256 alternative models will be reviewed and compared.

The plan of the paper is as follows. In section 2, maximal data information prior densities will be discussed. Then in section 3, posterior odds used in model selection problems will be compared to several non-Bayesian model selection criteria. Section 4 is devoted to results of applying Bayesian and non-Bayesian techniques to select a forecasting model. A summary of results and concluding remarks are presented in section 5.

2. Maximal Data Information Prior (MDIP) Distributions

In Zellner (1971, 1977, 1990), MDIPs' rationale, derivation and explicit forms for many problems have been discussed. As pointed out earlier, the negative entropy of a joint density function $p(y, \theta)$ for observations and parameters is given by $-H = \int \int p(y, \theta) \ell n \ p(y, \theta) dy d\theta$, which on using $p(y, \theta) = \pi(\theta) f(y|\theta)$ can be expressed as

$$-H = \int I(\theta) \pi(\theta) d\theta + \int \pi(\theta) \ell n \ \pi(\theta) d\theta \qquad (1)$$

with

$$I(\theta) = \int f(y|\theta) \ell n \ f(y|\theta) dy \qquad (2)$$

the negative entropy associated with $f(y|\theta)$, the data density. In (1), $-H$, a measure of the information in $p(y, \theta)$, breaks up into two parts, $\int I(\theta) \pi(\theta) d\theta$, the average information in the data density, $f(y|\theta)$, and $\int \pi(\theta) \ell n \ \pi(\theta) d\theta$, the information in the prior density $\pi(\theta)$.

The criterion functional which is maximized with respect to choice of $\pi(\theta)$ subject to side conditions to yield a MDIP, denoted by $\pi_*(\theta)$ is:

$$G[\pi(\theta)] = \int I(\theta) \pi(\theta) d\theta - \int \pi(\theta) \ell n \ \pi(\theta) d\theta \qquad (3)$$

which is just the difference of the two terms on the r.h.s. of (1). Thus $G[\pi(\theta)]$ is equal to the average information in the data density, $I(\theta)$ given in (2) minus the information in the prior density, $\pi(\theta)$. Maximizing G with respect to the choice of $\pi(\theta)$, subject to side conditions reflecting initial information, e.g. $\int \pi(\theta) d\theta = 1$, etc. provides a unique prior density which gives maximal data information relative to the information in the prior; that is the expression on the r.h.s. of (3) is maximized. Thus the information in the data density is featured by use of a MDIP.[1]

An alternative expression for $G[\pi(\theta)]$ in (3) is:

$$G[\pi(\theta)] = \int \left[\int g(\theta|y) \ell n \{\ell(\theta|y)/\pi(\theta)\} d\theta \right] h(y) dy \qquad (4)$$

---

[1] In cases in which y is a vector $y' = (y_1, y_2, ... y_n)$ and the y_i's are independently and identically distributed, $f(y|\theta) = \prod_{i=1}^{n} f(y_i|\theta)$; then (2) is $I_n(\theta) = nI(\theta)$, where $I(\theta) = \int f(y_i|\theta) \ell n \ f(y_i|\theta) dy_i$, $i = 1, 2, ..., n$, and maximization of G in (3) yields the prior $\pi_*(\theta)$ that makes the average information in the data density $I_n(\theta)/n$ as large as possible relative to the information in the prior density.

where $p(y, \theta) = h(y)g(\theta|y)$, with $g(\theta|y)$ the posterior density of θ, $h(y)$ the marginal density of y and $\ell(\theta|y) \equiv f(y|\theta)$ is the well-known likelihood function. Thus (4) indicates that $G[\pi(\theta)]$ is an average of the logarithm of the likelihood-prior ratio, $\ell(\theta|y)/\pi(\theta)$, taken w.r.t. $g(\theta|y)$, the posterior density and $h(y)$, the marginal density of the observations. It follows that maximizing G w.r.t. choice of $\pi(\theta)$ **makes the likelihood function $\ell(\theta|y)$ as high as possible relative to the prior, $\pi(\theta)$.** This is indeed a satisfying feature of a MDIP.

As shown in Zellner (1971, 1977, 1990), on maximizing $G[\pi(\theta)]$ in (3) with respect to $\pi(\theta)$ subject to $\int \pi(\theta)d\theta = 1$, that is that $\pi(\theta)$ is a proper density, the solution is:

$$\pi_*(\theta) = c \exp\{I(\theta)\} \qquad a < \theta < b \tag{5}$$

with $I(\theta)$ given in (2) and $c^{-1} = \int_a^b e^{I(\theta)}d\theta$ is a normalizing constant. Thus if $I(\theta) = $ constant, $\pi_*(\theta)$ is a uniform density. Also for the class of densities $f[(y - \theta)/\sigma], \pi_*(\theta, \sigma) \propto 1/\sigma$ with $a_1 < \theta < b_1$ and $0 < a_2 < \sigma < b_2$. See Zellner (1977) for many other MDIPs.

Above, we have used just the side condition, $\int \pi(\theta)d\theta = 1$ in the maximization problem. If other side conditions are imposed reflecting given initial information, e.g. $\int \theta^i \pi(\theta)d\theta = \mu_i$, $i = 1,2,...,$m, then the MDIP is given by $\pi_*(\theta) \propto \exp\{I(\theta) + \lambda_1\theta + \lambda_2\theta^2 + ... + \lambda_m\theta^m\}$, where the λ_i are Lagrange multipliers which can be evaluated given values of the prior moments, μ_i, $i = 1,2,...,$m. If, e.g. m = 2 and $I(\theta) = $ const., $\pi_*(\theta)$ has a normal form. If m = 4 and $I(\theta) = $ const., $\pi_*(\theta)$ has an exponential quartic form. We see that side conditions reflecting initial information that may be available can readily be utilized in deriving MDIPs. Since all of these MDIPs are proper, they can be used in computing posterior odds. However, before discussing this topic, we shall consider some MDIPs for time series models.

For a **stationary** autoregressive model, $y_t - \mu = \rho(y_{t-1} - \mu) + \varepsilon_t$, where $\varepsilon_t \sim N(0,\sigma^2)$, $-1 < \rho < 1$, $Ey_t = \mu$, $\sigma_{y_t}^2 = \sigma^2/(1-\rho^2)$, and $f(y_t|\mu, \rho, \sigma) \propto (1-\rho^2)^{1/2}\sigma^{-1}\exp\{-(1-\rho^2)(y_t-\mu)^2/2\sigma^2\}$. Then $\ell n\, f(y_t|\mu, \rho, \sigma) = $ const. $+ \frac{1}{2}\ell n(1-\rho^2) - (1-\rho^2)(y_t-\mu)^2/2\sigma^2$ and $I(\theta) = E\, \ell n\, f(y_t|\mu, \rho, \sigma) = $ const. $+ \frac{1}{2}\ell n(1-\rho^2) - \frac{1}{2}$ and thus $\pi_*(\mu, \rho, \sigma) \propto \exp\{I(\theta)\} = (1-\rho^2)^{\frac{1}{2}}/\sigma$ as pointed out in Zellner (1977). Thus μ, $\ell n\, \sigma$ and ρ are independently distributed with μ and $\ell n\, \sigma$ uniform and ρ having a density $\pi_*(\rho) \propto (1-\rho^2)^{\frac{1}{2}}$, $-1 < \rho < 1$, a form of a beta density symmetric about a mode at $\rho = 0$ and with a value equal to zero for $\rho = 1$ and $\rho = -1$.

Now for an AR(2) stationary process, $y_t - \mu = \rho_1(y_{t-1} - \mu) + \rho_2(y_{t-2} - \mu) + \varepsilon_t$, where $\varepsilon_t \sim N(0,\sigma^2)$, $Ey_t = \mu$, $\sigma_{y_t}^2 = \rho_1^2\sigma_{y_t}^2 + \rho_2^2\sigma_{y_t}^2 + 2\rho_1\rho_2\gamma_1 + \sigma^2$, with $E(y_{t-1}-\mu)(y_{t-2}-\mu) = \gamma_1$. Since $\gamma_1 = \rho_1\sigma_{y_t}^2 + \rho_2\gamma_1$, $\gamma_1 = \sigma_{y_t}^2\rho_1/(1-\rho_2)$ and thus $\sigma_{y_2}^2 = \sigma^2/[1-\rho_1^2-\rho_2^2-2\rho_1^2\rho_2/(1-\rho_2)]$. For this problem, the MDIP is,

$$\pi_*(\mu, \rho_1, \rho_2, \sigma) \propto [1 - \rho_1^2 - \rho_2^2 - 2\rho_1^2\rho_2/(1 - \rho_2)]^{1/2}/\sigma \tag{6}$$

with ρ_1 and ρ_2 constrained to the region of stationarity in the ρ_1, ρ_2 plane, see Zellner (1971,p.196) for a graphical display of this region and Fig. 1 for a plot of the contours of this prior for ρ_1 and ρ_2. For higher order stationary autoregressive processes, similar procedures can be employed to obtain MDIP priors.

It is often the case that one may not be sure that an autoregressive process is stationary. For example in terms of an AR(1) process $y_t = \rho y_{t-1} + \varepsilon_t$, it may be that $\rho \geq 1$ which implies that the process is not stationary. If such a process begins at $t = 0$, $y_t = \rho^t y_0 + \varepsilon_t +$

$\rho\varepsilon_{t-1} + ... + \rho^{t-1}\varepsilon_1$ and if y_0 is fixed, $Ey_t = \rho^t y_0$, and $\sigma^2_{y_t} = [1 + \rho^2 + \rho^4 + \rho^6 + ... + \rho^{2(t-1)}]\sigma^2$ under the assumption that the ε_t's are uncorrelated with zero means and common variance σ^2. If it is further assumed that the ε_t's are NID$(0, \sigma^2)$, then y_t has a normal density with mean $\rho^t y_0$ and variance $\sigma^2_{y_t}$, $f(y_t|y_0, \rho, \sigma) \propto \sigma^{-1}_{y_1} \exp\{-(y_t - \rho^t y_0)^2 / 2\sigma^2_{y_t}\}$ and the MDIP is $\pi_*(\sigma_{y_t}) \propto 1/\sigma_{y_t}$ or

$$\pi_*(\rho, \sigma) \propto 1/\sigma[1 + \rho^2 + \rho^4 + ... + p^{2(t-1)}]^{1/2} \qquad (7)$$

In this case, the MDIP depends on t, as well as σ and ρ where σ and ρ are confined to finite, possibly very large, sub-regions of the parameter space, $0 < \sigma < \infty$ and $-\infty < \rho < \infty$. For $t = 1$, $\pi_*(\rho)$ is uniform while for $t > 1$, it has a mode at $\rho = 0$ and falls off symmetrically for $\rho \neq 0$. Given a MDIP relative to $f(y_t|\rho, \sigma, y_0)$, on observing y_t we can use the prior in (7) to obtain a posterior density for ρ and σ which can be used as a prior in the analysis of the next observation y_{t+1} and so on. If on the other hand we want the MDIP relative to the data density for $(y_t, y_{t+1}, ..., y_{t+n}) = (y_t, \mathbf{y}')$, then $f(y_t, \mathbf{y}|y_0, \rho, \sigma) = f(y_t|y_0, \rho, \sigma)f(\mathbf{y}|y_t, y_0, \rho, \sigma)$ and application of the MDIP approach to $f(\mathbf{y}|y_t, \rho, \sigma)$ alone yields $\pi_*(\rho, \sigma) \propto 1/\sigma$. On the other hand when the complete joint density is considered, the information in the $n + 1$ observations $(y_t \mathbf{y})$ is $I_{n+1} = E[\ell n\ f(y_t|y_0, \rho, \sigma) + \ell n\ f(\mathbf{y}|y_t y_0, \rho, \sigma)] = -[\ell n\ \sigma_{y_t} + n\ \ell n\ \sigma]$, with $\sigma^2_{y_t} = \sigma^2[1 + \rho^2 + ... + \rho^{2(t-1)}]$. Then the information per observation, $I = I_{n+1}/(n+1)$ and $\pi_*(\rho, \sigma) \propto \exp\{I\} = 1/\sigma[1 + \rho^2 + \rho^4 + ... + \rho^{2(t-1)}]^{1/(n+1)}$. This prior density will tend to become uniform in ρ as the sample size n becomes large and/or t approaches 1.

The above analysis rests on the presumption that y_0 and t, the time the process has been operative, are both known. If one or the other of these quantities has an unknown value, then the density for y_t cannot be implemented. In such a case it is possible to formulate $f(y_{t+1}|y_t, \rho, \sigma)$ and use it to obtain the following MDIP, $\pi_*(\rho, \sigma) \propto 1/\sigma$, $0 < a_1 < \sigma < b_1$ and $-a_2 < \rho < b_2$. Thus, if it is thought that ρ might possibly be slightly larger than 1 but surely not less than -1, the density of ρ could be taken uniform with $-1 \leq \rho \leq 1.1$, as was done in Anton (1986) who used a beta prior for ρ defined over this interval.[2] Similarly, the MDIP associated with $f(y_{t+1}, y_{t+2}, ... y_{t+n}|y_t, \rho, \sigma)$ is $\pi_*(\rho, \sigma) \propto 1/\sigma$ with $0 < a_1 < \sigma < b_1$ and $a_2 < \rho < b_2$.

For higher order AR processes say an AR(3), $y_t = \alpha + \beta_1 y_{t-1} + \beta_2 y_{t-2} + \beta_3 y_{t-3} + \varepsilon_t$, with the ε_t's NID$(0, \sigma^2)$, if y_0, y_{-1} and y_{-2} have known values and if it is not known when the process started, we can consider the conditional density $f(y_1, y_2, ... y_n|y_0, y_{-1}, y_{-2}\boldsymbol{\theta})$, where $\boldsymbol{\theta}' = (\alpha_1, \beta_1, \beta_2, \beta_3)$ and the MDIP for the parameters is $\pi_*(\alpha, \beta_1, \beta_2, \beta_3, \sigma) \propto 1/\sigma$, defined over a finite region of the parameter space. See Hong (1989) for an analysis of this prior distribution's implications for the properties of the roots of the AR(3) process and for the amplitude and period of periodic components associated with complex conjugate roots. Of course if other side conditions are utilized in obtaining the MDIP, its form will not be that shown above. For example if we use side conditions for the first and second moments of α, β_1, β_2, and β_3, the MDIP for these parameters will be normally shaped and not uniform. While any of the above proper prior densities can be employed in computation of

---

[2] The interval $-1 \leq \rho \leq 1.1$ is equivalent to the interval $0 \leq \frac{1+\rho}{2.1} \leq 1$. Letting $\theta = (1 + \rho)/2.1$, Anton (1986) used a proper beta density for θ such that the associated density for ρ had a modal value=1. In the MDIP approach, a beta density will be obtained as a solution if $I(\theta)$=const. and one uses the side conditions $\int_0^1 \pi(\theta)d(\theta) = 1$, $\int_0^1 \pi(\theta)\ell n\ \theta d\theta = c_1$ and $\int_0^1 \pi(\theta)\ell n\ (1 - \theta)d\theta = c_2$.

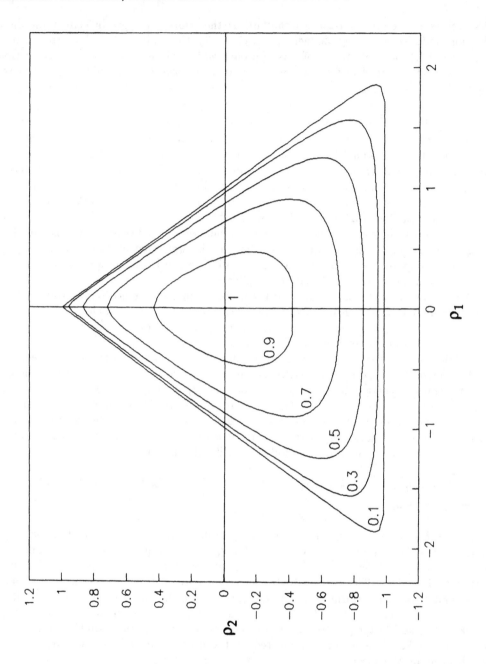

Fig. 1. Contours of the maximal data information prior density for parameters ρ_1 and ρ_2 of a stationary AR(2) model (The MDIP prior for ρ_1 and ρ_2, shown in (6) in the text is

$$\pi_*(\rho_1, \rho_2) \propto \left[1 - \rho_1^2 - \rho_2^2 - \frac{2\rho_1^2 \rho_2}{1 - \rho_2} \right]^{1/2}$$ with $-2 < \rho_1 < 2$, $\rho_2 < \rho_1 + 1$ for $-2 < \rho_1 \leq 0$ and $\rho_2 < -\rho_1 + 1$ for $0 \leq \rho_1 < 2$.)

Bayes' factors, in the present paper we shall utilize the priors developed in Zellner and Siow (1980) for model selection–see Zellner (1984, Ch. 3.7), Connolly (1991), Moulton (1991), and Koop (1992) for applications of this approach which will also be employed in section 4 after a consideration of alternative models and model selection techniques in the next section.

3. Models and Model Selection Techniques

In considering models and model selection, it is relevant to consider the origins of models. First, in situations in which little is known, it may be wise to assume, as Jeffreys and others have suggested, that all variation is random unless shown otherwise. Use of simple "benchmark" models, e.g. random walk models may be a good starting point as was done in early analyses of stock prices and in other areas. In our empirical work to follow, we shall evaluate a random walk model along with others.

Second, some suggest starting with a broad model or a broad class of models and then testing to determine whether a member of the class is supported by the data. Of course this approach requires us to have enough information to specify a relevant class of models that is not too heavily parameterized or complicated.

Third, if one can enumerate all possible models under consideration, as will be done below, there may be a certain subclass, e.g. models with four input variables, that has more members than other subclasses and this fact, given no other relevant information may lead us to have an a priori preference for this particular subclass of models.

Fourth, many utilize intuition and background theoretical and empirical knowledge to formulate models. For example, in modeling the rate of growth of annual output denoted by y_t, it seemed important to us to have a model incorporating an oscillatory component to represent business cycles. Thus in Garcia-Ferrer et al. (1987), we considered an autoregression (AR) of order three, $y_t = \alpha + \beta_1 y_{t-1} + \beta_2 y_{t-2} + \beta_3 y_{t-3} + \varepsilon_t$. On finding that this model did not fit the data well in the vicinity of turning points, we added some leading indicator (LI) variables, to obtain the following AR(3)LI model:

$$
\begin{aligned}
y_t &= \beta_0 + \beta_1 y_{t-1} + \beta_2 y_{t-2} + \beta_3 y_{t-3} + \beta_4 S_{t-1} + \beta_5 S_{t-2} + \beta_6 M_{t-1} + \beta_7 W_{t-1} + u_t \\
&= \beta_0 x_t' \beta + u_t
\end{aligned} \tag{8}
$$

where the subscript t denotes the value of a variable in the t'th year and the variables, y_t, S_t, M_t, and W_t are growth rates of real GNP, real stock prices, real money and real world stock prices, respectively and u_t is a non-autocorrelated zero-mean error term with variance σ^2 for all t. The relation in (8) has been derived from simple macroeconomic theories and shown to perform better in forecasting than random walk models and AR(3) models–see Garcia-Ferrer, et al. (1987) and Zellner and Hong (1989). Still we can ask, as Mittnik (1990) and Otter (1990) did, would models containing other combinations of the input variables be more strongly supported by the data? We shall present some results below bearing on this issue.

Finally, is it possible to produce a model such as that in (8) by a formal method, say by maximizing entropy subject to side conditions? The answer to this question is yes. That is, with $f(y_t|\beta_0 + x_t'\beta, \sigma^2)$ the conditional density of y_t given $\beta_0 + x_t'\beta$ and σ^2, consider maximizing $H = -\int f(y_t|\cdot) \ln f(y_t|\cdot) dy_t$ subject to $\int f(y_t|\cdot) dy_t = 1$, $\int y_t f(y_t|\cdot) dy_t = \beta_0 + x_t'\beta$, and $\int (y_t - \beta_0 - x_t'\beta)^2 f(y_t|\cdot) dy_t = \sigma^2$. The solution to this problem is $f_*(y_t|\beta_0 +$

$x'_t\beta, \sigma^2$), a normal density function for y_t given $\beta_0 + x'_t\beta$ and σ^2 with mean $\beta_0 + x'_t\beta$ and variance σ^2. Thus maxent can be used to obtain the model in (8) and others as shown in Ryu (1991). However, the side conditions employed, for example, the variables to include in x_t can be varied and thus a model selection problem arises.

In approaching the problem of which variables to include in an equation like (8), we broadened the set to include the seven variables shown in (8) plus M_{t-2} to give a total of 8 input variables denoted by $(y_1, y_2, y_3, S_1, S_2, M_1, M_2, W_1)$ and the intercept β_0 which will be included in all 256 alternative models. That is, we have $N_r = \begin{pmatrix} n \\ r \end{pmatrix} = n!/(n-r)!r!$ models containing r variables. Thus, $N_0 = 1$, $N_1 = 8$, $N_2 = 28$, $N_3 = 56$, $N_4 = 70$, $N_5 = 56$, $N_6 = 28$, $N_7 = 8$, and $N_8 = 1$, a total of 256 distinct models. It is to these 256 possible models that we shall apply model selection procedures.

For the purpose of deriving posterior odds on a particular model versus the broadest model, that containing all 8 input variables and an intercept, let us write

$$y_t = \beta_0 + x'_{1t}\,\beta_1 + x'_{2t}\,\beta_2 + u_t$$
$$1 \times k_1 k_1 \times 1 \quad 1 \times k_2 k_2 \times 1 \tag{9}$$

with $x'_t = (x'_{1t}, x'_{2t})$, the vector of all $k_1 + k_2 = 8$ input variables. Then consider

$$H_A : \beta_2 = 0 \text{ versus } H_B : \beta_2 \neq 0 \tag{10}$$

H_A implies that a particular model with k_1 input variables, x_{1t}, is appropriate, while H_B implies that the model containing all 8 input variables is appropriate. Given a set of observations on y_t and x_t, the posterior odds relating to H_A vs. H_B, denoted by K_{AB}, is given by

$$K_{AB} = \frac{\pi_A}{\pi_B} \frac{\int p_A(y|\beta_1, \sigma)\pi_A(\beta_1, \sigma)d\beta_1 d\sigma}{\int p_B(y|\beta_1, \beta_2\sigma)\pi_B(\beta_1, \beta_2, \sigma)d\beta_2 d\beta_2 d\sigma} \tag{11}$$

where π_A/π_B is the prior odds on H_A vs. H_B and the second factor on the r.h.s. of (11) is the Bayes Factor. If we take $\pi_A/\pi_B = 1$, use normal likelihood functions and the prior densities employed in Zellner and Siow (1980, p.593), then as shown in this last reference

$$K_{AB} \doteq b(\nu_B/2)^{k_2/2}\{(1 - R_B^2)/(1 - R_A^2)\}^{(\nu_B - 1)/2} \tag{12}$$

where $b = \pi^{1/2}/\Gamma[(k_2 + 1)/2]$, R_A^2 and R_B^2 are the sample squared multiple correlation coefficients under H_A and H_B, respectively and $\nu_B = n - k_1 - k_2 - 1$. Alternatively, the posterior odds can be expressed as

$$K_{AB} \doteq b(\nu_B/2)^{k_2/2}(SSE_B/SSE_A)^{(\nu_B - 1)/2} \tag{13}$$

where SSE_B and SSE_A are the **sums of squared least squares residuals** under hypotheses B and A, respectively. We see that the better is the fit of the restricted model, M_A, as measured by R_A^2 or SSE_A, the higher the odds in favor of M_A vs. M_B, the broadest model. Also note that the larger is k_2, the number of input variables added under H_B, the greater the odds in favor of M_A vs. M_B, all else held constant. Of particular importance here is to note that K_{AB} in (13) is a monotonic function of SSE_A, all else held constant.

As will be seen, this is a characteristic of alternative model selection criteria considered below.

Some well known alternative model selection criteria are, with $p \equiv k_1 + 1$,

(i)   Mean squared error: $MSE = SSE_A/(n-p)$
(ii)   Schwarz's criterion: $SC = n\, \ell n(SSE_A/n) + p\, \ell n\, n$
(iii)   Akaike's information criterion: $AIC = n\, \ell n\, SSE_A/n + 2p$
(iv)   Mallows' $C(p)$: $C(p) = SSE_A/\hat{\sigma}^2 + 2p - n$
(v)   Sawa's Bayesian information criterion: $BIC = n\, \ell n(SSE_A/n) + 2(p+2)q - 2q^2$

where $\hat{\sigma}^2 = SSE_B/(n - k_1 - k_2 - 1)$ and $q = \hat{\sigma}^2/(SSE_A/n)$. It is seen that the criteria (i)-(v) are all monotonic functions of SSE_A and **thus for given k_1 and n will give an identical ordering of models**. However, the "penalties" imposed for including additional input variables are different in the alternative model selection rules.

Now if we were to select the model, say M_A^*, with the highest posterior odds greater than one relative to the broadest model, this would be optimal given a zero-one loss function. However, it must be recognized that on screening 256 models, we may get one that appears to be a good one just by chance, a so-called "data-mining process." Jeffreys (1967, p.254) refers to this as a "selection problem". He suggests the following solution. If we put prior probability 1/2 on M_A and 1/2 on the $m = 255$ alternative models, each having prior probability q, then $(1 - q)^m = 1/2$ is the probability that the m alternative models are false. Thus, Jeffreys obtains, $q = 1 - (2)^{-1/m} \doteq 1/m \log 2$ for large m. In our case, $m = 255$ and the prior odds $\pi_A/\pi_B = 1/2q \doteq m/2\log 2 = .7m = (.7)255 = 178.5$ rather than 1. Thus Jeffreys' correction for selection is quite large in our case.

Another way to think about the selection problem is to assign a probability, P_A to a null model, M_{A_0}, say the ARLI model in (8) and assign probability P_B to the remaining 255 models with $P_B/255$ assigned to each. Then the prior odds for M_{A_0} vs. any other model is equal to $P_A/(P_B/255) = 255P_A/P_B$. If, for example $P_A = 1/16$ and $P_B = 5/16$, then the prior odds for M_{A_0} versus any other model of the remaining 255 models is 51 to 1, again a large number. Note also that $P_A + P_B < 1$ so as to allow for the possibility that models not included in the 256 under consideration, for example nonlinear or time-varying parameter models, may be appropriate. As these considerations indicate, it is not always appropriate to use 1:1 prior odds in computing posterior odds. In what follows, we shall show the results of computing posterior odds based on 1:1 prior odds and then consider effects of adjustments for selection on results.

4. Model Selection and Prediction Results

Here we relate an annual output growth rate variable for the U.S., y_t, to y_{t-1}, y_{t-2}, y_{t-3}, S_{t-1}, S_{t-2}, M_{t-1}, M_{t-2}, and W_{t-1}, that is to 8 input variables which we denote by $c\, y_1 y_2 y_3 S_1 S_2 M_1 M_2 W_1$, where c denotes a constant or intercept term. Thus there are 8 input variables in the broadest model. As indicated above, there is a total of 256 models, that is $\sum_{r=0}^{8} \binom{8}{r}$. Using the odds expression in (12), we computed the odds on each sub-model versus the broadest model, the model with 8 input variables and an intercept term for two periods, 1954-73 and 1954-87. Also for these periods and models we computed within sample mean squared errors, MSEs. The results for our autoregressive-leading indicator (AR(3)LI) model are shown in line 1 of Table 1. It is seen that for the periods 1954-73 and 1954-87, the posterior odds for the AR(3)LI model versus the broadest model, given in the last line of the table are 2.80 and 6.19, respectively. Note also that the within sample MSEs for the AR(3)LI model are 1.79 and 2.70 versus 1.81 and 2.80 for the broadest model. Finally, the root mean-squared error of forecast for 14 one year ahead forecasts, 1974-87 is 2.31 for the AR(3)LI model versus 2.47 for the broadest model. Thus in this comparison, the AR(3)LI model is preferred on all counts. The AR(3)LI model also dominates the AR(3) and random walk benchmark models in lines 8 and 9 of Table 1. The same can be said with respect to the models in lines 6 and 7 that were selected by Mittnik (1990) and Otter (1990) using Hankel matrix, state-space model identification techniques.

Table 1

Model Selection Using U.S. Annual Data, Posterior Odds,
Mean Squared Error and Out-of-Sample Forecast RMSEs

Model[a]	1954-73 Odds[b]	1954-73 MSE[c]	1954-87 Odds[b]	1954-87 MSE[c]	Out-of-Sample, 1974-87, RMSE of Forecast[d]
1. AR(3)LI $cy_1y_2y_3S_1S_2M_1W_1$	2.80	1.79	6.19	2.70	2.31
2. $cy_3M_1M_2W_1$	17.42	1.57	2.74	3.46	2.74
3. $cy_3S_1M_1M_2W_1$	15.67	1.53	14.29	2.88	2.44
4. $cy_3S_1M_1W_1$	8.61	1.81	26.34	2.87	2.30
5. $cy_3S_1S_2M_1W_1$	12.88	1.59	59.92	2.56	2.22
6. $cS_1S_2M_1M_2W_1$[e]	2.18	2.27	1.68	3.44	2.59
7. $cy_1S_1M_1W_1$[f]	2.11	2.39	2.55	3.48	2.54
8. $AR(3), cy_1y_2y_3$	.04	5.41	.01	5.76	2.69
9. Random Walk for $\ln$ GNP with Drift, c	.06	4.89	.02	5.75	2.68
10. General Model, $cy_1y_2y_3S_1S_2M_1M_2W_1$	1.00	1.81	1.00	2.80	2.47

[a]The general model in line 10 is given by $y_t = c + \beta_1 y_{t-1} \beta_2 y_{t-2} + \beta_3 y_{t-3} + \beta_4 S_{t-1} + \beta_5 S_{t-2} + \beta_6 M_{t-1} + \beta_7 M_{t-2} + \beta_8 W_{t-1} + \epsilon_t$ and denoted by $cy_1y_2y_3S_1S_2M_1M_2W_1$ where y_t, S_t, M_t, and W_t denote the annual growth rates of real GNP, real stock prices, real money and real world stock prices, respectively. Other models are particular cases of the general model.

[b]Posterior odds are computed for each specific model versus the general model in line 10 with prior odds in each case taken 1:1. See equations (12) and (13) in the text and Zellner and Siow (1980), and Zellner (1984, Ch.3.7) for expressions for odds.

[c]Within sample mean squared error $= \sum_{t=1}^{N} \hat{\epsilon}_t^2 / \nu$, where $\nu =$ degrees of freedom, or number of observations minus number of estimated parameters.

[d]One-step-ahead forecast errors, Δ_t were employed to compute RMSEs given by RMSE $= \sqrt{\sum_1^{14} \Delta_t^2 / 14}$.

[e]This is the model selected by Mittnik (1990), based on data 1953-73 for the U.S.

[f]This is the model selected by Otter (1990), based on data 1954-73 for the U.S.

As regards the models in lines 2-5, these were the ones most favored by the posterior odds and several other model selection criteria. While the model selection procedures favor these models, it has to be remembered that they were selected from 255 alternatives to the AR(3)LI model in line 1. There is a danger that the models so selected may fit the data by chance. If that is the case, the out-of-sample forecasting performance of these models with new data will usually not be as satisfactory as the within sample fits might imply. In the present instance, note that while the models in lines 2 to 5 appear to be favored relative to the AR(3)LI model in line 1 by posterior odds of from about 3 to 6 to 1 for the 1954-73 data, these models' out-of-sample RMSEs of forecast are either nearly the same as or higher than that of the AR(3)LI model. It would appear that use of 1:1 prior odds is not entirely satisfactory. If a Jeffreys-like correction for selection is introduced resulting in a prior odds favoring the AR(3)LI model, we would have some protection from effects of data-mining or of getting a model that happens to fit by chance. On the other hand, note that all the models in lines 2 to 5 have a special AR(3) structure–only y_{t-3} appears in these equations with an estimated coefficient that is negative. On examining the roots of $y_t = \beta_3 y_{t-3}$ with $\beta_3 < 0$, it is the case that there are two complex conjugate roots and a real root giving rise to an oscillatory component with a period of six years. Thus this particular formulation is a parsimonious way of obtaining a cycle.

These calculations indicate that the AR(3)LI model which we have employed in our past work–see e.g. Garcia-Ferrer et al. (1987) and Zellner and Hong (1989) is not grossly dominated by any of the 255 alternative models considered using data for the U.S. However, this is not to say that other types of models cannot dominate the AR(3)LI. Calculation of posterior odds for fixed versus time-varying parameter models for 18 countries in Min and Zellner (1990) shows that for many countries models with time-varying parameters drawn from hyper-distributions are favored and perform better in forecasting out-of-sample.

5. Summary and Concluding Remarks

In this paper, we have provided additional analysis of maximal data information prior densities. We have shown that they are the prior densities for which the posterior expectation of the logarithm of the ratio of a likelihood function to a prior is maximized. Thus the information in the likelihood function is emphasized in this sense by use of MDIPs. Also, we derived and discussed MDIPs for stationary and non-stationary autoregressive processes which are sensible and operational. Then we turned to a discussion of posterior odds and other model selection criteria. In analyzing alternative criteria, it was found that they all provided the same ordering of models with the same number of input variables since they are monotonic functions of the sum of squared least squares residuals. Further, attention was given to the problem of "selection effects" or "data-mining." In our application of posterior odds and other model selection techniques to the problem of evaluating 256 alternative models, it was found that a selection problem was probably present and that a Jeffreys-like correction to the prior odds was needed. Finally, it was found that the posterior odds approach is quite operational and yielded sensible results.

REFERENCES

Anton, L.A.M. (1986), "Empirical Regularities in Short-Run Exchange Rate Behavior," doctoral dissertation, Department of Economics, U. of Chicago.

Bernardo, J. (1979), "Reference Posterior Distributions for Bayesian Inference," J. of Royal Statistical Society, B, 41, 113-147.

Box, G.E.P. and G.C. Tiao (1973), Bayesian Inference in Statistical Analysis, Reading, MA: Addison-Wesley Publishing Co.

Connolly, R.A. (1991), "A Posterior Odds Analysis of the Weekend Effect," J. of Econometrics, 49, 51-104.

Diebold, F.X. (1990), "A Note on Bayesian Forecast Combination," ms., Department of Economics, U. of Pennsylvania, Philadelphia, PA.

Garcia-Ferrer, A., R.A. Highfield, F. Palm, and A. Zellner (1987), "Macroeconomic Forecasting Using Pooled International Data," J. of Business and Economic Statistics, 5, 53-67.

Geisel, M.S. (1975), "Bayesian Comparisons of Simple Macroeconomic Models," in S.E. Fienberg and A. Zellner, editors, Studies in Bayesian Econometrics and Statistics in Honor of Leonard J. Savage, Amsterdam, The Netherlands: North-Holland Publishing Co., 227-256.

Hartigan, J. (1964), "Invariant Prior Distributions," Annals of Mathematical Statistics, 35, 836-845.

Hong, C. (1989), "Forecasting Real Output Growth Rates and Cyclical Properties of Models: A Bayesian Approach," doctoral dissertation, Department of Economics, U. of Chicago.

Jaynes, E.T. (1968), "Prior Probabilities," IEEE Transactions on Systems Science and Cybernetics, SSC-4, 227-241.

_____ (1980), "Discussion," in J.M. Bernardo, M.H. DeGroot, D.V. Lindley, and A.F.M. Smith, editors, Bayesian Statistics, Valencia, Spain: University Press, 618-629.

Jeffreys, H. (1967), Theory of Probability (3rd rev. ed.; 1st ed., 1939), London: Oxford U. Press.

Koop, G. (1992), " 'Objective' Unit Root Tests," J. Applied Econometrics, 7, 65-82.

Leamer, E. (1978), Specification Searches, New York: J. Wiley and Sons, Inc.

Lindley, D.V. (1956), "On a Measure of Information Provided by an Experiment," Annals of Mathematical Statistics, 27, 986-1005.

Min, C.K. and A. Zellner (1990), "Bayesian and Non-Bayesian Methods for Combining Models and Forecasts with Applications to Forecasting International Growth Rates," ms., H.G.B. Alexander Research Foundation, U. of Chicago, to appear in J. of Econometrics.

Mittnik, S. (1990), "Macroeconomic Forecasting Using Pooled International Data," J. of Business and Economic Statistics, 8, 205-208.

Moulton, B.R. (1991), "A Bayesian Approach to Regression Selection and Estimation with Application to a Price Index for Radio Services," J. of Econometrics, 49, 169-193.

Novick, M. and W. Hall (1965), "A Bayesian Indifference Procedure," J. of the American Statistical Association, 60, 1104-1117.

Otter, P.W. (1990), "Canonical Correlation in Multivariate Time Series Analysis with an Application to One-Year-Ahead and Multiyear-Ahead Macroeconomic Forecasting," J. of Business and Economic Statistics, 8, 453-457.

Palm, F.C. and A. Zellner (1991), "To Combine or Not to Combine? Issues of Combining Forecasts," ms., H.G.B. Alexander Research Foundation, U. of Chicago, to appear in the J. of Forecasting.

Phillips, P.C.B. (1991), "To Criticize the Critics: An Objective Bayesian Analysis of Stochastic Trends," J. of Applied Econometrics, 6, 333-364.

Poirier, D.J. (1991), Bayesian Empirical Studies in Economics and Finance, editor, Journal of Econometrics, 49, 1-304.

Ryu, H.K. (1991), "Orthogonal Basis and Maximum Entropy Estimation of Probability Density and Regression Functions," doctoral dissertation, Department of Economics, U. of Chicago.

Savage, L.J. (1961), The Subjective Basis of Statistical Practice, ms., Department of Statistics, U. of Michigan, Ann Arbor, MI.

Zellner, A. (1971), An Introduction to Bayesian Inference in Econometrics, New York: J. Wiley and Sons, Inc. (reprinted by Krieger Publishing Co., Malabar, FL, 1987).

_____ (1977), "Maximal Data Information Prior Distributions," in A. Aykac and C. Brumat, eds., New Developments in the Applications of Bayesian Methods, Amsterdam: North-Holland Publishing Co., 211-232 and reprinted in A. Zellner (1984).

_____ (1984), Basic Issues in Econometrics, Chicago: U. of Chicago Press.

Zellner, A. (1990), "Bayesian Methods and Entropy in Economics and Econometrics," in W.T. Grandy, Jr. and L.H. Schick (eds.), Maximum Entropy and Bayesian Methods, Dordrecht, Netherlands: Kluwer Academic Publishers, 17-31.

Zellner, A. and C. Hong (1989), "Forecasting International Growth Rates Using Bayesian Shrinkage and Other Procedures," J. of Econometrics, 40, 183-202.

_____ and A. Siow (1980), "Posterior Odds Ratios for Selected Regression Hypotheses," in J.M. Bernardo, M.H. DeGroot, D.V. Lindley, and A.F.M. Smith, editors, Bayesian Statistics, Valencia, Spain: University Press, 585-603.7

BAYESIAN NUMERICAL ANALYSIS

John Skilling
Cavendish Laboratory
Madingley Road
Cambridge CB3 0HE, UK

ABSTRACT. Algorithms are presented which are intended for practical computation with probability distributions in many dimensions. The central feature of these methods is full use of the second-order curvatures which define the distances and volumes that underlie proper quantification. The second-order matrix operations are, however, simulated in such a way that the computation costs grow only linearly with the dimensionality.

1. Introduction

The greatest challenge facing Bayesians today is the sheer technical difficulty of doing the sums. Due in no small part to clear expositions by Jaynes, the status of probability calculus as the required calculus of consistent inference (Cox, 1946) is widely understood. There is ever-increasing application to a remarkable variety of increasingly difficult problems. These involve manipulating ever-larger and more awkward hypothesis spaces, hence the practical challenge. For, if Bayesians are unable to solve the problems properly, other methods will gain credit for improper solutions. Interestingly, it turns out that the Bayesian mode of thought itself repeatedly guides the practical numerical analysis we need. So, to the beginning....

$$\text{Given } \mathcal{I}: \quad \text{Joint } \Pr(\mathbf{f}, \mathbf{D}) = \Pr(\mathbf{f}) \, \Pr(\mathbf{D}|\mathbf{f}) \equiv \text{prior} \times \text{likelihood}$$
$$= \Pr(\mathbf{D}) \Pr(\mathbf{f}|\mathbf{D}) \equiv \text{evidence} \times \text{inference} \tag{1}$$

Subject to any particular background information $\mathcal{I}$, we have data $\mathbf{D}$ (perhaps many thousands), from which we wish to infer parameters $\mathbf{f}$ (perhaps millions). The author's motivation for this derives from quantified maximum entropy data analysis, where the data are noisy experimental measurements of a quantity f distributed across N cells (perhaps millions). However, the aim of this work is not to justify or use any particular prior, but to develop state-of-the-art ideas about computing deductions from the joint distribution in many dimensions, and thus to extend the boundaries of practical Bayesian methods.

After a particular dataset has been obtained, the joint distribution is a function of $\mathbf{f}$ only, not normalised as such, which we write as

$$\Pr(\mathbf{f}, \mathbf{D}) = P(\mathbf{f}) = \exp(-E(\mathbf{f})) \tag{2}$$

in direct form (P) or in logarithmic form as a potential (E).

The first task, at least logically, is to compute the prior predictive

$$Z = \Pr(\mathbf{D}) = \int d\mathbf{f} \, \Pr(\mathbf{f}, \mathbf{D}) \tag{3}$$

known colloquially as the "evidence". Technically, this integral over $d\mathbf{f}$ should be carried out with respect to measure rather than volume, but for simplicity we shall assume uniform measure so that the distinction disappears. The second task is to obtain the posterior

$$\Pr(\mathbf{f}|\mathbf{D}) = \Pr(\mathbf{f}, \mathbf{D}) / \Pr(\mathbf{D}) \tag{4}$$

known colloquially as the "inference".

The ground rules for our computations will be that the only allowed operations with the joint distribution are as follows:
1. For any $\mathbf{f}$, the joint distribution P (or its logarithm E) can be evaluated. In many applications, this may take $\mathcal{O}(N)$ arithmetical operations, using fast transforms.
2. For any $\mathbf{f}$, assuming its components f_i to be continuous variables, the gradient ∇E with components $\partial E / \partial f_i$ can also be evaluated. This might take $\mathcal{O}(2N)$ operations.
3. For any $\mathbf{f}$, and any second vector $\mathbf{u}$, the Hessian matrix $\mathbf{B} = \nabla\nabla E$ with components $\partial^2 E / \partial f_i \partial f_j$ can be applied to $\mathbf{u}$ to give $\mathbf{Bu}$. This might also take $\mathcal{O}(2N)$ operations.

No operation more expensive than these is allowed. For example, it would take $\mathcal{O}(N^2)$ operations to list all the individual Hessian components B_{ij}, and $\mathcal{O}(N^3)$ to invert $\mathbf{B}$ component-wise. These operations are forbidden.

2. Gaussian Approximation

The megadimensional "evidence" integral (3) is the centrepiece of Bayesian analysis, used for assessing choices of prior, and for normalising the posterior inference. Yet it will never be possible to evaluate any such integral by brute force summation over all $\mathbf{f}$. The only way of proceeding is to use algebraic manipulation to give some equivalent expression which can be computed. One formula is here pre-eminent.

Suppose P is, at least approximately, a multivariate Gaussian about its mode $\widehat{\mathbf{f}}$.

$$P(\mathbf{f}) = \exp\left(-\widehat{E} - \tfrac{1}{2}(\mathbf{f} - \widehat{\mathbf{f}})^T \mathbf{B}(\mathbf{f} - \widehat{\mathbf{f}})\right) \tag{5}$$

Then the evidence integral is

$$Z = \int d\mathbf{f} \, P(\mathbf{f}) = \exp(-\widehat{E}) \left(\det(\mathbf{B}/2\pi)\right)^{-\frac{1}{2}} \tag{6}$$

Hence we need to find $\widehat{\mathbf{f}}$ and $\det \mathbf{B}$. Commonly, $\widehat{\mathbf{f}}$ is found iteratively, by some variant of the basic algorithm

$$\mathbf{f} \Leftarrow \mathbf{f} - \mathbf{B}^{-1}\nabla E \tag{7}$$

so that we need to apply the inverse as well as evaluate the determinant of the (perhaps million $\times$ million) matrix $\mathbf{B}$. Although too expensive to carry out directly, these operations can be simulated in $\mathcal{O}(\text{several} \times N)$ operations by using conjugate gradient techniques.

Conjugate Gradient Inversion

The conjugate gradient algorithm (Hestenes and Stiefel, 1952) takes a vector $\mathbf{u}$ and applies $\mathbf{B}$ repeatedly to span progressively wider subspaces $S_1, S_2, \ldots, S_R$ defined by $\{\mathbf{u}\}$, $\{\mathbf{u}, \mathbf{Bu}\}, \ldots, \{\mathbf{u}, \mathbf{Bu}, \ldots, \mathbf{B}^{R-1}\mathbf{u}\}$. At the r'th stage, S_r has dimension r and the vector $\mathbf{y}_r \in S_r$ which maximises

$$Q(\mathbf{y}) = \mathbf{y}^T\mathbf{u} - \tfrac{1}{2}\mathbf{y}^T\mathbf{By} \tag{8}$$

within S_r is found, along with the value and gradient there.

$$Q_r = Q(\mathbf{y}_r), \qquad \nabla Q(\mathbf{y}_r) = \mathbf{u} - \mathbf{By}_r \tag{9}$$

As the subspace expands with r, the values Q_r necessarily increase towards the final maximum

$$Q_{\max} = \tfrac{1}{2}\mathbf{u}^T\mathbf{B}^{-1}\mathbf{u} \tag{10}$$

and the $\mathbf{y}_r$ tend towards the maximising

$$\hat{\mathbf{y}} = \mathbf{B}^{-1}\mathbf{u} \tag{11}$$

Thus $\mathbf{B}$ is effectively inverted by repeated forward application.

With $\mathbf{y}_r$ lying within S_r, the gradient $\nabla Q(\mathbf{y}_r)$ lies by construction in S_{r+1}, but it is also orthogonal to all vectors in S_r (because $\mathbf{y}_r$ maximised Q in S_r), and is in particular orthogonal to all earlier gradients. Hence the normalised gradients

$$\mathbf{g}_{r+1} = \nabla Q(\mathbf{y}_r)/|\nabla Q(\mathbf{y}_r)|, \qquad (r \geq 0 \text{ with } \mathbf{y}_0 = \mathbf{0}) \tag{12}$$

form an orthonormal set of base vectors for $S_1, S_2, S_3, \ldots$ which are more useful than the naive vectors $\mathbf{B}^r\mathbf{u}$. Moreover, (9) shows that $\mathbf{By}_r$ lies in S_{r+1} when $\mathbf{y}$ lies in S_r, and is hence orthogonal to all $\mathbf{g}_{r+s}$ for $s \geq 2$, showing that

$$\mathbf{g}_{r+s}^T\mathbf{Bg}_r = 0, \qquad (s \geq 2) \tag{13}$$

Hence the symmetric matrix $\mathbf{B}$ is tridiagonal in the orthonormal R-dimensional basis $\{\mathbf{g}\}$. In this form, $\mathbf{B}$ is correspondingly easy to invert, so that the $\mathbf{y}_r$ are easy to obtain. The tridiagonal form is also straightforward to diagonalise as $\mathbf{LDL}^T$, where $\mathbf{L}$ is orthogonal and $\mathbf{D}$ is diagonal, giving

$$\mathbf{Bu} \approx \mathbf{GLDL}^T\mathbf{G}^T\mathbf{u} \quad \text{and} \quad \phi(\mathbf{B})\mathbf{u} \approx \mathbf{GL}\,\phi(\mathbf{D})\mathbf{L}^T\mathbf{G}^T\mathbf{u} \tag{14}$$

where $\mathbf{G}$ is the $N \times R$ matrix of gradients $\mathbf{g}$ derived from $\mathbf{u}$, and ϕ is any function.

In brief, conjugate gradient is a neat and optimal way of approaching $\mathbf{B}^{-1}\mathbf{u}$ and allowing related approximations for $\phi(\mathbf{B})\mathbf{u}$, while obtaining successive lower bounds

$$0 = Q_0 \leq Q_1 \leq Q_2 \leq \cdots \leq Q_{\max} = \tfrac{1}{2}\mathbf{u}^T\mathbf{B}^{-1}\mathbf{u} \tag{15}$$

Unfortunately, it is not clear in general how many steps of the procedure should be taken.

In our Bayesian application, we seek to maximise a joint distribution comprising both a prior and a likelihood, for which the Hessian $\mathbf{B}$ is a sum of two parts:

$$\mathbf{B} = \nabla\nabla\left(-\log \Pr(\mathbf{f})\right) + \nabla\nabla\left(-\log \Pr(\mathbf{D}|\mathbf{f})\right) \tag{16}$$

A body of practical Bayesian experience favours gentle priors smoothly diminishing away from some global maximum, and for which $\nabla\nabla\left(-\log \Pr(\mathbf{f})\right)$ is positive-definite. Suppose for simplicity that this is just the identity $\mathbf{I}$ in appropriately scaled units.

$$\nabla\nabla\left(-\log \Pr(\mathbf{f})\right) = \mathbf{I} \tag{17}$$

This form is correct for a simple Gaussian prior on $\mathbf{f}$, and for an entropic prior under its correct metric, and some other generalisations are straightforward. Suppose also that, as is common, the log-likelihood is a concave function of $\mathbf{f}$, so that, within the Gaussian approximation

$$\nabla\nabla\left(-\log \Pr(\mathbf{D}|\mathbf{f})\right) = \mathbf{A} \tag{18}$$

with $\mathbf{A}$ non-negative. Then

$$\mathbf{B} = \mathbf{I} + \mathbf{A} \tag{19}$$

This common special form allows a termination condition to be set.

Consider the conjugate gradient maximisation of

$$Q^*(\mathbf{y}) = \mathbf{y}^T \mathbf{A} \mathbf{u} - \tfrac{1}{2}\mathbf{y}^T \mathbf{B} \mathbf{A} \mathbf{y} \tag{20}$$

instead of Q. A maximising $\mathbf{y}$ remains $\mathbf{B}^{-1}\mathbf{u}$ as in (11), though this is not unique if $\mathbf{A}$ is singular. If this maximisation is taken to be over $\mathbf{A}^{\frac{1}{2}}\mathbf{y}$, the conjugate gradient calculation then mimics that for Q, save that $\mathbf{u}$ is similarly replaced by $\mathbf{A}^{\frac{1}{2}}\mathbf{u}$. The subspaces $\mathcal{S}_r$ for $\mathbf{y}$ are then the same, so that the basis vectors $\mathbf{g}$ are unchanged. Only the scalar products which control the construction of the minimising $\mathbf{y}_r$ are different, incorporating extra factors of $\mathbf{A}$. The new sequence of lower bounds is

$$0 = Q_0^* \leq Q_1^* \leq Q_2^* \leq \dots \leq Q_{\max}^* = \tfrac{1}{2}\mathbf{u}^T \mathbf{B}^{-1} \mathbf{A} \mathbf{u} \tag{21}$$

But

$$Q_{\max} + Q_{\max}^* = \tfrac{1}{2}\mathbf{u}^T \mathbf{B}^{-1}(\mathbf{I} + \mathbf{A})\mathbf{u} = \tfrac{1}{2}\mathbf{u}^T \mathbf{u} \tag{22}$$

which is available at the outset. Thus the bounds Q_r^* give a progressively *decreasing* set of *upper* bounds to $Q_{\max}$, giving convergent constraints

$$Q_r \leq Q_{\max} \leq \tfrac{1}{2}\mathbf{u}^T \mathbf{u} - Q_r^* \tag{23}$$

Hence $Q_{\max}$ can be determined to any assigned precision. For many Bayesian purposes, it is thus possible to define a proper termination criterion for conjugate gradient, which makes the calculation definitive, to specified accuracy.

Bayesian Iteration

In fact, nothing in Bayesian calculus instructs us to maximise anything. The inference from the Bayesian calculation is not just the mode $\hat{\mathbf{f}}$ of maximum probability, but includes the surrounding probability distribution, here taken to be multivariate Gaussian. The proper criterion for terminating an iterative calculation is when the correct distribution such as (5) becomes sufficiently well approximated by the current distribution, such as

$$\bar{P}(\mathbf{f}) = \exp\left(-\bar{E} - \tfrac{1}{2}(\mathbf{f} - \bar{\mathbf{f}})^T \mathbf{B}(\mathbf{f} - \bar{\mathbf{f}})\right) \qquad (24)$$

surrounding the algorithm's current iterate $\bar{\mathbf{f}}$.

In order to assign an acceptable degree of approximation error, we need to consider the purpose of the analysis. The purpose is not usually to infer $\mathbf{f}$ as an object in its own right, but to infer various **properties** of $\mathbf{f}$. The inference about a completely arbitrary property may be arbitrarily peculiar, so we here restrict our consideration to *linear* properties. Within the Gaussian approximation, any linear property

$$\gamma = \mathbf{c}^T \mathbf{f} \qquad (25)$$

of $\mathbf{f}$ will be estimated from the current iterate as multivariate Gaussian with mean and variance

$$\bar{\gamma} = \mathbf{c}^T \bar{\mathbf{f}}, \qquad \sigma^2 = \mathbf{c}^T \mathbf{B}^{-1} \mathbf{c} \qquad (26)$$

Because $\bar{\mathbf{f}}$ differs from the true mode $\hat{\mathbf{f}}$, the estimate $\bar{\gamma}$ will be in error by

$$\delta\bar{\gamma} = \mathbf{c}^T(\bar{\mathbf{f}} - \hat{\mathbf{f}}) = \mathbf{c}^T \mathbf{B}^{-1} \nabla E \qquad (27)$$

which may be compared with the standard deviation σ.

We now introduce the idea of a *typical relevant* property. The vector $\mathbf{c}$ representing property γ could in principle be arbitrary (whatever that means to a Bayesian), but the data will only be relevant insofar as they have an influence on that property. Presumably we do not intend to request inferences about properties regarding which the data are silent. Because the precision of the influence of the data is mediated through the log-likelihood curvature $\mathbf{A}$, the most natural way of defining the distribution of relevant properties is to assign to them Gaussian statistics with

$$\langle \mathbf{c} \rangle = \mathbf{0}, \qquad \langle \mathbf{c}\mathbf{c}^T \rangle = \mathbf{A} \qquad (28)$$

We then have the expectations

$$\begin{aligned}
\langle (\delta\bar{\gamma})^2 \rangle &= \nabla E^T \mathbf{B}^{-1} \mathbf{A} \mathbf{B}^{-1} \nabla E < \nabla E^T \mathbf{B}^{-1} \nabla E \\
\langle \sigma^2 \rangle &= \operatorname{trace}(\mathbf{B}^{-1}\mathbf{A}) = G
\end{aligned} \qquad (29)$$

where G also represents the number of "good" measurements (Gull, 1989).

A convenient termination criterion is

$$\nabla E^T \mathbf{B}^{-1} \nabla E < \epsilon G, \qquad \epsilon \ll 1 \qquad (30)$$

at which typical estimate errors $\langle (\delta\bar{\gamma})^2 \rangle$ are a suitably small (ϵ) fraction of typical variances $\langle \sigma^2 \rangle$. The worst case condition for any individual property requires $\nabla E^T \mathbf{B}^{-1} \nabla E < \epsilon$, but in large-scale problems this degree of accuracy may not be necessary.

CONJUGATE GRADIENT DETERMINANT AND TRACE

Within our ground rules, the matrix $\mathbf{B}$ can be used only by applying it to some seed vector $\mathbf{r}$. When evaluating a determinant or a trace, there is no specified vector to use, and indeed $\det \mathbf{B}$ is invariant under coordinate rotation. But that does not mean that $\mathbf{r}$ itself is rotationally invariant, and hence null.

In Bayesian terms, it is our knowledge of $\mathbf{r}$, expressed as its probability distribution $\Pr(\mathbf{r})$, which is rotationally invariant, hence spherically symmetric. A convenient choice is

$$\Pr(\mathbf{r}) = (2\pi)^{-\frac{N}{2}} \exp(-\tfrac{1}{2}\mathbf{r}^T\mathbf{r}) , \qquad r_i \Leftarrow \mathcal{N}(0,1) \tag{31}$$

from which $\mathbf{r}$ is sampled by taking each component independently from the unit normal distribution. This vector has statistics (up to fourth order)

$$\langle r_i \rangle = 0, \quad \langle r_i r_j \rangle = \delta_{ij}, \quad \langle r_i r_j r_k \rangle = 0, \quad \langle r_i r_j r_k r_l \rangle = \delta_{ij}\delta_{kl} + \delta_{ik}\delta_{jl} + \delta_{il}\delta_{jk} \tag{32}$$

Consider

$$x = \mathbf{r}^T(\log \mathbf{B})\mathbf{r} \tag{33}$$

where $\log \mathbf{B}$ is defined by letting ϕ be the logarithm in (14). The expectation value of x is

$$\langle x \rangle = \operatorname{trace}(\log \mathbf{B}) = \log(\det \mathbf{B}) \tag{34}$$

Hence x is an unbiased estimator of $\log \det \mathbf{B}$. Furthermore, the variance is

$$\langle (x - \langle x \rangle)^2 \rangle = 2\operatorname{trace}\big((\log \mathbf{B})^2\big) \tag{35}$$

which can be estimated from

$$y = 2\,\mathbf{r}^T(\log \mathbf{B})^2\mathbf{r} \tag{36}$$

As the complexity and dimensionality N of problems become larger, the expectation (34) and variance (35) both tend to increase in proportion to N, because both are matrix traces. Accordingly, the relative accuracy of a single estimate x improves roughly as $\sqrt{N}$. For large problems, a *single* random vector $\mathbf{r}$ can easily suffice to give an adequately precise estimate of $\det \mathbf{B}$! In any case, the expected error is available from (36), and some extra accuracy can be obtained by averaging the results from more than one seed vector $\mathbf{r}$.

Moreover, conjugate gradient can be used to give converging upper and lower bounds on the values of x. The identity

$$x = \mathbf{r}^T(\log \mathbf{B})\mathbf{r} = \int_0^1 d\theta\, \mathbf{r}^T(\mathbf{I} + \theta\mathbf{A})^{-1}\mathbf{A}\mathbf{r} \tag{37}$$

shows that x is a positively-weighted sum of inversions each based on $\mathbf{A}$ and seeded with $\mathbf{r}$. All these inversions, parameterised by θ, share the same subspaces $\mathcal{S}_1, \mathcal{S}_2, \mathcal{S}_3, \ldots$ The conjugate gradient calculation of x gives progressively increasing lower bounds on each individual inversion $\mathbf{r}^T(\mathbf{I} + \theta\mathbf{A})^{-1}\mathbf{A}\mathbf{r}$, and by implication gives progressively increasing lower bounds on x also. Likewise, the identity

$$x^* = \mathbf{r}^T(\mathbf{A} - \log \mathbf{B})\mathbf{r} = \int_0^1 d\theta\, \mathbf{r}^T(\mathbf{I} + \theta\mathbf{A})^{-1}\theta\mathbf{A}^2\mathbf{r} \tag{38}$$

shows that x^* too is a positively-weighted sum of inversions each based on $\mathbf{A}$ and seeded with $\mathbf{r}$. The conjugate gradient calculation of x^* (which has the same subspaces as that of x) gives progressively increasing lower bounds on each individual inversion $\mathbf{r}^T(\mathbf{I} + \theta\mathbf{A})^{-1}\mathbf{A}^2\mathbf{r}$, and by implication on x^* also. But

$$x + x^* = \mathbf{r}^T\mathbf{A}\mathbf{r} \tag{39}$$

which is available at the outset. Hence the lower limits on x^* become upper limits on x, so that x can be determined to any desired accuracy.

By this means, the calculation of the determinant involved in Bayesian inference is made firm. The quantity x in (33) can be computed to any precision, and it bears a defined relationship (34,35) to the determinant. A simpler version of the above, based on

$$z = \mathbf{r}^T\mathbf{B}^{-1}\mathbf{A}\mathbf{r} \tag{40}$$

enables the matrix trace G (29) to be evaluated to any desired accuracy.

Conjugate Gradient Samples

We have already seen (26) that any linear property γ of $\mathbf{f}$ has a variance which takes the form of a standard inversion. Convergent upper and lower bounds to this are available from conjugate gradient. Thus the standard statistics of γ can be computed to any desired accuracy, albeit at the cost of performing a separate conjugate gradient inversion for each desired property. This can become expensive. Also, nonlinear properties (such as the location of recognisable features) cannot be found in this way.

A more general method of finding properties is afforded by taking several, M, Monte Carlo samples $\mathbf{f}_1, \mathbf{f}_2, \ldots, \mathbf{f}_M$ from the posterior inference

$$\Pr(\mathbf{f}|\mathbf{D}) = \left(\det(\mathbf{B}/2\pi)\right)^{\frac{1}{2}} \exp{-\tfrac{1}{2}(\mathbf{f} - \widehat{\mathbf{f}})^T\mathbf{B}(\mathbf{f} - \widehat{\mathbf{f}})} \tag{41}$$

Remarkably few samples suffice, even though they span only a tiny fraction of the full space $\mathbb{R}^N$. For any linear property, the mean calculated from

$$\bar{\gamma} = (\gamma_1 + \gamma_2 + \ldots + \gamma_M)/M , \qquad \gamma_i = \mathbf{c}^T\mathbf{f}_i \tag{42}$$

is expected to be in error by only $1/\sqrt{M}$ standard deviations of γ. The chance of a serious error of more than one standard deviation decreases exponentially with M, being less than 0.01% for $M = 16$. Similarly, the standard deviation σ, calculated from

$$\sigma^2 = \left((\gamma_1 - \bar{\gamma})^2 + (\gamma_2 - \bar{\gamma})^2 + \ldots + (\gamma_M - \bar{\gamma})^2\right)/(M - 1) \tag{43}$$

is expected to be in error by only 1 part in $1/\sqrt{2M}$.

Properties of $\mathbf{f}$ can thus be inferred easily and with fair accuracy by using random samples. Such samples ought to have statistics

$$\langle\mathbf{f}_i\rangle = \widehat{\mathbf{f}} , \qquad \langle(\mathbf{f}_i - \widehat{\mathbf{f}})(\mathbf{f}_i - \widehat{\mathbf{f}})^T\rangle = \mathbf{B}^{-1} \tag{44}$$

in accordance with (41). This can be achieved by acquiring random normal vectors $\mathbf{r}_i$ as in (31), and applying conjugate gradient to reach

$$\mathbf{f}_i = \widehat{\mathbf{f}} + \mathbf{B}^{-\frac{1}{2}}\mathbf{r}_i \tag{45}$$

as in (14) with the function ϕ being the inverse square root.

SUMMARY OF GAUSSIAN APPROXIMATION

Provided that the joint distribution $\Pr(\mathbf{f}, \mathbf{D})$ can be adequately approximated as a Gaussian around a maximum, then the location of this maximum, the evaluation of the evidence $\Pr(\mathbf{D})$, and the generation of samples of $\mathbf{f}$ for inferring arbitrary properties, can all be accomplished by using the Hessian matrix $\mathbf{B}$ of the Gaussian. Moreover, all the required matrix operations can be simulated to any required accuracy with conjugate gradients, so that the cost of a computation need increase no faster than linearly with the dimension N. Such techniques of numerical analysis are the key to reliable quantified probabilistic analysis in many dimensions.

However, the Gaussian approximation may be less than perfect. Figure 1 (top left) shows a small-scale (64-cell) simulation of a blurred and noisy dataset $\mathbf{D}$. At top right is shown the corresponding $\hat{\mathbf{f}}$ from quantified maximum entropy (Skilling, 1989). Because the entropic prior is not Gaussian, the Hessian matrix changes with $\mathbf{f}$, and the Gaussian approximation involves using its value at $\hat{\mathbf{f}}$. Four "typical" random samples of $\mathbf{f}$ in this approximation are shown in the lower part of Figure 1. All four show pronounced variability around zero, with several negative excursions. Although the entropic prior forbids negative values, implicitly assigning them zero prior probability, the Gaussian approximation has no such definitive edge and allows negative values. In this example, almost all typical samples from the Gaussian turn out to be impossible!

To some extent, this direct display of typical $\mathbf{f}$ accentuates the difficulty, and macroscopic averages over several cells are less prone to be negative. Even so, the lax behaviour of the Gaussian near the orthant boundaries must be damaging the quality of the inferences drawn from it. There is, indeed, some empirical evidence suggesting that the variances inferred in such examples tend to be too large. We need to go beyond the Gaussian approximation, but preferably without losing the power of the associated mathematics.

3. "CLOUDS": Direct Probabilistic Sampling

The lesson from algorithms for the Gaussian approximation is that the Hessian matrix $\mathbf{B}$ plays a critical rôle. Although it may vary with position, local curvature is vital in performing the volumetric integration over $\mathbf{f}$. Curvature is also important in iterating $\mathbf{f}$ to approach the mode $\hat{\mathbf{f}}$. On this point, the inefficiency of simple steepest ascent algorithms, and of methods which merely use function values, is well known. From a fundamental viewpoint, it is always a second-order metric tensor which defines lengths, angles, and volumes, so it is no accident that the second-order Hessian is central to any powerful numerical method.

If, however, the Hessian is to be allowed to vary, then a single sample such as $\hat{\mathbf{f}}$ can no longer suffice to describe the full inference $\Pr(\mathbf{f}|\mathbf{D})$. It seems necessary to use a "cloud" of at least several samples $\mathbf{f}_1, \mathbf{f}_2, \mathbf{f}_3, \ldots$. Often thought of as "particles" in $\mathbb{R}^N$, these samples are to be evolved towards "equilibrium", at which each is independently distributed according to the posterior inference. Also, the form (2) of the required distribution suggests thermal equilibrium of particles in a potential energy field $E(\mathbf{f})$.

This idea has a long history, dating back at least to Metropolis *et al.* (1953). Metropolis algorithms evolve a sample by moving it in accordance with some Markov process, and accepting or rejecting the change on the basis of the gain or loss of probability. If the step is in a purely arbitrary direction, the systematic part of the net motion corresponds to steepest ascents, and is anyway small because most random offsets are almost orthogonal to this direction. Thus the simple Metropolis algorithm is very slow, though provably convergent (eventually). Importance samplers have been used in attempts to anticipate the local form of the inference, and to sample appropriately. But, insofar as the importance weights of the samples differ from each other, computer memory is then being wasted on samples of small weight. In spaces of high dimension, it becomes increasingly more difficult to find importance samplers which are adequately faithful when faced with exponentially large probability factors, possibly reaching thousands of orders of magnitude. The Gibbs sampler (due to Geman and Geman (1984), not to J.W. Gibbs, and developed further by Gelfand and Smith (1990)) has become popular, but it relies upon moving along pre-ordained directions such as coordinate axes. When N is large, almost all directions in $\mathbb{R}^N$ are almost orthogonal to all the axes, so that motion along axes alone is usually seriously inefficient. Possibly methods such as the above pay insufficient attention to local curvature.

The hybrid Monte Carlo method (Duane, Kennedy, Pendleton and Roweth, 1987) combines the convergence characteristics of Metropolis algorithms with the "molecular dynamics" approach of Alder and Wainwright (1959). In this method, samples are evolved dynamically by sending them in orbits, often with relatively long mean free path, through the energy field $E(\mathbf{f})$. Although this method effectively uses curvature by repeatedly evaluating ∇E in order to follow the orbits, there is an arbitrariness about the dynamics (why not, say, relativistic dynamics?) which suggests that this approach too may be sub-optimal.

LOCAL EQUILIBRATION: CONVECTION-DIFFUSION

At the location of any sample or "particle", the potential E can be evaluated, along with its gradient. The conjugate gradient method also allows us to use the local Hessian matrix $\mathbf{B}$ for a variety of purposes, without fatal penalty in computation time. It does not appear to be useful to attempt any analysis of higher order than second. However, if $\mathbf{B}$ varies with position, calculations based on second-order analysis will progressively lose accuracy beyond some loosely defined "trust region" around the particle.

Within this trust region, E is taken to be quadratic in $\mathbf{f}$, and we want the particle to equilibrate as effectively as possible. This means that it should undergo as many thermalising interactions as possible before it leaves the region. In other words, it should evolve as if it were embedded in a collision-dominated fluid. Given any initial information about the location, the probability $\psi(\mathbf{f}, t)$ that the particle will be found at location $\mathbf{f}$ at later time t obeys the convection-diffusion equation

$$\frac{\partial \psi}{\partial t} = \nabla \cdot (\psi \nabla E + \nabla \psi) \qquad (46)$$

If the matrix $\mathbf{B}$ has the special form (19), in order to take advantage of the full power of conjugate gradient for appropriate Bayesian problems, then the gradient operator ∇ should be defined with respect to whatever metric enabled that special form. The first term on the right represents the systematic effect of the force ∇E on the particle convecting it towards regions of lower energy, while the second term $\nabla^2 \psi$ represents the random diffusion due to collisions. Whatever the initial conditions, the solution of (46) tends towards $\psi \nabla E + \nabla \psi = 0$, so that $\psi \propto \exp(-E)$, which is the required solution for thermal equilibrium.

At time t after release of the i'th sample from its initial position $\mathbf{f}_i(0)$ in the quadratic energy field E, the probability density for its location is a Gaussian, sampled by

$$\mathbf{f}_i(t) = \mathbf{f}_i(0) - (\mathbf{I} - e^{-\mathbf{B}t})\mathbf{B}^{-1}\nabla E \pm (\mathbf{I} - e^{-2\mathbf{B}t})^{\frac{1}{2}}\mathbf{B}^{-\frac{1}{2}}\mathbf{r} \qquad (47)$$

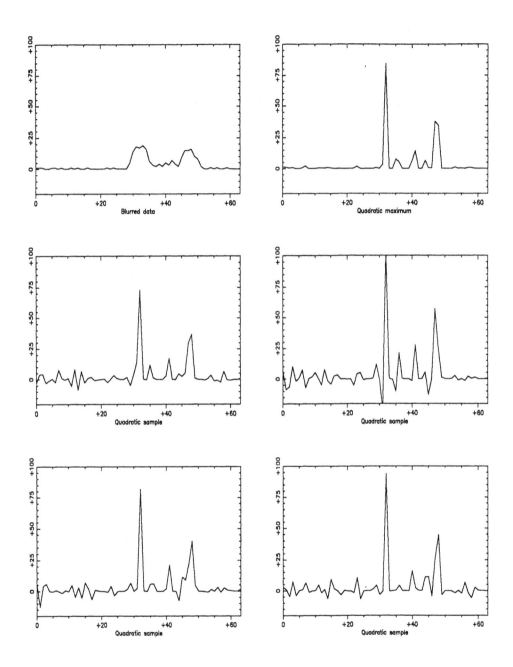

Fig. 1. Maximum entropy in the Gaussian approximation. Top left: blurred and noisy data. Top right: MaxEnt reconstruction. Middle and bottom: four samples from its Gaussian cloud.

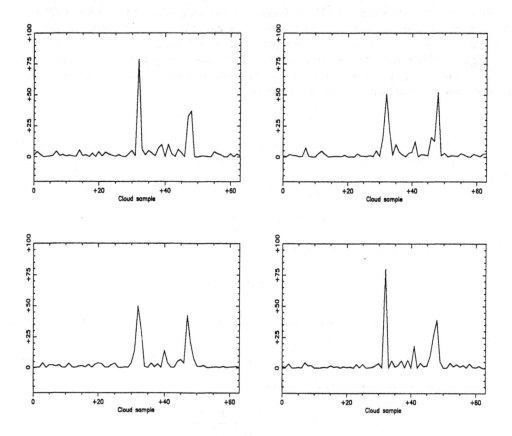

Fig. 2. Four samples from the MaxEnt cloud, for the same problem as in Fig. 1.

Here ∇E and $\mathbf{B}$ are evaluated at the original location, and $\mathbf{r}$ is a random normal vector as in (37). The same effect could be produced much more expensively by direct computation of very many correspondingly tiny dynamical or Metropolis steps, but such details are washed away by the macroscopic convection-diffusion equation. For large times, the exponential factors in (47) disappear, leaving

$$\mathbf{f}_i(\infty) \; = \; \mathbf{f}_i(0) \; - \; \mathbf{B}^{-1} \nabla E \; \pm \; \mathbf{B}^{-\frac{1}{2}} \mathbf{r} \tag{48}$$

as an equilibrium sample – obtained to any desired accuracy in a single conjugate-gradient step. When E is not exactly quadratic, the imposition of a finite time step t can prevent the sample from moving too far. Early time steps in an algorithm need not be particularly short: only the last steps may need to be short in order to converge towards accurate equilibrium.

Termination can no longer be assured by requiring $\nabla E^T \mathbf{B}^{-1} \nabla E$ to be much less than G, as in equation (30). As they fill their cloud of probability, the individual particles move away from the centre where $\nabla E = \mathbf{0}$. In the absence of any more formal criterion, it may be adequate to terminate when $\nabla E^T \mathbf{B}^{-1} \nabla E$ settles down to a value of the order of G. When this happens, convection towards the centre of the cloud is balanced overall by diffusion, which is a plausible indication (but not a proof) of convergence.

Figure 2 shows four samples from the cloud, for the same problem as Figure 1. Each sample is necessarily non-negative in this more accurate representation, and this leads to markedly reduced variability in the background.

Evaluating the Evidence

By themselves, a set of point samples cannot give the evidence integral (3), because they lack the volume property. One method, exact in principle, is to use "thermodynamic integration".

Generalise the joint distribution by "heating" the likelihood to a temperature $1/\lambda$

$$P_\lambda(\mathbf{f}) = Pr(\mathbf{f})\, Pr(\mathbf{D}|\mathbf{f})^\lambda\,, \qquad (\lambda \le 1) \tag{49}$$

so that the evidence integral becomes generalised to

$$Z(\lambda) = \int d\mathbf{f}\, P_\lambda(\mathbf{f}) \tag{50}$$

The desired value is $Z(1)$, and we also have $Z(0) = 1$ by normalisation. Its logarithmic derivative is

$$\frac{d\log Z(\lambda)}{d\lambda} = \left\langle \log Pr(\mathbf{D}|\mathbf{f}) \right\rangle_\lambda \tag{51}$$

where the λ-average of a quantity ξ is defined as

$$\langle \xi \rangle_\lambda = \frac{\int d\mathbf{f}\, \xi(\mathbf{f})\, P_\lambda(\mathbf{f})}{\int d\mathbf{f}\, P_\lambda(\mathbf{f})} \tag{52}$$

This is an average over the generalised joint distribution, and it can be estimated as usual from samples $\mathbf{f}_j^{(\lambda)}$ drawn randomly from it.

$$\langle \xi \rangle_\lambda \approx \left(\xi(\mathbf{f}_1^{(\lambda)}) + \xi(\mathbf{f}_2^{(\lambda)}) + \ldots + \xi(\mathbf{f}_M^{(\lambda)}) \right)/M \tag{53}$$

Derivatives of $\langle \xi \rangle_\lambda$ with respect to λ are also available from the samples, though they become subject to increasing sampling noise.

Hence we can estimate the desired evidence numerically by integrating the derivative (51) to

$$\log Z(1) = \log Z(0) + \int_0^1 d\lambda \left\langle \log Pr(\mathbf{D}|\mathbf{f}) \right\rangle_\lambda \tag{54}$$

provided we have an equilibrated set of samples at several inverse temperatures λ. The obvious procedure here is to use "simulated annealing" (Kirkpatrick, Gelatt and Vecchi, 1983), starting at $\lambda = 0$ and gradually cooling to $\lambda = 1$. The disadvantage of this procedure is that repeated equilibration is computationally expensive, and usually involves sampling over large volumes which the full likelihood factor effectively excises from the evidence integral.

At equilibrium, though, most of the sample particles usually have much the same energy, to within a range of relative order $N^{-\frac{1}{2}}$. Hence there would be little error in forcing all the particles to have the mean energy $\langle E \rangle$.

$$Z = \int d\mathbf{f}\, e^{-E(\mathbf{f})} \approx e^{-\langle E \rangle} \int d\mathbf{f}\, \delta(E(\mathbf{f}) - \langle E \rangle) \qquad (55)$$

The single energy constraint only affects one degree of freedom out of N, and $\log Z$ remains accurate to about 1 part in N. In thermodynamics, the analogous operation is the passage from canonical to microcanonical ensemble, where the microscopic approximation involved usually goes unremarked.

Once more neglecting corrections of 1 part in N, we have

$$\log Z \approx -\langle E \rangle + \log \mathcal{A} \qquad (56)$$

where $\mathcal{A}$ is the area of the energy surface defined by $E(\mathbf{f}) = \langle E \rangle$. This area is a geometrical constant, which it should usually be possible to evaluate from the surface's local properties. Thermodynamic integration involves following this surface as it compresses from its prior position, and in difficult problems this could be an expensive procedure. After all, $\langle \log \Pr(\mathbf{D}|\mathbf{f}) \rangle_\lambda$ can be a somewhat erratic function of λ, involving sharp changes in value which in thermodynamics would be called first order phase changes.

Currently, the author's preferred method of estimation is to use

$$\log \mathcal{A} \approx \tfrac{N}{2} - \tfrac{1}{2}\langle \log \det(\mathbf{B}/2\pi) \rangle \qquad (57)$$

or, more directly for our purposes,

$$\log Z \approx \tfrac{N}{2} - \tfrac{1}{2}\langle \log \det(\mathbf{B}/2\pi) \rangle - \langle E \rangle \qquad (58)$$

Formula (58) is exact for any Gaussian distribution. Moreover, its error is only second order in any deviation from a Gaussian. The formula is also easy to evaluate, relying as it does purely upon samples drawn from the "correct" unit-temperature cloud. Hence it is of practical value.

GLOBAL EQUILIBRATION: BIRTH AND DEATH

For difficult problems, particle-following algorithms can be seriously inefficient. If the required valley in the potential energy E is difficult to find, an algorithm may have to wait until all the particles have fallen into it before it can approach equilibrium. Worse, if the problem is multi-modal with several valleys, particles may become exponentially well trapped in the wrong ones. This problem is especially serious because the volumes of the basins which entrap the particles may be hugely different from the volumes of the valleys they lead to, so deductions from wrongly entrapped clouds may be grossly in error. Moreover, this difficulty cannot even be detected, let alone corrected, unless some attention is paid to volume factors.

Fortunately, there is a solution. Physical particles can equilibrate between different volumes if they can exchange identity, possibly via an external reservoir. In thermodynamic terms, the sample particles should be drawn from a grand canonical ensemble and thus able to exchange identity directly, instead of from a canonical ensemble which merely allows energy to be exchanged with the reservoir.

The spatial density $\rho(\mathbf{f})$ of particles ought to be proportional to $\exp(-E(\mathbf{f}))$ at equilibrium. From knowledge of the locations of particles in a cloud of samples, we can estimate the actual local density around each particle by simply adding up its neighbours, duly weighted down for distance. The estimate of ρ need not even be particularly accurate, because its main purpose is to catch gross trapping errors. This is fortunate because with few samples in many dimensions, the estimate will be subject to substantial sampling errors. In order to approach equilibrium, particles in over-populated domains should disappear (into the grand canonical reservoir), whereas particles should appear (from the reservoir) in under-populated domains. To accomplish this, introduce

$$\begin{aligned} \text{death rate} &\propto (\rho\, e^{E})^{\text{any positive power}} \\ \text{birth rate} &= \text{constant} \end{aligned} \tag{59}$$

The positive power in the death rate can be chosen so that there is a reasonably wide choice of particles to kill: a particle which dies is removed from computer memory. The constant birth rate can conveniently be chosen to ensure a constant number of particles in the cloud. This scheme tends to drive $\rho\, e^{E}$ towards the constant value it should take at equilibrium, irrespective of the distance between birth and death sites. Even in simple problems, the inclusion of birth and death can accelerate convergence towards the gross features of equilibrium. For final approach, though, birth and death should probably be switched off, lest errors in the estimates of ρ bias the samples.

The birth location should, of course, be sampled as well as may be from the required distribution $P(\mathbf{f})$. To this end, note the "$\pm$" sign in the convection-diffusion solution (47). Normally, when a particle propagates, one sign is chosen, and the particle's original location is over-written by the new. However, the alternative sign remains available for use at negligible cost. This alternative destination serves as a suitable birth site, associated with but different from the currently propagating particle.

4. Conclusions

The "CLOUDS" algorithm (Conjugate-gradient Local Optimisation Upon Diffusing Samples) offers some prospect of significant improvement in the power of numerical analysis applied to large-scale Bayesian inference. The algorithm uses conjugate gradient methods to simulate a variety of matrix operations to any desired accuracy, without requiring $\mathcal{O}(N^2)$ or $\mathcal{O}(N^3)$ operations. At all stages, the algorithm takes proper account of the volume factors inherent in Bayesian integrals, and by these methods calculations with Gaussian probabilities become fast but still accurate.

Beyond the Gaussian case, it is necessary to compute a "cloud" of samples drawn from the posterior inference. Locally, this cloud evolves to its required equilibrium through a convection-diffusion equation, equivalent to embedding the sample in a collision-dominated gas designed for rapid equilibration. Globally, the cloud evolves as a member of a grand canonical ensemble, with particles free to appear and disappear. This allows different domains to equilibrate, irrespective of the distance between them, which should substantially ameliorate difficulties caused by particle-trapping.

In the development of "CLOUDS", it is remarkable how techniques from physics have guided the development of the numerical analysis required to perform ever-more-detailed probabilistic calculations. The author hopes that methods such as these will continue to push back the frontiers of the practical use of probability calculus, and that they may even feed back into computational physics to further the synergy between physics and probability.

FINALE

In this volume honouring the distinguished contributions of E.T. Jaynes, it may be fitting to offer a paean to logic, extending the words of Jaynes (1985) to emphasise the art of practical application.

Ed Jaynes Has Shown The Way

From astronomy to zoology, applications will be found,
And with wailing lamentation will the sceptics all be drowned,
For the principles of logic are the same in every field,
And regardless of your ignorance you'll always know they yield
What your information indicates and (whether good or bad)
The best predictions one could make from data that you had.
Your problem may be huge – you may think you need a Cray,
But if the sums are done aright, straight Bayes will win the day!

REFERENCES

Alder, B.J. and Wainwright, T.E.: 1959, 'Studies in Molecular Dynamics. I. General Method', *J. Chem. Phys.*, **31**, 459.

Cox, R.T.: 1946, 'Probability, Frequency and Reasonable Expectation', *Amer. J. Phys.* **14**, 1.

Duane, S., Kennedy, A.D., Pendleton, B.J. and Roweth, D.: 1987, 'Hybrid Monte Carlo', *Phys. Lett. B*, **195**, 216.

Gelfand, A.E. and Smith, A.F.M.: 1990, 'Sampling-based Approaches to Calculating Marginal Densities', *J. Amer. Statistical Association*, **85**, 398.

Geman, S. and Geman, D.: 1984, 'Stochastic Relaxation, Gibbs Distributions and the Bayesian Restoration of Images', *IEEE Trans. Pattern Analysis and Machine Intelligence*, **6**, 721.

Gull, S.F.: 1989, 'Developments in Maximum Entropy Data Analysis', in J. Skilling (ed.), *Maximum Entropy and Bayesian Methods*, Kluwer, Dordrecht.

Hestenes, M.R. and Stiefel, E.: 1952, 'Methods of conjugate gradients for solving linear systems', *J. Res. Nat. Bur. Standards*, **49**, 409.

Jaynes, E.T.: 1985, 'Where Do We Go From Here?', in C.R. Smith and W.T. Grandy, Jr. (eds.), *Maximum Entropy and Bayesian Methods in Science and Engineering*, Kluwer, Dordrecht.

Kirkpatrick, S., Gelatt, C.D. and Vecchi, M.P.: 1983, 'Optimization by Simulated Annealing', *Science*, **220**, 671.

Metropolis, N., Rosenbluth, A.W., Rosenbluth, M.N., Teller, A.H. and Teller. E.: 1953, 'Equations of State Calculations by Fast Computing Machines', *J. Chem. Phys.* **21**, 1087.

Skilling, J.: 1989, 'Classic Maximum Entropy', in J. Skilling (ed.), *Maximum Entropy and Bayesian Methods*, Kluwer, Dordrecht.

QUANTUM STATISTICAL INFERENCE

R. N. Silver
Theoretical Division
MS B262 Los Alamos National Laboratory
Los Alamos, New Mexico 87545

ABSTRACT. Can quantum probability theory be applied, beyond the microscopic scale of atoms and quarks, to the human problem of reasoning from incomplete and uncertain data? A unified theory of quantum statistical mechanics and Bayesian statistical inference is proposed. QSI is applied to ordinary data analysis problems such as the interpolation and deconvolution of continuous density functions from both exact and noisy data. The information measure has a classical limit of negative entropy and a quantum limit of Fisher information (kinetic energy). A smoothing parameter analogous to a de Broglie wavelength is determined by Bayesian methods. There is no statistical regularization parameter. A priori criteria are developed for good and bad measurements in an experimental design. The optimal image is estimated along with statistical and incompleteness errors. QSI yields significantly better images than the maximum entropy method, because it explicitly accounts for image continuity.

1. Introduction

Jaynes has been an eloquent advocate for a compelling hypothesis: *Probability Theory as Logic*. That is, probabilities represent degrees of belief; probability theory develops and applies universal principles of logical inference from incomplete information. Two of his primary interests have been Bayesian probability theory and the interpretation of quantum mechanics. Bayesian probability theory yields inferences by systematically and consistently combining new data with prior knowledge. Jaynes pioneered the maximum entropy (ME) class of Bayesian methods for density function estimation (Jaynes, 1983), which has been applied successfully to numerous data analysis problems.

Quantum mechanics is a probability theory which provides an empirically flawless description of physical phenomena on the microscopic scale of atoms and quarks. The Copenhagen interpretation of quantum theory is very different from Bayesian inference. Nevertheless, the probability-theory-as-logic hypothesis requires that the application of quantum theory to physics should be one example of universal principles of inference (Jaynes, 1989). Perhaps quantum physics can provide a big clue toward finding these, as yet incompletely formulated, principles. As a second example, this paper proposes a unified application of Bayesian inference and quantum mechanics to the same class of non-parametric density function estimation problems previously addressed by ME.

The justifications for ME (see, e.g., Tikochinsky, 1984) require no prior correlations between points in an image. Despite its many successes, experience has shown that ME

223

images often exhibit spurious artifacts and overly sharp structure (Skilling, 1989). Most applications of ME are to *continuous* density functions, which means that the first derivatives of the density function are finite. This prior knowledge can only be incorporated by additional image smoothing. The developers of the leading ME code have proposed this smoothing to be a user-chosen *pre-blur* of a *hidden ME image*.

An alternative approach to image smoothing using quantum probability theory is motivated by a series of observations. There exists a one-to-one mathematical analogy between ME and classical statistical mechanics. Quantum statistical mechanics is also a successful theory of inference. It has classical statistical mechanics as a limit. Quantum density functions are smoother than classical density functions. The quantum smoothing parameter is the de Broglie wavelength. The Schrödinger equation at the heart of quantum mechanics may be derived from a variational principle on Fisher information (kinetic energy) (Frieden, 1989). Variational principles are preferable to *ad hoc* formulations.

These observations suggest a natural way to improve upon ME: formulate a Bayesian density-function estimation method in one-to-one mathematical analogy with quantum statistical mechanics. *Quantum Statistical Inference* (QSI) is the realization of this approach. It introduces a broad new class of prior probabilities incorporating correlations between points in an image.

2. The Data Analysis Problem

The generic data analysis problem is to solve a Fredholm equation of the first kind,

$$\hat{D}_k = \int dy \hat{R}_k(y)\hat{f}(y) + \hat{N}_k \quad . \tag{1}$$

Here $\hat{D}_k$ are N_d data which are typically incomplete, $\hat{R}_k(y)$ is an integral transform which is often an instrumental resolution function, and $\hat{N}_k$ represents noise. The true $\hat{f}(y)$ which generated the data is usually termed the *object*. Finding a normalized density function, $\hat{f}(y)$, is an ill-posed inverse problem. The goal of a statistical inference method is to provide a best estimate of $\hat{f}(y)$, termed the *image*, and its reliability. These estimates should be based on both the data and any prior knowledge available. Success is measured by the fidelity of the image to the object.

In ME, the image is derived from a variational principle: maximize the Shannon-Jaynes entropy,

$$S_c \equiv - \int dy \hat{f}(y) \ln\left(\frac{\hat{f}(y)}{\hat{m}(y)}\right) \quad , \tag{2}$$

subject to the constraints of the data. Here, $\hat{m}(y)$ is a *default model* for $\hat{f}(y)$, so-named because it is the answer ME returns in the absence of data. It is most convenient to remove the explicit reference to the default model by introducing renormalized variables $x \equiv \int_{-\infty}^{y} dy' \hat{m}(y')$, $f(x) \equiv \hat{f}(y)/\hat{m}(y)$, and $R_k(x) \equiv \hat{R}(y)$. Since $\hat{m}(y)$ must also be a normalized density function, clearly $0 \le x \le 1$. The data analysis problem is then rewritten in vector notation,

$$\vec{D} = \int_0^1 dx \vec{R}(x)f(x) + \vec{N} \quad . \tag{3}$$

In the special case of noiseless data ($\vec{N} = \vec{0}$), the solution by the method of Lagrange multipliers is

$$f_C(x) = \frac{1}{Z_C} \exp\left(-\vec{\lambda}_C \cdot \vec{R}(x)\right) \qquad Z_C \equiv \int_0^1 dx \exp\left(-\vec{\lambda}_C \cdot \vec{R}(x)\right) \qquad (4)$$

The N_d Lagrange multipliers, $\vec{\lambda}_C$, are to be determined by the fits to the data. This is formally analogous to a density function in *classical* statistical mechanics (subscript "C") with the identifications $\vec{\lambda}_C \cdot \vec{R}(x) \Leftrightarrow V(x)/T$, where $V(x)$ is potential energy and T is temperature. Z_C is a classical partition function.

ME works best for problems, such as deconvolution, where $\vec{R}(x)$ is a broad function and $f(x)$ contains comparatively sharp features. However, for problems where $\vec{R}(x)$ is sharp and $f(x)$ broad, ME tends to produce spurious structure. An extreme example is the data interpolation problem, in which the goal is to infer a density function from knowledge of its values at a finite number of points. Then $\vec{R}(x)$ consists of a set of δ-functions, and the ME solution is nonsense. Figure 1 shows a test density function and three data analysis problems corresponding to interpolation (I), Gaussian deconvolution (G), and exponential deconvolution (E). The data are taken to be exact and measured at 32 equally spaced data points. The corresponding ME images are shown in Fig. 2 displayed in 128 pixels. The ME image for the G data is credible, but it exhibits overly sharp and occasionally spurious structure. The ME image for the E data exhibits sharp edges reflecting the sharp feature in the resolution function. The ME image for the I data equals the data for measured pixels and equals the default model value of 1.0 for unmeasured pixels. The I and E ME images are not credible because they clearly reflect how the data were measured.

3. Quantum Statistical Mechanics & Fisher Information

To apply quantum probability theory to the data analysis problem defined by Eq. (3), a variational principle is needed which leads to a wave equation analogous to the Schrödinger equation. Consider the concept of *Fisher information* originally introduced (Fisher, 1925) in statistics as a measure of the inverse uncertainty in determining a position parameter by maximum likelihood estimation. In information theory, Fisher information and entropy have been proven to be related by derivative and metrical relations (Dembo, 1991). Recently Frieden observed that Fisher information is analogous to the kinetic energy operator in quantum mechanics, and he has proposed to derive the Schrödinger equation from a principle of minimum Fisher information (MFI) subject to constraints on the average potential energy and the normalization (Frieden, 1989). Such a proposed relation between statistics, information theory and quantum mechanics concepts is very controversial among physicists. Nevertheless, MFI provides a convenient derivation of a Schrödinger-like wave equation for density function estimation.

Applying MFI to the data analysis problem defined by Eq. (3), extremize

$$Q_1 \equiv \frac{1}{4} \int_0^1 dx \frac{1}{f(x)} \left(\frac{\partial f(x)}{\partial x}\right)^2 + \vec{\lambda} \cdot \int_0^1 dx \vec{R}(x) f(x) - E \int_0^1 dx f(x) \quad . \qquad (5)$$

The first term is Fisher information, the second term imposes the constraints due to the data with Lagrange multipliers $\vec{\lambda}$, and the third term is the normalization constraint with

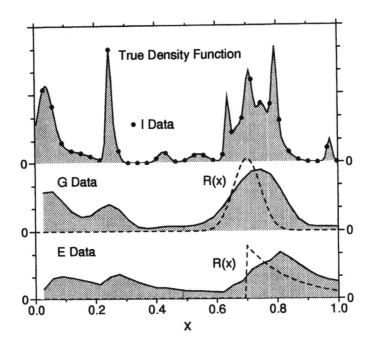

Fig. 1. Exact Data Example - Top curve is the test density function, and dots are the interpolation (I) data. Middle curves are the Gaussian (G) data (solid) and resolution function $R(x)$ (dashed ×0.35). Bottom curves are the exponential (E) data and resolution function. Data are exact and measured at 32 points. The Gaussian has a standard deviation of 0.044 and the exponential has a decay constant of 0.15.

Lagrange multiplier E. Defining a *wave function* by $\psi(x) \equiv \pm\sqrt{f(x)}$, the Euler-Lagrange equation is

$$\frac{\partial Q_1}{\partial \psi(x)} - \frac{d}{dx}\left(\frac{\partial Q_1}{\partial(d\psi(x)/dx)}\right) = 0 \quad . \tag{6}$$

This results in a Schrödinger-like wave equation,

$$-\frac{d^2\psi(x)}{dx^2} + \vec{\lambda} \cdot \vec{R}(x)\psi(x) = E\psi(x) \quad . \tag{7}$$

The data constraints $\vec{\lambda} \cdot \vec{R}(x)$ are analogous to potential energy $V(x)$, and E is analogous to total energy. Eq. (7) may also be written in matrix form as $\mathbf{H} \mid \psi >= E \mid \psi >$ using the *Heisenberg representation* $\psi(x) \equiv < x \mid \psi >$. The matrix $\mathbf{H}(x, x')$ is a *Hamiltonian*.

Solving the Schrödinger equation is an eigenvalue problem subject to a requirement that the solutions form a Hilbert space in $0 \leq x \leq 1$. For example, for $V(x) = 0$ a complete orthonormal set of solutions is $\psi(x) = \sqrt{2}\cos(k_n x); \sqrt{2}\sin(k_n x)$ where wave vectors are $k_n = 2\pi n$ and eigenenergies are $E_n^o = k_n^2$. This paper does not consider the possibility of complex solutions to the Schrödinger equation. For $V(x) \neq 0$, the $\psi(x)$ for the ground state is nodeless and would provide an image corresponding to the MFI principle.

However, the ME principle is certainly successful for many data analysis problems. Fortunately, it is not necessary to make a choice between these two variational principles,

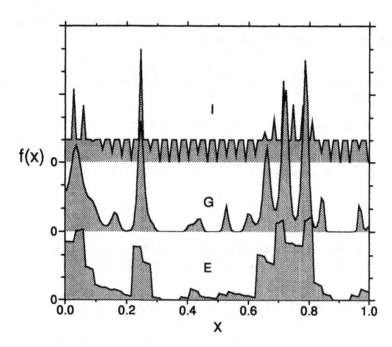

Fig. 2. Maximum Entropy Images - For exact data example in Fig. 1, displayed in 128 pixels.

since quantum statistical mechanics (see, e.g., McQuarrie, 1976) provides a seamless interpolation between the MFI principle in the quantum limit and the ME principle in the classical limit. The image is taken to be a weighted sum of eigensolutions,

$$f(x) = \sum_{n=0}^{\infty} w_n \psi_n^2(x) \; . \tag{8}$$

The weights, w_n, are to be interpreted as the probabilities for the n'th eigensolutions, and they should be determined by maximizing the quantum entropy in the Hilbert space subject to any applicable constraints. One constraint is that the image should be normalized to unity. Higher energy solutions have increasing numbers of nodes enabling them to describe increasingly sharp structure in $f(x)$. A second constraint on the average energy would, therefore, act as a low-pass spatial filter resulting in smoothing of the image.

Therefore, maximize

$$Q_2 \equiv \sum_{n=0}^{\infty} \left[-w_n \ln(w_n) - \frac{1}{T} E_n w_n + (\ln Z_Q + 1) w_n \right] \; . \tag{9}$$

The first term is the quantum entropy, $S_Q \equiv -\sum_{n=0}^{\infty} w_n \ln(w_n)$. The Lagrange multiplier for average energy is $1/T$, where T is temperature. It is related to a *de Broglie wavelength*, Λ, which is the characteristic scale for smoothing of the image, by $\Lambda \equiv \sqrt{4\pi/T}$. (In ordinary quantum mechanics, $\Lambda = \sqrt{2\pi\hbar^2/mk_BT}$, so that QSI corresponds to setting

$\hbar^2/2mk_B \to 1$.) The Lagrange multiplier for the normalization of the image is $\ln Z_Q + 1$. The result is the *canonical ensemble* of statistical mechanics,

$$w_n = \frac{1}{Z_Q} \exp\left(-\frac{E_n}{T}\right) \quad Z_Q \equiv \sum_{n=0}^{\infty} \exp\left(-\frac{E_n}{T}\right) = \exp\left(-\frac{F}{T}\right) \quad . \tag{10}$$

Z_Q is the quantum partition function, and F is the *Gibbs free energy*. The form chosen for the constraints was partially motivated by the many attractive mathematical properties of the canonical ensemble.

This theory may be written more compactly in the Heisenberg representation,

$$f(x) = \frac{1}{Z_Q} < x \mid e^{-\mathbf{H}/T} \mid x > \quad Z_Q = Tr\left(e^{-\mathbf{H}/T}\right) \quad . \tag{11}$$

The expectation value for any integral function, $O(x)$, of the image is termed an *observable*, and it is equivalent to a thermodynamic expectation value in statistical mechanics, i.e.

$$\ll O \gg \equiv \int_0^1 dx O(x) f(x) = \frac{1}{Z_Q} Tr\left(O e^{-\mathbf{H}/T}\right) \quad . \tag{12}$$

For example, setting $O(x) \to \delta(x - x_o)$ gives the image $f(x_o)$, and setting $O(x) \to R_k(x)$ gives the k'th fit to the data, $\ll R_k \gg$. The Lagrange multipliers, $\vec{\lambda}$, and the fits, $\ll \vec{R} \gg$, are conjugate variables in the sense of a Legendre transformation in statistical mechanics. The image is an implicit function of $\vec{\lambda}$ and Λ. In the classical limit ($\Lambda \to 0, T \to \infty$), the images reduce to the ME results given by Eq. (4) with the identifications $\vec{\lambda}_C \leftrightarrow \vec{\lambda}/T$ and $Z_Q \leftrightarrow Z_C/\Lambda$. In the quantum limit ($\Lambda \to \infty, T \to 0$), they reduce to the MFI images.

4. Bayesian Statistical Inference

The optimal $\vec{\lambda}$ and Λ are to be determined from the data by the application of Bayesian statistical inference (see, e.g., Loredo, 1990). Bayes theorem is

$$P[\vec{\lambda}, \vec{D}, \Lambda] = P[\vec{\lambda} \mid \vec{D}, \Lambda] \times P[\vec{D}, \Lambda] = P[\vec{D} \mid \vec{\lambda}, \Lambda] \times P[\vec{\lambda}, \Lambda] \quad . \tag{13}$$

In Bayesian terminology $P[\vec{\lambda} \mid \vec{D}, \Lambda]$ is termed the *posterior probability*, $P[\vec{D}, \Lambda]$ is the *evidence*, $P[\vec{D} \mid \vec{\lambda}, \Lambda]$ is the *likelihood function*, and $P[\vec{\lambda}, \Lambda]$ is the *prior probability*. The most probable $\vec{\lambda}$ is determined from the maximum of the posterior probability, and the most probable Λ is determined from the maximum of the evidence. The data are embodied in the likelihood function.

The prior probability is determined by an analogy between quantum statistical mechanics and statistical inference. In statistical mechanics, the system probability is proportional to the partition function, $Z_Q = \exp(-F/T)$, which counts the number of different eigensolutions the system can have within the constraints. In Bayesian inference the system probability is proportional to the product of the likelihood function and the prior probability. Since the data constraints have already been identified with the potential energy, the likelihood function may be identified as $\propto \exp(- \ll V \gg /T)$. That leaves a prior probability $P[\vec{\lambda}, \Lambda] \propto \exp(-I_Q - F_o/T)$. A subscript "$o$" will be used to denote a quantity evaluated at the default model value of $\vec{\lambda} = \vec{0}$. The quantum generalization of an information measure is

$$I_Q \equiv \delta \left[\frac{F}{T} - \frac{\ll V \gg}{T} \right] = \delta \left[\frac{1}{T} \ll -\frac{\partial^2}{\partial x^2} \gg -S_Q \right] \quad . \tag{14}$$

Here "δ" denotes a change in this quantity from the $\vec{\lambda} = \vec{0}$ value, so that $I_Q^o = 0$. The right hand side is a simple sum of the expectation values for Fisher information (kinetic energy) and quantum negentropy. In the classical ($\Lambda \to 0$) limit, I_Q reduces to the information measure in ME, $-S_C$.

To establish the properties of I_Q, consider the *susceptibility* (or *Hessian*) matrix,

$$\mathbf{K} \equiv -T\vec{\nabla}\vec{\nabla}F = -T\frac{\partial \ll \vec{R} \gg}{\partial \vec{\lambda}} \quad . \tag{15}$$

Here $(\vec{\nabla})_k \equiv \partial/\partial \lambda_k$. A susceptibility in statistical mechanics is the derivative of an observable (e.g. $\ll R_k \gg$) with respect to a *field* (e.g. $\lambda_{k'}$). The *fluctuation-dissipation theorem* (Kubo, 1959) of quantum statistical mechanics relates $\mathbf{K}$ to a correlation function, which may be written in two ways,

$$\mathbf{K}_{k,l} = \int_0^1 d\tau \ll \delta R_k e^{-\mathbf{H}\tau/T} \delta R_l e^{+\mathbf{H}\tau/T} \gg = \int_0^1 dx \int_0^1 dx' \delta R_k(x) \delta R_l(x') \Theta(x, x') \quad . \tag{16}$$

Here $\delta R_k(x) \equiv R_k(x) - \ll R_k \gg$.. Eq. (16) may be derived by second-order quantum perturbation theory. The first expression for $\mathbf{K}$ is the same as a susceptibility in the Kubo theory for linear response (see, e.g., Fick, 1990). The second expression introduces a *spatial correlation function* $\Theta(x, x')$, whose width characterizes prior correlations between points in the image. It is a strictly positive real symmetric function satisfying $\int_0^1 dx \Theta(x, x') = f(x')$. In the classical limit of $\Lambda \to 0$, $\Theta(x, x') \to f(x)\delta(x - x')$. The peak in $\Theta(x, x')$ broadens with increasing $\Lambda \neq 0$. In the default model limit of $\vec{\lambda} = \vec{0}$, the variance of the peak is given by $\Lambda/\sqrt{12\pi}$. In the quantum limit of $\Lambda \to \infty$, $\Theta(x, x') \to f(x)f(x')$. It follows that $\mathbf{K}$ must be a positive definite matrix. At fixed $\ll \vec{R} \gg$, the eigenvalues of $\mathbf{K}$ are maximal at $\Lambda = 0$, decrease monotonically with increasing Λ, and go to zero at $\Lambda = \infty$. This behavior of the eigenvalues will be important to determining the optimal Λ from the evidence.

Derivatives of I_Q with respect to observables may be evaluated using the chain rule,

$$\frac{\partial I_Q}{\partial \ll \vec{R} \gg} = -\frac{\vec{\lambda}}{T} \quad , \tag{17}$$

and

$$\frac{\partial^2 I_Q}{\partial \ll R_k \gg \partial \ll R_{k'} \gg} = \mathbf{K}_{k,k'}^{-1} \quad . \tag{18}$$

Since $\mathbf{K}$ is positive definite, Eq. (18) implies that I_Q must be a strictly positive convex function of the fits to the data, $\ll \vec{R} \gg$ and, therefore, $\vec{\lambda}$. In this paper, Gaussian approximations will be used to evaluate all integrals over these parameters. The normalized prior probability is then

$$P[\vec{\lambda}, \Lambda] = \frac{1}{T^{N_d}} \sqrt{\frac{\det(\mathbf{K})}{(2\pi)^{N_d}}} \exp\left(-I_Q - \frac{F_o}{T}\right) \qquad P[\ll \vec{R} \gg, \Lambda] = \frac{T^{N_d}}{\det(\mathbf{K})} P[\vec{\lambda}, \Lambda] \quad . \tag{19}$$

In the first relation, $\sqrt{\det(\mathbf{K})}$ is the effective *metric* for integrals over $\vec{\lambda}$. The second relation follows from Eq. (15), which states that $\det(\mathbf{K})/T^{N_d}$ is the Jacobian for the transformation from an integral over $\vec{\lambda}$ to an integral over $\ll \vec{R} \gg$. More generally, QSI satisfies geometrical properties related to those of classical statistical thermodynamics (Levine, 1986).

Consider the application of QSI to the exact data example considered in Figures 1,2. Then, the likelihood function is

$$P[\vec{D} \mid \vec{\lambda}, \Lambda] = \prod_{k=1}^{N_d} \delta(D_k - \ll R_k \gg) \quad . \tag{20}$$

The posterior probability is non-zero only for the $\vec{\lambda}$ such that $\vec{D} = \ll \vec{R} \gg$, which can be found by non-linear optimization methods such as Newton-Raphson. In principle, the image should be obtained by marginalizing Bayes theorem over Λ. In practice, a single Λ_{opt} may be used instead of marginalizing if the evidence, $P[\vec{D}, \Lambda]$, has a single sharp peak. The evidence for exact data is

$$P[\vec{D}, \Lambda] = \frac{1}{\sqrt{(2\pi)^{N_d} \det(\mathbf{K})}} \exp\left(-I_Q - \frac{F_o}{T}\right) \quad . \tag{21}$$

This separates into a product of three factors whose behavior may be understood using the dependence of the eigenvalues of $\mathbf{K}$ on Λ. The *Occam factor*, $1/\sqrt{\det(\mathbf{K})}$ (Bretthorst, 1988) favors the simpler model of larger Λ (or fewer quantum states contributing to the image). Eq. (18) implies that the *data factor*, $\exp(-I_Q)$, favors the more complex model of smaller Λ (or more quantum states contributing to the image). The factor $\exp(-F_o/T)$ may be regarded as a slowly varying Jeffrey's prior for Λ which is independent of $\ll \vec{R} \gg$.

Figure 3 shows the evidence, Occam factor and data factor for the interpolation problem as a function of Λ. The $\Lambda_{opt} \approx 0.6$, corresponding to a variance of the spatial correlation function given by $\Lambda_{opt}/\sqrt{12\pi} \approx 0.1$. Comparable values of Λ_{opt} are found for all three data analysis problems in Figure 1. Figure 4 shows the optimal QSI images for the data in Figure 1. They are clearly more credible than the corresponding ME images shown in Figure 2. Credibility is quantitatively expressed by the many orders of magnitude larger Bayesian evidences for the QSI images relative to the ME images.

5. Noisy Data

Consider the extension of QSI to data subject to Gaussian independent noise defined by $E(\hat{N}_k) = 0$ and $Cov(\hat{N}_k \hat{N}_{k'}) = \delta_{k,k'} \hat{\sigma}_k^2$ in Eq. (1). It is most convenient to renormalize to new variables $D_k \equiv \hat{D}_k/\hat{\sigma}_k$ and $R_k(x) \equiv \hat{R}_k(y)/\hat{\sigma}_k$. The likelihood function is then

$$P[\vec{D} \mid \vec{\lambda}, \Lambda] = \frac{1}{\sqrt{(2\pi)^{N_d}}} \exp\left(-\frac{\chi^2}{2}\right) \qquad \chi^2 = (\vec{D} - \ll \vec{R} \gg)^\dagger \cdot (\vec{D} - \ll \vec{R} \gg) \quad . \tag{22}$$

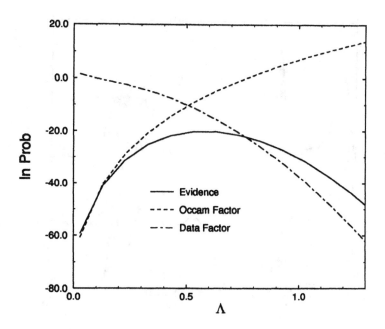

Fig. 3. Optimization of de Broglie Wavelength - The evidence, Occam factor, and data factor for the interpolation problem in Fig. 1 as a function of Λ.

Bayes theorem for noisy data becomes

$$P[\ll \vec{R} \gg | \vec{D}, \Lambda] \times P[\vec{D}, \Lambda] = \frac{1}{(2\pi)^{N_d} \sqrt{\det(\mathbf{K})}} \exp\left(-I_Q - \frac{\chi^2}{2} - \frac{F_o}{T}\right) \quad . \quad (23)$$

Regarding $1/\sqrt{\det(\mathbf{K})}$ as a metric for integrals over $\ll \vec{R} \gg$, the most probable image is obtained from the minimum of

$$Q_3 \equiv I_Q + \frac{\chi^2}{2} \quad . \quad (24)$$

In contrast to the traditional literature (Titterington, 1985) on regularizing ill-posed inverse problems including previous implementations of ME (Skilling, 1989), there is no statistical regularization parameter multiplying the regularizing functional I_Q, even in the classical limit where $I_Q \rightarrow -S_C$. Equivalently, the regularization parameter in quantum statistical mechanics is known *a priori* to be one when the density function is normalized to unity. The minimum is obtained from the first derivative of Q_3,

$$\frac{\partial Q_3}{\partial \ll \vec{R} \gg} = -\frac{\vec{\lambda}}{T} - \vec{D} + \ll \vec{R} \gg = \vec{0} \quad . \quad (25)$$

A sub(super)script "f" denotes a quantity evaluated at the *final* solution, $\vec{\lambda}^f$, to Eq. (25). The second derivative

$$\frac{\partial^2 Q_3}{\partial \ll \vec{R} \gg \partial \ll \vec{R} \gg} = \mathbf{K}^{-1} + 1 \quad , \quad (26)$$

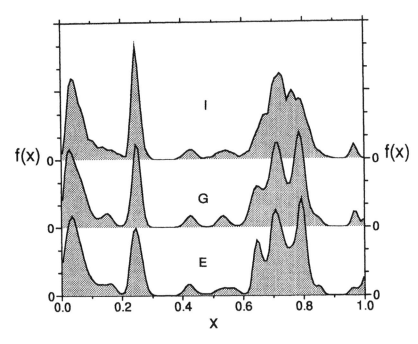

Fig. 4. Quantum Statistical Inference Images - For the exact data example in Fig. 1, displayed in 128 pixels.

is strictly positive implying that a unique solution to Eq. (25) exists.

For typical data analysis problems, the eigenvalue spectrum of $\mathbf{K}^f$ is a very steep function varying over many orders of magnitude. Independent measurements can be defined as eigenvectors of $\mathbf{K}^f$, and their quality can be ranked according to the size of their eigenvalues. *Good measurements* may be defined as eigenvectors of $\mathbf{K}^f$ whose eigenvalues are much greater than 1. The data for good measurements are dominated by signal, so that the typical values of $|\ll R_k \gg| \approx | D_k |$ are large compared to one. *Bad measurements* may be defined by eigenvectors of $\mathbf{K}^f$ whose eigenvalues are much less than 1. The data for bad measurements are dominated by noise so that the typical values of $| D_k |$ are of order 1 and the typical values for $|\ll R_k \gg|$ are much less than one. In other words, QSI fits the good measurements and ignores the bad ones. The *number of good measurements* may be defined by

$$N_g^f \equiv Tr\left[\mathbf{K}^f \cdot (1 + \mathbf{K}^f)^{-1}\right] \quad . \tag{27}$$

An error scaling consistency argument suggests that

$$\chi_f^2 \approx N_d - N_g^f \quad , \tag{28}$$

so that the χ^2 for the most probable image is less than the number of data. Numerical experiments suggest that N_g satisfies inequalities,

$$N_g^{\vec{\lambda}\neq\vec{0},\Lambda\neq0} \leq N_g^{o,\Lambda\neq0} \leq N_g^{o,\Lambda=0} \quad . \tag{29}$$

Therefore, the initial good measurements can be identified from the eigenvectors of the classical susceptibility, $\mathbf{K}^{o,\Lambda=0}$. They can often be calculated *a priori* from the resolution functions and noise, before any data are acquired. Throwing out the initial bad measurements can dramatically reduce the computational and data acquisition tasks without affecting the results. Relations similar to Eqs. (27,28) were first found in the Classic MaxEnt formulation (Skilling, 1989), but they depended on the values of a statistical regularization parameter. Since the initial regularization parameter in Classic MaxEnt is infinite, the initial N_g in Classic MaxEnt is zero.

The evidence is given by

$$P[\vec{D}, \Lambda] = \frac{1}{(2\pi)^{N_d} \sqrt{\det(\mathbf{K}^f + 1)}} \exp\left(-I_Q^f - \frac{\chi_f^2}{2} - \frac{F_o}{T}\right) \quad . \tag{30}$$

Only good measurements contribute to the Occam factor $\propto 1/\sqrt{\det(\mathbf{K}^f + 1)}$. The data factor is now $\propto \exp(-I_Q^f - \chi_f^2/2)$. Again, the Occam factor favors small Λ, while the data factor favors large Λ.

Figure 5 shows noisy data sets for the test density function in Fig. 1 and the same Gaussian (G) and exponential (E) resolution functions. The data are measured at 64 equally spaced points with the noise levels indicated. Figure 6 shows the corresponding ME images obtained by setting $\Lambda = 0$ in QSI, although similar results would be obtained from using the MEMSYS routines (Skilling, 1989). The ME images clearly show excessive overfitting and noise artifacts. Figure 7 shows the corresponding QSI images for the same data, which are clearly superior to the ME images. Although χ^2 is larger for the QSI images than for the ME images, the evidence for them is greater because they correspond to the simpler model of larger Λ. The data sets are placed in order of increasing information content, as measured by the value of I_Q^f. This ranking also corresponds to the visual quality of the images in Figure 7 and to the values of N_g^f. Note that the lineshape can be more important than the statistical errors in determining the information content of the data, with sharper lineshapes generally yielding more information (everything else being equal).

The behavior of the evidence can be seen most readily by evaluating it in a quadratic approximation,

$$P[\vec{D}, \Lambda] \approx \frac{1}{(2\pi)^N \sqrt{\det(\mathbf{K}^o + 1)}} \exp\left(-\frac{1}{2}\delta\vec{D} \cdot (\mathbf{K}^o + 1)^{-1} \cdot \delta\vec{D} - \frac{F_o}{T}\right) \quad , \tag{31}$$

where the *discrepancy* between the data and the default model predictions is given by $\delta\vec{D} \equiv \vec{D} - \ll \vec{R} \gg_o$. Since the eigenvalues of $\mathbf{K}^o$ are monotonically decreasing functions of Λ, the Occam factor again prefers large Λ and the data factor again prefers small Λ. In this quadratic approximation, the image is given by a *quantum filter*,

$$f(x) \approx \int_0^1 dx' \Theta_o(x, x') \left[1 + \delta\vec{R}(x')^\dagger \cdot (\mathbf{K}^o + 1)^{-1} \cdot \delta\vec{D}\right] \quad . \tag{32}$$

This can be a computationally effecient approximation which is adequate for many data analysis problems, although it will not enforce the positivity of the image or achieve the full resolution of the non-linear QSI equations.

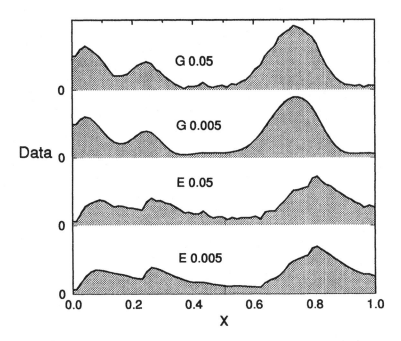

Fig. 5. Noisy Data Example - The test density function is convolved with the same resolution functions in Fig. 1 at **64** equally spaced data points, and then Gaussian independent noise is added. The notation "E 0.05" stands for the exponential resolution function with a noise standard deviation of 0.05.

6. Statistical and Incompleteness Errors

The goal of a statistical inference procedure is to use the data and prior knowledge to make predictions for unmeasured observables. This should include estimates of the reliability of those predictions. For any observable, $\ll O \gg$, these may be calculated from the conditional probability, $P[\ll O \gg| \vec{D}]$, using Bayes theorem. In the absence of data on that observable, the most probable value for $\ll O \gg$ is obtained at $\vec{\lambda}^f$ and $\lambda_O = 0$. This section summarizes the results for the errors on these estimates.

For two observables, $\ll A \gg$ and $\ll B \gg$, define generalized susceptibilities,

$$K_{A,B}^f \equiv \int_0^1 d\tau \ll \delta A e^{-\mathbf{H}\tau/T} \delta B e^{+\mathbf{H}\tau/T} \gg^f = \int_0^1 dx \int_0^1 dx' \delta A(x) \delta B(x') \Theta_f(x,x') \quad , \quad (33)$$

with $\delta A(x) \equiv A(x) - \ll A \gg_f$. Then the composite covariance $C[\delta \ll A \gg \delta \ll B \gg]$ is given by

$$C = K_{A,B}^f - \vec{K}_{A,R}^{f\dagger} \cdot (1 + \mathbf{K}^f)^{-1} \cdot \vec{K}_{R,B}^f \quad . \quad (34)$$

In the absence of measurements the covariance is maximal, given by $K_{A,B}^o$. The second term in Eq. (34) says that good measurements reduce the covariance for unmeasured quantities, as should be required of any valid statistical inference procedure.

Insight can be gained by separating the covariance into the sum of a *statistical covariance*,

$$C_S \equiv \vec{K}_{A,R}^{f\dagger} \cdot (\mathbf{K}^f + \mathbf{K}^f \cdot \mathbf{K}^f)^{-1} \cdot \vec{K}_{R,B}^f \quad , \quad (35)$$

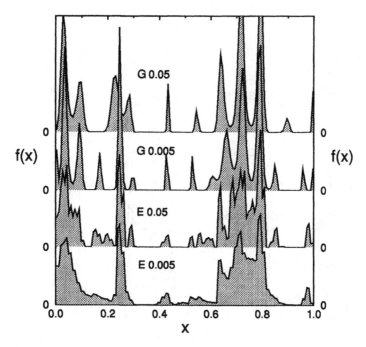

Fig. 6. Maximum Entropy Images - For the noisy data example in Fig. 5.

and an *incompleteness covariance*,

$$C_I \equiv K_{A,B} - \vec{K}_{A,R}^{f\dagger} \cdot (\mathbf{K}^f)^{-1} \cdot \vec{K}_{R,B}^f \quad . \tag{36}$$

These definitions make sense in several respects. If the statistical errors on the data are zero, C_I remains finite while C_S goes to zero. If $\ll A \gg$ is one of the measurements, C_I goes to zero while C_S remains finite. Bad measurements are the dominant contribution to C_S, which cancel the reduction in C_I.

An example is the covariance on the fits to the data, $\delta \ll \vec{R} \gg^\dagger \cdot \delta \ll \vec{R} \gg$. Then $C_I = 0$ and $C_S = N_g^f$. Combined with Eq. (28), it implies that the average fit to the data satisfies $E(\chi^2) \approx N_d$, which corresponds to the intuitive expectation.

Another example is the covariance on points in the image. Define

$$\vec{\Gamma}(x) \equiv \int_0^1 dx' \Theta(x, x') \delta \vec{R}(x') \quad . \tag{37}$$

Then the covariance of two points in the image is

$$C[\delta f(x_1) \delta f(x_2)] = \Theta_f(x_1, x_2) - f(x_1) f(x_2) - \vec{\Gamma}(x_1)^\dagger \cdot (\mathbf{1} + \mathbf{K}^f)^{-1} \cdot \vec{\Gamma}(x_2) \quad . \tag{38}$$

The case $x_1 = x_2$ would give the errors on points. The first term on the r.h.s. is the spatial correlation function. In the absence of data it indeed characterizes prior correlations between points in the image. In the classical ME limit of zero Λ, it approaches a δ- function,

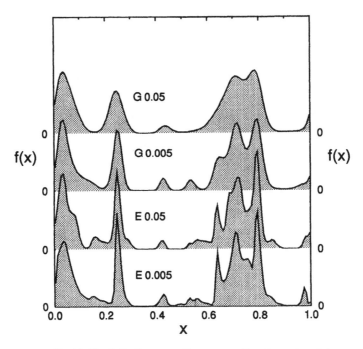

Fig. 7. Quantum Statistical Inference Images - For the noisy data example in Fig. 5.

so that the errors on individual points diverge. For $\Lambda \neq 0$, the errors on points are finite and they decrease with increasing Λ. The second term preserves the normalization of the image. The third term states how good measurements reduce the covariance of points in the image.

Generally, the optimal QSI images are smoother and have smaller errors than the ME images for the same data. If the object consists of a small number of isolated δ-functions, $\Lambda_{opt} \approx 0$ (T large) so that QSI may reduce to ME with maximal errors. If the object is smooth but has sharper structure than the resolution function, the QSI image may have $\Lambda_{opt} \neq 0$ even though it may resemble a ME image at $\Lambda_{opt} = 0$; however, the errors on the QSI image will be smaller than errors on the ME image. If the data are close to the default model predictions, Λ_{opt} will be large ($T \approx 0$) and the errors will be small.

7. Conclusions and Discussion

QSI is the first explicit application of quantum theory to human reasoning. The compelling motivations for QSI include the unquestioned success of quantum statistical mechanics in the physical domain, the *probability-theory-as-logic* hypothesis, and the inadequacies of the maximum entropy method for a broad class of density function estimation problems. This paper has proposed that quantum mechanics is a theory of statistical inference for *continuous* density functions, it has demonstrated applications to the interpolation and deconvolution of data, and the results are superior to the maximum entropy method.

QSI can best be tested and developed by experience with diverse data analysis problems. So far QSI has used only a small fraction of the rich structure of quantum theory. Other quantum concepts may also apply to information theory and statistics. These may include pure quantum states, complex wave functions, path integrals, gauge fields, etc. A practical issue now is how to code QSI efficiently. The dimension of the Hamiltonian is the number of pixels in the image. Diagonalizing the Hamiltonian should be amenable to sparse matrix and stochastic methods. The number of Lagrange multipliers is the number of data. The optimization problem can often be drastically reduced by dropping initial bad measurements. Conjugate gradient methods can avoid the necessity to calculate the susceptibility matrix at each iteration step. A vast reservoir of experience in quantum calculations may be applicable to the development of a robust QSI code. They can also motivate controlled approximations to QSI. Bayesian methods can be used to optimize any *hyperparameters* of the data analysis procedure such as noise scaling, density function normalization, etc.

Did the founders of quantum mechanics, in solving the riddles of the atom, serendipitously discover how to reason from incomplete and uncertain information? The successful validation of QSI would do much more than provide a powerful new tool for data analysis. Lifting quantum theory out of its original physical context would aid in the development of a more rigorous mathematical approach. It would also support the probability-theory-as-logic hypothesis and, thereby, provide new perspectives on the interpretation of quantum mechanics.

ACKNOWLEDGMENTS. This research was supported by the U. S. Dept. of Energy.

REFERENCES

Bretthorst, G. L.: 1988, *Bayesian Spectrum Analysis and Parameter Estimation*, Springer-Verlag, Berlin..

Dembo, A.; Cover, T. M.; Thomas, J. A.: 1991, 'Information Theoretic Inequalities', *IEEE Trans. Info. Theory* **37**, 1501.

Fick, E.; Sauermann, G.: 1990, *The Quantum Statistics of Dynamic Processes*, Springer-Verlag, Berlin

Fisher, R. A.: 1925,'Theory of Statistical Estimation', *Proc. Cambr. Phil. Soc.* **22**, 700.

Frieden, B. R.: 1989, 'Fisher Information as the basis for the Schrödinger wave equation', *Am. J. Phys.* **57**, 1004.

Jaynes, E. T.: 1983, in R. D. Rosenkrantz, (ed.), *E. T. Jaynes: Papers on Probability, Statistics and Statistical Physics*, D. Reidel Publishing Co., Dordrecht.

Jaynes, E.T.: 1989, 'Clearing Up Mysteries – The Original Goal', in J. Skilling (ed.), *Maximum Entropy and Bayesian Methods*, Klüwer, Dordrecht.

Kubo, R.: 1959, 'Some Aspects of the Statistical-Mechanical Theory of Irreversible Processes', in Britten & Dunham, (eds.), *Lectures in Theoretical Physics*, **1**.

Levine, R. D.: 1986, 'Geometry in Classical Statistical Thermodynamics', *J. Chem. Phys.* **84**, 910.

Loredo, T.: 1990, 'From Laplace to Supernova SN 1987A: Bayesian Inference in Astrophysics',in P. Fougere (ed.), *Maximum Entropy and Bayesian Methods*, Kluwer, Dordrecht., 81.

McQuarrie, D. A.: 1976, *Statistical Mechanics*, Harper & Row, New York.

Skilling, J.; Gull, S.: 1989, 'Classic MaxEnt', in J. Skilling (ed.), *Maximum Entropy and Bayesian Methods*, Kluwer, Dordrecht, 45-71.

Tikochinsky, Y.; Tishby, N. Z.; Levine, R. D.: 1984, 'Consistent Inference of Probabilities for Reproducible Experiments', *Phys. Rev. Letts.* **52**, 1357.

Titterington, D. M.: 1985, 'Common Structure of Smoothing Techniques in Statistics', *Int. Statist. Rev.*, **53**, 141.

APPLICATION OF THE MAXIMUM ENTROPY PRINCIPLE TO NONLINEAR SYSTEMS FAR FROM EQUILIBRIUM

H. Haken
Institut für Theoretische Physik und Synergetik
Pfaffenwaldring 57/4
D-7000 Stuttgart 80 (Vaihingen)
Germany

ABSTRACT. In this paper it is shown how Jaynes' maximum entropy principle or, more generally, his maximum calibre principle can be cast in such a form that the stochastic process that underlies observed data can be determined under the assumption that the process is Markovian. Under suitable constraints it becomes possible to derive the Fokker-Planck equation and the Îto-Langevin equation of that process.

1. Introduction

Jaynes' maximum entropy principle [1] has found wide-spread applications not only in systems in thermoequilibrium but also in systems far from it. In addition, Jaynes treated time-dependent processes by means of his maximum calibre principle [2]. In the present paper we show how his principles can be cast in such a form that the underlying process that is observed by certain correlation functions can be represented by a Fokker-Planck equation and an Îto-Langevin equation. It will be assumed throughout our paper that the process is Markovian. At the end of this contribution explicit examples treated recently by Lisa Borland and the present author will be presented.

2. Derivation of the Fokker-Planck Equation

In order to express the definition of a Markov process in a rigorous mathematical form, we choose a time sequence $t_0, t_1, \ldots t_M$. We may attribute a probability distribution to the path taken by the state vectors at the corresponding times. The joint probability distribution for this path (trajectory) is given by

$$P(\boldsymbol{q}_M, t_M; \boldsymbol{q}_{M-1}, t_{M-1}; \ldots; \boldsymbol{q}_0, t_0). \tag{1}$$

If the process is Markovian, the P of (1) can be split into a product

$$\prod_{i=0}^{M-1} P(\boldsymbol{q}_{i+1}, t_{i+1} | \boldsymbol{q}_i, t_i) \, P(\boldsymbol{q}_0, t_0) \tag{2}$$

where $P(\boldsymbol{q}_0, t_0)$ is the initially given probability distribution, whereas $P(\boldsymbol{q}_{i+1}, t_{i+1} | \boldsymbol{q}'_i, t_i)$ is the conditional probability of finding the state vector $\boldsymbol{q}$ at the position $\boldsymbol{q}_{i+1}$ at time t_{i+1} given that the system was in the state $\boldsymbol{q}_i$ at time t_i.

In the following we shall allow time t to change continuously so that we replace the index $i+1$ by $i+\tau$ or, occasionally, by $t+\tau$, where t corresponds to i. Furthermore it will be necessary to exhibit the individual components of the state vector q. For these reasons we shall adopt the following notation: We replace the index $i+1$ of the state vector by the argument $i+\tau$ and we reserve the index to denote the individual components. Thus we make the following replacement:

$$q_{i+1} \rightarrow q(i+\tau) \equiv [q_1(i+\tau), \quad q_2(i+\tau), ..., q_N(i+\tau)]. \tag{3}$$

Since the argument of q clearly indicates the time at which the variables of the conditional probability are to be taken, and since we shall treat stationary processes, the arguments t_{i+1} and t_i in (2) will be dropped altogether. Thus the conditional probability will be represented in the abbreviated form

$$P[q(i+\tau)|q(i)]. \tag{4}$$

Since it is known from the theory of Markov processes that the Fokker-Planck equation describing a continuous and stationary Markov process is determined by the first and second moments, we shall introduce the following constraints:

$$f_{1,\ell} = \langle q_\ell(i+\tau) \rangle_{q(i)} = \int q_\ell(i+\tau) P[q(i+\tau)|q(i)] d^N q(i+\tau) \tag{5}$$

and

$$f_{2,\ell,k} = \langle q_\ell(i+\tau) q_k(i+\tau) \rangle_{q(i)} \tag{6}$$

where (6) is defined in complete analogy to (5). The subscript $q(i)$ at the bracket in (5,6) means that the average values are taken under the assumption that the state vector is precisely known at time $t = i$. This is also exhibited explicitly by the right-hand side of (5) where the conditional probability P depends on $q(i)$. We now consider $q(i)$ as a fixed parameter, whereas the $q_\ell(i+\tau)$ represent a set of variables. We may then apply the maximum entropy principle. The application of this principle is straightforward and we obtain the result

$$P[q(i+\tau)|q(i)] = \exp\left[-\lambda - \sum_\ell \lambda_\ell q_\ell(i+\tau) - \sum_{k\ell} \lambda_{k\ell} q_k(i+\tau) q_\ell(i+\tau)\right]. \tag{7}$$

We note that the Lagrange multipliers λ, λ_ℓ, and $\lambda_{k\ell}$ are still functions of the initial value of the state vector q_i:

$$\lambda(q_i), \lambda_\ell(q_i), \lambda_{k\ell}(q_i). \tag{8}$$

Our final goal will it be to derive a Fokker-Planck equation for the conditional probability (7). To this end we make a few transformations and introduce the vector of the Lagrange parameters λ_ℓ by means of

$$\lambda^T = (\lambda_1 \lambda_2 ... \lambda_N) \tag{9}$$

and the matrix

$$\Delta = (\lambda_{k\ell}). \tag{10}$$

In the following we shall assume that the determinant of Δ does not vanish

$$\det\Delta \neq 0. \tag{11}$$

We then define a new vector $h\prime\prime$ by means of

$$h\prime\prime = \frac{1}{2}\Delta^{-1}\lambda' \tag{12}$$

Using this newly defined vector $h\prime\prime$ we may write P (7) in the form

$$P = \exp\left\{-\tilde{\lambda} - \sum_{k\ell}[q_k(i+\tau) - h_k]\lambda_{k\ell}[q_\ell(i+\tau) - h\prime\prime_\ell]\right\} \tag{13}$$

where we have used the abbreviation

$$\tilde{\lambda} = \lambda - \sum_{k\ell} h_k \lambda_{k\ell} h_\ell. \tag{14}$$

In the limit $\tau \to 0$, the state vector $q(i+\tau)$ must approach $q(i)$, i.e. P must meet the requirement

$$\tau \to 0: \quad P \overset{!}{\underset{\to}{}} \delta[q(i+\tau) - q(i)]. \tag{15}$$

Thus P becomes a singular function, namely the δ function. As is well-known, the δ function can be approximated by a Gaussian. This leads us to the idea that the matrix $\lambda_{k\ell}$ becomes singular as $\tau \to 0$. We thus put

$$\lambda_{k\ell} = \frac{1}{\tau}G_{k\ell}[q(i)]. \tag{16}$$

Furthermore we may read off from (15) that h_k, which depends on $q(i)$, approaches $q_k(i)$ for $\tau \to 0$. To satisfy this requirement, we make the hypothesis

$$h_k = q_k(i) + \tau K_k[q(i)]. \tag{17}$$

Under the assumptions (16, 17) the conditional probability P is transformed into

$$\begin{aligned}
P(q_{i+\tau}|q_i) = \exp[-\tilde{\lambda}(q_i)]\exp\Big\{&-\frac{1}{\tau}\sum_{k\ell}[q_k(i+\tau) - q_k(i) - \tau K_k]G_{k\ell} \\
&\times [q_\ell(i+\tau) - q_\ell(i) - \tau K_\ell]\Big\}.
\end{aligned} \tag{18}$$

It will be now our goal to derive a Fokker-Planck equation for the distribution function belonging to (18). To this end we start from an equation defining the probability distribution function P which depends on $q(i+\tau)$ alone. It obeys the Chapman-Kolmogorov equation

$$P(q_{i+\tau}) = \int P(q_{i+\tau}|q_i)P(q_i)d^N q_i, \tag{19}$$

where the conditional probability is given by (18). We wish to derive a differential equation for $P(\boldsymbol{q}_{i+\tau})$. To this end we multiply (19) by an arbitrary function $g(\boldsymbol{q}_{i+\tau})$ and integrate over $d^N q_{i+\tau}$ to obtain

$$
\begin{aligned}
I &\equiv \int g(\boldsymbol{q}_{i+\tau})\, P(\boldsymbol{q}_{i+\tau}) d^N q_{i+\tau} \\
&= \int d^N q_i\, P(\boldsymbol{q}_i)\, F(\boldsymbol{q}_i)
\end{aligned}
\tag{20}
$$

where $F(\boldsymbol{q}_i) = \int d^N q_{i+\tau}\, g(\boldsymbol{q}_{i+\tau})\, P(\boldsymbol{q}_{i+\tau}|\boldsymbol{q}_i)$.

To evaluate $F(\boldsymbol{q}_i)$ we introduce a new variable vector

$$
\boldsymbol{q}_{i+\tau} - \boldsymbol{q}_i - \tau K'(q_i') = \xi,
\tag{21}
$$

which yields

$$
F = \int d^N \xi N(\boldsymbol{q}_i) \exp\left(-\xi \frac{G}{\tau}\xi\right) g\left[\boldsymbol{q}_i + \tau K'(\boldsymbol{q}_i) + \xi\right]
\tag{22}
$$

where $N(\boldsymbol{q}_i) = \exp[-\tilde{\lambda}(\boldsymbol{q}_i)]$, and G is the matrix $(G_{k\ell})$.

In order to evaluate the integral up to terms linear in τ, we expand g in (22) into a power series in ξ up to second order and into a power series in τ up to first order. Using the normalization condition

$$
\int N(\boldsymbol{q}_i) \exp\left(-\xi \frac{G}{\tau}\xi\right) d^N \xi = 1,
\tag{23}
$$

the properties of Gaussian integrals,

$$
\int N(\boldsymbol{q}_i) \exp\left(-\xi \frac{G}{\tau}\xi\right) \xi_k \xi_\ell d^N \xi = \frac{\tau}{2}(G^{-1})_{k\ell},
\tag{24}
$$

and knowing that integrals over odd powers of ξ vanish, we readily obtain

$$
\begin{aligned}
F(\boldsymbol{q}_i) &= g(\boldsymbol{q}_i) + \tau K'(\boldsymbol{q}_i)\, \nabla_{q_i}\, g(\boldsymbol{q}_i) \\
&\quad + \tau \frac{1}{4} \sum_{k\ell} (G^{-1})_{k\ell} \partial^2 g/\partial q_{ik}\partial q_{i\ell}
\end{aligned}
\tag{25}
$$

where ∇ is the gradient operator.

Inserting (25) into (20) and performing a partial integration we obtain in place of (20) the relation

$$
I = \int d^N q_i g(\boldsymbol{q}_i) \left\{ 1 - \tau \nabla q_i K'(\boldsymbol{q}_i) + \frac{\tau}{4} \sum_{k\ell} \left(\frac{\partial^2}{\partial q_{ik}\partial q_{i\ell}}\right) [G^{-1}(\boldsymbol{q}_i)]_{k\ell} \right\} P(\boldsymbol{q}_i).
\tag{26}
$$

We now introduce the notation

$$
P(\boldsymbol{q}_{i+\tau}) \to f(\boldsymbol{q}; t + \tau),
\tag{27}
$$

$$
P(\boldsymbol{q}_i) \to f(\boldsymbol{q}; t).
\tag{28}
$$

Because g, which occurs in the middle part of (20) and on the r.h.s. of (20, 26) is an arbitrary function, the corresponding expressions must be equal even without integration. This leads us to the relation

$$f(\boldsymbol{q}; t + \tau) = f(\boldsymbol{q}; t) + \tau L f(\boldsymbol{q}; t) \tag{29}$$

where L is the Fokker-Planck operator

$$Lf = -\nabla_q[K'(\boldsymbol{q})f] + \frac{1}{4} \sum_{k\ell} \left(\frac{\partial^2}{\partial q_k \partial q_\ell} \right) \{[G^{-1}(q)]_{k\ell} f\} \tag{30}$$

which occurs in the Îto calculus. Bringing $f(\boldsymbol{q}_i; t)$ in (29) to the l.h.s., dividing both sides of the resulting equation by τ and letting $\tau \to 0$, we obtain the Fokker-Planck equation

$$\frac{\partial f}{\partial t} = Lf. \tag{31}$$

3. Derivation of the Îto-Langevin Equation

We now derive the Îto-Langevin equation corresponding to the Fokker-Planck equation (31). (A brief introduction to the formalism of the Îto-Langevin equation can be found in Haken [3].) We write the Îto-Langevin equation in the form

$$dq_\ell(t) = K_\ell[\boldsymbol{q}(t)]dt + \sum_m g_{\ell m}[\boldsymbol{q}(t)]dw_m(t) \tag{32}$$

where the stochastic process is defined by

$$\langle dw_m \rangle = 0 \tag{33}$$

and

$$\langle dw_m(t)dw_\ell(t) \rangle = \delta_{\ell m} dt. \tag{34}$$

The diffusion coefficients $G_{k\ell}$ are connected with the functions $g_{\ell m}$ by the formula

$$\frac{1}{2} G_{k\ell} = \sum_m g_{km} g_{\ell m}. \tag{35}$$

We introduce the matrices G and g by means of

$$(G_{k\ell}) = G; \quad (g_{km}) = g \tag{36}$$

so that (35) can be written in the form

$$\frac{1}{2} G = g\tilde{g} \tag{37}$$

where $\tilde{g}$ means the transposed matrix. For the solution of (35) it will be sufficient to assume that g is a square matrix and also that g is symmetric as is G. Equation (37) then acquires the form

$$\frac{1}{2}G = g^2. \tag{38}$$

The solution matrix g can be easily determined by assuming that G, which is positive definite, can be diagonalized by means of the orthogonal matrix U. We denote the corresponding diagonal matrix by D and its matrix elements by D_ℓ. We thus obtain

$$D = \frac{1}{2}UG\tilde{U} = (Ug\tilde{U})^2 \tag{39}$$

where

$$. \quad U\tilde{U} = 1. \tag{40}$$

We then find

$$(Ug\tilde{U})_{k\ell} = \delta_{k\ell}D_\ell^{1/2} \tag{41}$$

from which we may calculate g. To this end we define the diagonal matrix $D^{1/2}$ by means of

$$(D^{1/2})_{k\ell} = \delta_{k\ell}D_\ell^{1/2}. \tag{42}$$

This allows us to cast (41) into the form

$$g = \tilde{U}D^{1/2}U \tag{43}$$

so that g is explicitly determined. Thus both the quantities occurring in (32), i.e. K_ℓ and $g_{\ell g}$, are now determined. Equation (32) can be solved on a serial computer which uses a random noise generator to realize the terms $dw_m(t)$. Equation (32) can also be interpreted as a parallel network with elements ("neurons") labeled by the index ℓ [4]. The element ℓ receives inputs from all other elements and processes these by means of $K_\ell(q)$ and the sum over m in (32), where random noise generators are required.

4. Taking Care of a Reduced Information

In a number of practical applications the network may not be able to measure

$$\langle q_\ell(i+\tau)\rangle_{q(i)} \text{ and } \langle q_\ell(i+\tau)q_k(i+\tau)\rangle_{q(i)}$$

for all the values of the state vector $q(i)$ but rather only some moments of $q(i)$ determined by samples of $q(i)$. We assume that the measurements are made under steady-state conditions. In this case we can express the joint probability by a product of the conditional probability and the steady-state probability distribution

$$P(q_{i+\tau}, q_i) = P(q_{i+\tau}|q_i)P_{st}(q_i). \tag{44}$$

In order to define moments or other correlation functions, we shall introduce the functions $U_j(q_i)$ which may be assumed for instance in the form

$$U_j(q_i) \equiv U_{j,i} = q_{1,i}^{\mu_1}q_{2,i}^{\mu_2}...q_{N,i}^{\mu_N}; \quad \mu_1 + \mu_2 + ... + \mu_N = R \tag{45}$$

Note that the index i refers to the time index whereas the other indices 1,2,... refer to components of a vector. We then introduce the following constraints

$$\langle U_{j,i}^{(1)} \rangle = \int U_j^{(1)}(\boldsymbol{q}_i) P_{st}(\boldsymbol{q}_i) d^N q_i, \quad j = 1, 2, \ldots \tag{46}$$

$$\langle q_{k,i+\tau} U_{m,i}^{(2)} \rangle = \int q_{k,i+\tau} U_m^{(2)}(\boldsymbol{q}_i) P(\boldsymbol{q}_{i+\tau}, \boldsymbol{q}_i) d^N q_{i+\tau} d^N q_i,$$

$$k = 1, \ldots N_j, \quad m = 1, 2, \ldots \tag{47}$$

and

$$\langle q_{k,i+\tau} q_{\ell,i+\tau} U_{u,i}^{(3)} \rangle, k = 1, \ldots, N$$

$$\ell = 1, \ldots, N \tag{48}$$

$$n = 1, 2, \ldots$$

which is defined in complete analogy to (47).

Using our previous results we may immediately determine the steady-state distribution in the form

$$P_{st}(\boldsymbol{q}_i) = \exp\left(- \lambda_{st} - \sum_j \lambda_{j,st} U_{j,i}^{(1)} \right). \tag{49}$$

We now make use of a generalization of Jaynes' maximum entropy principle to the space-time domain as made by Jaynes himself; *i.e.*, we may now subject a multi-time joint probability to the maximum information (entropy) principle. For our present approach it is sufficient to consider the two-time joint probability, which, by means of the maximum information principle, acquires the form

$$P(\boldsymbol{q}_{i+\tau}, \boldsymbol{q}_i) = \exp[-\lambda_0 + A(\boldsymbol{q}_i) + B'(\boldsymbol{q}_i) q_{i+\tau} + \bar{q}_{i+\tau} C(\boldsymbol{q}_i) q_{i+\tau}] \tag{50}$$

where we have used the following abbreviations

$$A(\boldsymbol{q}_i) = - \sum_j \lambda_j^{(1)} U_j^{(1)}(\boldsymbol{q}_i) \tag{51}$$

$$B_k(\boldsymbol{q}_i) = - \sum_m \lambda_{km}^{(2)} U_j^{(2)}(\boldsymbol{q}_i) \tag{52}$$

$$C_{k\ell} = - \sum_n \lambda_{k\ell n} U_n^{(3)}(\boldsymbol{q}_i) \tag{53}$$

where the λ's are the Lagrange multipliers.

Using (54,49,50) we readily obtain the conditional probability in the form

$$P(\boldsymbol{q}_{i+\tau}|\boldsymbol{q}_i) = \exp[-\hat{\lambda} + \hat{A}(\boldsymbol{q}_i) + B'(\boldsymbol{q}_i) q_{i+\tau} + \bar{q}_{i+\tau} C(\boldsymbol{q}_i) q_{i+\tau}] \tag{54}$$

where we have used the abbreviations

$$\hat{\lambda} = \lambda_0 - \lambda_{st} \tag{55}$$

and

$$\hat{A} = A + \sum_j \lambda_{j,st} U_j^{(1)}(\boldsymbol{q}_i). \tag{56}$$

A comparison of the result (54) with (7) reveals that we may now make the following identifications between the Lagrange parameters which appeared previously and those which appear now:

$$\lambda(\boldsymbol{q}_i) \leftrightarrow \hat{\lambda} - \hat{A}(\boldsymbol{q}_i), \tag{57}$$

$$\lambda_\ell(\boldsymbol{q}_i) \leftrightarrow -B_\ell(\boldsymbol{q}_i), \tag{58}$$

and

$$\lambda_{k\ell}(\boldsymbol{q}_i) \leftrightarrow -C_{k\ell}(\boldsymbol{q}_i). \tag{59}$$

In this way we may again derive an Îto-Fokker-Planck equation in complete analogy to Sect. 2. If we use the constraints in the form of the functions (45) we have polynomials in the drift coefficients K' and the diffusion coefficients $G_{k\ell}$.

5. Learning of the Lagrange parameters

In this section, we present results obtained by Lisa Borland and this author. A major problem is the explicit determination of the Lagrange multipliers. Now we want to determine the values of the Lagrange multipliers occurring in the above expressions. To this end, we treat the example of a single variable and introduce a time-dependent form of the Kullback information as a measurement of distance between two joint distribution functions $\tilde{P}(q_{i+\tau}, q_i)$ and $P(q_{i+\tau}, q_i)$:

$$k = \int\int \tilde{P}(q_{i+\tau}, q_i) \ln \left(\frac{\tilde{P}(q_{i+\tau}, q_i)}{P(q_{i+\tau}, q_i)} \right) dq_{i+\tau} dq_i \geq 0 \tag{60}$$

where the constraints

$$\int\int P(q_{i+\tau}, q_i) dq_{i+\tau} dq_i = 1 \tag{61}$$

and

$$\int\int \tilde{P}(q_{i+\tau}, q_i) dq_{i+\tau} dq_i = 1 \tag{62}$$

must hold.

We assume that our system is offered known measurements of the type

$$\langle \Omega(q_i, q_{i+\tau}) \rangle_{\tilde{P}} = \int\int \Omega(q_i, q_{i+\tau}) \, \tilde{P}(q_{i+\tau}, q_i) dq_{i+\tau} dq_i \tag{63}$$

where Ω represents some function of $q_i, q_{i+\tau}$ and $\tilde{P}$ is the joint probability function according to which the input values of $q_{i+\tau}$ and q_i are distributed. We do not know $\tilde{P}$, but we consider it a fixed quantity. The idea now is to vary the internal distribution P created by our system until it matches $\tilde{P}$ as well as possible. This match is the better the smaller k is, and optimal for $k = 0$. A simple study shows that to minimize k it suffices to minimize the quantity

$$W = -\int\int \tilde{P}(q_{i+\tau}, q_i) \ln P(q_{i+\tau}, q_i) dq_{i+\tau} dq_i. \tag{64}$$

Inserting the explicit form of P from (50) above and using (62), this expression becomes

$$W = \lambda^{(0)} - \int \int \tilde{P}(q_i, q_{i+\tau}) \left(- \sum_{\mu=1}^{M} \lambda_\mu^{(1)} q_i^\mu + \sum_{\nu=0}^{N} \lambda_\nu^{(2)} q_i^\mu q_{i+\tau} - \sum_{\kappa=0}^{L} \lambda_\kappa^{(3)} q_i^\kappa q_{i+\tau}^2 \right) dq_{i+\tau} dq_i. \quad (65)$$

Normalization of P (eq. 61) further gives

$$\exp(\lambda^{(0)}) = \int \int \exp\left(- \sum_{\mu=1}^{M} \lambda_\mu^{(1)} q_i^\mu + \sum_{\mu=0}^{N} \lambda_\mu^{(2)} q_i^\mu q_{i+\tau} - \sum_{\kappa=0}^{L} \lambda_\kappa^{(3)} q_i^\kappa q_{i+\tau}^2 \right) dq_{i+\tau} dq_i \quad (66)$$

so that the final expression for W becomes

$$W = \ln\left[\int \int \exp\left(- \sum_{\mu=1}^{M} \lambda_\mu^{(1)} q_i^\mu + \sum_{\mu=0}^{N} \lambda_\mu^{(2)} q_i^\mu q_{i+\tau} - \sum_{\kappa=0}^{K} \lambda_\kappa^{(3)} q_i^\kappa q_{i+\tau}^2 \right) dq_{i+\tau} dq_i \right]$$

$$- \int \int \tilde{P}(q_i, q_{i+\tau}) \left(- \sum_{\mu=1}^{M} \lambda_\mu^{(1)} q_i^\mu + \sum_{\nu=0}^{N} \lambda_\nu^{(2)} q_i^\nu q_{i+\tau} - \sum_{\kappa=0}^{K} \lambda_\kappa^{(3)} q_i^\kappa q_{i+\tau}^2 \right) dq_{i+\tau} dq_i. \quad (67)$$

W can be seen as a potential function to be minimized. Alternatively, it can be interpreted as a network of "neurons" q_i and $q_{i+\tau}$, connected by "synapses" λ_ν^i. One way of minimizing W is to subject the connectivities to fluctuations, which can be done by submitting the λ's to a gradient strategy:

$$\dot{\lambda}_\mu^{(i)} = \gamma_i \frac{\partial W}{\partial \lambda_\mu^i}. \quad (68)$$

The result of these dynamics is that the λ's adjust themselves so as to optimally reproduce $\tilde{P}$, or in other words, our network "learns" the input probability distribution $\tilde{P}$. We shall therefore refer to the dynamics of eq. (68) as learning dynamics.

Explicit evaluation of the derivative in (68) gives, for $i = 1$:

$$\frac{\partial W}{\partial \lambda_\mu^{(1)}} = \left(\int \int \exp(...) dq_{i+\tau} dq_i \right)^{-1} \int \int q_i^\mu \exp(...) dq_{i+\tau} dq_i$$

$$- \int \int \tilde{P}(q_{i+\tau}, q_i) q_i^\mu dq_i dq_{i+\tau} \quad (69)$$

$$= (\langle q_i^\mu \rangle)_P - (\langle q_i^\mu \rangle)_{\tilde{P}}.$$

and similarly for $\lambda_\nu^{(2)}$ and $\lambda_\kappa^{(3)}$.

The learning dynamics for the Lagrange multipliers λ in the joint probability distribution thus become:

$$\dot{\lambda}_\mu^{(1)} = \gamma_1 \left(\langle q_i^\mu \rangle_P - \langle q_i^\mu \rangle_{\tilde{P}} \right) \quad (70)$$

$$\dot{\lambda}_\nu^{(2)} = -\gamma_2 \left(\langle q_i^\nu q_{i+\tau} \rangle_P - \langle q_i^\nu q_{i+\tau} \rangle_{\tilde{P}} \right) \quad (71)$$

$$\dot{\lambda}_\nu^{(3)} = \gamma_3 \left(\langle q_i^\kappa q_{i+\tau}^2 \rangle_P - \langle q_i^\kappa q_{i+\tau}^2 \rangle_{\tilde{P}} \right) \quad (72)$$

where the first term on the right-hand side represents moments of q_i and $q_{i+\tau}$ calculated according to the internal distribution function P, while the second term corresponds to fixed values of measurements offered to the system. The extension of our approach to the multivariate case is, of course, straight forward.

6. Examples

We have tested our method on several computer-generated data sets. Here we quote two "experiments", one concerning Brownian motion and the other treating motion in a 4th-order potential. The details of our approach will be published elsewhere [5].

6.1. Brownian Motion

Our approach was first tested on the simple case of Brownian motion, where the Îto-Langevin equation has the form

$$dq_i = -\alpha q_i dt + \tilde{\omega}\sqrt{Q}\sqrt{dt}. \tag{73}$$

This example was treated theoretically to check the validity of our results of section 4 and then simulated numerically in the computer. The system was offered input moments

$$\begin{aligned} \langle q_i^\nu \rangle &\quad \nu = 1, 2, 3 \\ \langle q_i^\kappa q_{i+\tau} \rangle &\quad \kappa = 0, 1, 2 \\ \langle q_i^\mu q_{i+\tau}^2 \rangle &\quad \mu = 0, 1, \end{aligned} \tag{74}$$

where for the sake of simplicity we choose $M = 2$, $L = 1$ and $N = 3$. These correlations are easily calculated using Gaussian integrals. Their numerical values for a situation with linear force $\alpha = 5.0$, constant diffusion term $Q = 2.0 = 0.14$ and sampling time $\tau = 0.01$ are shown in Table 1a. The learning dynamics for the Lagrange multipliers was started, using random initial values. Very soon, the system converges to give the values

$$K dt = \frac{-2.5q_i - 0.2q_i^2}{2(24.0 + 2.48q_i)} \approx -0.05q_i \tag{75}$$

and

$$g\sqrt{dt} = \frac{1}{\sqrt{2}(24.0 + 2.48q_i)} \approx 0.14. \tag{76}$$

Inserting the numerical value of τ then provides us with

$$dq_i = -5q_i dt + 1.4\tilde{\omega}\sqrt{dt}, \tag{77}$$

which shows that we really have determined the correct forces behind the process.

Table 1a - Steady state moments

$\langle q_i \rangle$	$\langle q_i^2 \rangle$	$\langle q_i^3 \rangle$
0.500	0.450	0.425

6.2. 4th-Order Potential

Our next experiment was the simulation of a process

$$dq = -\alpha q + \beta q^3 + \tilde{\omega}\sqrt{Q}\sqrt{dt}. \tag{78}$$

Again the input correlations were easily calculated and offered to our learning system. We chose to model a process with

$$\alpha = 6, \quad \beta = 19 \text{ and } \sqrt{Q} = 0.04.$$

The sampling time was $\tau = 0.01$. Numerical values of the correlation functions offered as input from the outer world, together with the internally produced values during the learning process are listed in Table 2. Numerical values of the λ's are listed in Table 3. From the learnt values of the Lagrange multipliers, we obtain

$$K(q_i) = \frac{1}{0.01}\frac{-30.43q_i + 0.002q_i^2 + 98.0q_i^3}{2(262.1 + 0.05q_i + 0.005q_i^2)} \approx -5.8q_i + 18.6q_i^3 \approx -6q_i + 19q_i^3$$

and

$$g(q_i) = \frac{1}{\sqrt{0.01}}\frac{1}{\sqrt{2(262.1 + 0.05q_i + 0.005q_i^2)}} \approx 0.043 \approx 0.04.$$

These values are again in very good agreement with the values of the forces underlying the process.

Table 2a - steady state moments

$\langle q_i \rangle$	$\langle q_i^2 \rangle$	$\langle q_i^3 \rangle$	$\langle q_i^4 \rangle$
0.0	0.270	0.0	0.0937

Table 2b - evolution of joint correlations

	$t=0$	$t=500$	$t=2000$	$t=4000$	$t=10000$	input values
$\langle q_{i+\tau} \rangle_P$	-3.10^{-6}	-7.10^{-6}	-2.10^{-5}	-2.10^{-5}	-2.10^{-5}	0.0
$\langle q_i q_{i+\tau} \rangle_P$	0.2491	0.2589	0.2680	0.2709	0.2717	0.27179
$\langle q_i^2 q_{i+\tau} \rangle_P$	-1.010^{-6}	-2.410^{-6}	-7.410^{-6}	-1.210^{-6}	-1.410^{-6}	0.0
$\langle q_i^3 q_{i+\tau} \rangle_P$	0.08802	0.09115	0.09410	0.09501	0.09528	0.09529
$\langle q_{i+\tau}^2 \rangle_P$	0.4274	0.2826	0.2777	0.2762	0.2757	0.27578
$\langle q_i q_{i+\tau} \rangle_P$	-2.10^{-6}	1.10^{-6}	4.10^{-6}	6.10^{-6}	7.10^{-6}	0.0
$\langle q_i^2 q_{i+\tau}^2 \rangle_P$	0.13613	0.09801	0.0977	0.09763	0.09758	0.09759

Table 3a - Final values of steady state dynamics

$\lambda_i^{(1)}$	$\lambda_2^{(1)}$	$\lambda_3^{(1)}$	$\lambda_4^{(1)}$
0.0	-15.0	0.0	24.0

Table 3b - "Learning" of the $\lambda^{(2)}$'s and $\lambda^{(3)}$'s

	t=10	t=500	t=1000	t=5000	t=10000	t=20000	$t \to \infty$
$\lambda_0^{(2)}$	0.15	0.16	0.20	0.02	0.00	0.00	0.00
$\lambda_1^{(2)}$	4.36	175.2	220.0	379.0	443.7	461.3	496.0
$\lambda_2^{(2)}$	1.02	0.99	0.43	0.05	0.00	0.00	0.00
$\lambda_3^{(2)}$	1.6	28.7	52.8	79.5	93.2	96.1	98.0
$\lambda_0^{(3)}$	4.0	92.0	111.2	202.3	235.0	245.1	262.0
$\lambda_1^{(3)}$	0.52	0.41	0.22	0.03	0.00	0.00	0.00
$\lambda_2^{(3)}$	1.19	1.12	0.93	0.71	1.04	0.00	0.00

7. Concluding Remarks

We have shown that it is possible to determine the deterministic and stochastic forces underlying a (Markovian) process, starting from measurements of correlation functions. After the analytical development of our approach, we tested it numerically on two computer-simulated processes. In both cases we obtained good confirmation of the theory. We have, however, yet to test the method on truly experimental data, which will be part of our future work.

REFERENCES

[1] Jaynes, E.T., Phys. Rev. **106**, 620 (1957)
[2] Jaynes, E.T., in *Complex Systems, Operational Approaches*, Springer Series in Synergetics, Haken, H. (ed.), Springer, Berlin (1985)
[3] Haken, H., *Advanced Synergetics*, Springer, Berlin (1983)
[4] Haken, H., *Synergetic Computers and Cognition*, Springer, Berlin (1990)
[5] Borland, L., and Haken, H., to be published

NONEQUILIBRIUM STATISTICAL MECHANICS

Baldwin Robertson
National Institute of Standards and Technology
Gaithersburg, Maryland 20899

ABSTRACT. The derivation of the equations of nonequilibrium statistical mechanics is reviewed. These exact closed equations are valid for nonlinear systems arbitrarily far from equilibrium. In the short-correlation-time and local limit they reduce to the equations of nonequilibrium thermodynamics with molecular expressions for the transport coefficients.

1. Introduction

The phenomenological equations of nonequilibrium thermodynamics have the form

$$\mathbf{J}_i = \mathbf{J}_i^0 + \sum_{j=1}^{m} L_{ij} \cdot \mathbf{X}_j, \qquad i = 1, \cdots, m, \tag{1.1}$$

where $\mathbf{J}_i$ is the total flux, $\mathbf{J}_i^0$ is its reversible (e.g. convective) part, the L_{ij} are phenomenological transport coefficients, and the $\mathbf{X}_j$ are thermodynamic forces. The $\mathbf{X}_j$ are gradients of the thermodynamic conjugate variables, such as the reciprocal of the temperature, etc. Examples of the diagonal coefficients L_{jj} are the heat conductivity, the ordinary diffusion coefficient, and the electrical conductivity, etc. The off-diagonal coefficients L_{ij} with $i \neq j$ are associated with the coupling of transport processes.

The thermodynamic conjugate variables, which are well known from equilibrium statistical thermodynamics, must be defined for a system that may be far from equilibrium. This can be done using Jaynes's information-theoretic approach to statistical mechanics,[1-4] which has been reviewed by Grandy.[5] One maximizes the information entropy or uncertainty subject to constraints on the expectations of the variables that are measured or controlled. The result is a generalization of Gibbs's canonical density, where the Lagrange multipliers are generalizations of the thermodynamic conjugate variables.

Jaynes[6,7] and Scalapino[8] have used this formalism in one way to analyze irreversible processes, and their work has been discussed by Zubarev and Kalashnikov[9-11] and by Grandy.[12] They maximize the information entropy just once for the entire nonequilibrium process, subject to constraints on the expectations over an interval of time. This gives an expression for the nonequilibrium statistical density, which they approximate and use to derive equations like (1.1) for systems close to (local) equilibrium.

In this paper we follow Jaynes's earlier work[1,2] in another way and maximize the information entropy repeatedly, subject to constraints only at the present instant of time. The generalized canonical density then yields time-dependent expressions for the entropy and the (reciprocal) temperature and other thermodynamic conjugate variables, with a substantial advantage. Even after a disturbance far from equilibrium, these expressions reduce to the usual thermodynamic expressions, whenever the system settles down to equilibrium. Furthermore, no matter how far the system is from equilibrium, the expressions for the entropy and the thermodynamic conjugates satisfy equations similar to those of equilibrium thermodynamics.[1,2]

The important difference between this formalism and equilibrium statistical thermodynamics is that here the constraints can involve operators other than constants of the motion. We use this formalism to define the thermodynamic conjugates and the entropy for a system that may be far from equilibrium.

We derive exact closed equations of motion for the expectations of the operators. These equations, like (1.1), consist of two kinds of terms: reversible (e.g. convective) terms $\mathbf{J}_i^0$ that do not change the entropy, and irreversible (i.e. dissipative) terms that do. The reversible terms are calculated using the generalized canonical density.

We use a generalization of the Zwanzig projection-operator method[13] to obtain an exact expression for the irreversible part of the statistical density. This immediately gives the irreversible part of the equations of motion.[14,15] We show how the exact equations reduce to the phenomenological equations[16] and obtain molecular expressions for the transport coefficients. Some references to related work and applications of the formalism have been given in a previous review.[17] Here we present the most economical derivation we can.

2. Statement of the Problem

Consider a system that for time $t < 0$ has a Hamiltonian H_0 and is in equilibrium. At $t = 0$ the Hamiltonian is suddenly changed to H, e.g., by changing an external field or by bringing subsystems into contact, etc. Then for $t > 0$ the system relaxes toward a new equilibrium. For clarity we restrict the present discussion to an isolated system with a time-independent Hamiltonian H, which does not commute with H_0. A more general problem has been considered previously,[15,16] and the analysis has also been extended to open systems.[18]

Let $F_1(\mathbf{r})$, $\cdots$, $F_m(\mathbf{r})$ be the quantum-mechanical operators (such as energy density, magnetic moment, etc.) whose expectations are variables that are observed or controlled in the experiment considered. Here $\mathbf{r}$ is an index denoting spatial position, and, as indicated, the operators are time-independent. The operators need not be constants of the motion, and they need not commute with each other.

The expectations of the operators are defined by

$$\langle F \rangle_t = \mathrm{Tr}[F\rho(t)], \tag{2.1}$$

where F is a column vector whose elements are the operators, $\rho(t)$ is the statistical density operator, and Tr denotes the trace of a matrix representation of the operators to its right. Although the expectations are defined here in terms of $\rho(t)$, they are ultimately to be calculated from the equations of motion to be derived.

The time dependence of $\rho(t)$ is given by the quantum Liouville equation

$$\partial\rho(t)/\partial t = -iL\rho(t), \tag{2.2}$$

where L is the Liouville operator.[19] The latter is a super operator that, when operating on any quantum-mechanical operator A, gives

$$LA \equiv \frac{1}{\hbar}[H, \ A], \tag{2.3}$$

where $[\ , \ ]$ is a commutator bracket. Since the Liouville operator L is linear, working with it is not any more difficult than working with the usual quantum-mechanical operators.

The initial condition for $\rho(t)$ can be the Gibbs canonical density

$$\rho(0) = \exp(-\beta H_0)/\text{Tr}[\exp(-\beta H_0)] \tag{2.4}$$

or some generalization of it. Equation (2.4) applies to a system initially at equilibrium at temperature $1/k\beta$. A possible generalization would be to have a number of noninteracting subsystems initially at equilibrium each at its own temperature. Then the different parts of the initial Hamiltonian would be multiplied by different reciprocal temperatures, and the initial condition would be a product of densities each of the form (2.4).

A formal solution to the Liouville equation (2.2) is

$$\rho(t) = \exp(-iLt)\,\rho(0). \tag{2.5}$$

This is equal to the more usual solution

$$\rho(t) = \exp\left(-\frac{i}{\hbar}Ht\right)\rho(0)\exp\left(\frac{i}{\hbar}Ht\right), \tag{2.6}$$

as can be shown by expanding both equations in powers of t. However, both solutions are of limited usefulness since they do not explicitly exhibit the separation into reversible and irreversible terms that is desired in the equations for $\langle F \rangle_t$ to be derived.

Furthermore, this $\rho(t)$ when used in

$$-k\,\text{Tr}[\rho(t)\ln\rho(t)] \tag{2.7}$$

yields an expression for entropy that is not suitable away from equilibrium. Consider a system that, after it is disturbed far from its initial equilibrium, settles down to a new equilibrium as the time t approaches infinity. At the initial equilibrium, the thermodynamic entropy will have one value, which is given by (2.7) with $\rho(t)$ replaced by (2.4). At the final equilibrium, the thermodynamic entropy will have another value, larger than the initial entropy, consistent with the second law of thermodynamics. A candidate for nonequilibrium entropy, at the very least, must equal the equilibrium entropy whenever the system is in equilibrium and so in this case must change in time.

However, (2.7) is independent of the time t because (2.6) is a unitary transformation and because of the cyclic invariance of the trace. Therefore, (2.7) can not be the nonequilibrium entropy.

3. Generalized Canonical Density

In order to obtain an expression that reduces to the thermodynamic entropy whenever the system settles down to equilibrium, we maximize the information entropy subject to the values of the expectations at a single instant of time as constraints. We use the formalism described by Jaynes.[1,2] The information entropy or uncertainty associated with a density operator $\sigma(t)$ is

$$S(t) = -k \operatorname{Tr}[\sigma(t) \ln \sigma(t)], \tag{3.1}$$

where we have included Boltzmann's constant k for interpretation later on. Maximizing the uncertainty subject to the constraint

$$\operatorname{Tr}[F\sigma(t)] = \langle F \rangle_t \tag{3.2}$$

incorporates information gathered at one instant of time t rather than over a time interval. Restricting information this way is necessary in order to obtain an equilibrium-like formalism because the thermodynamic state of a system at equilibrium depends only on the present values of the thermodynamic coordinates.

By introducing Lagrange multipliers $\lambda_1(\mathbf{r}, t), \cdots, \lambda_m(\mathbf{r}, t)$, one finds in a familiar way that (3.1) reaches a maximum subject to (3.2) when $\sigma(t)$ is the generalized canonical density

$$\sigma(t) = \exp[-\lambda(t) * F]. \tag{3.3}$$

Here $\lambda(t)$ is a row vector whose elements are $\lambda_1(\mathbf{r}, t), \cdots, \lambda_m(\mathbf{r}, t)$, and $*$ denotes a sum and integral

$$\lambda(t) * F \equiv \sum_{n=1}^{m} \int d^3r \, \lambda_n(\mathbf{r}, t) \, F_n(\mathbf{r}). \tag{3.4}$$

Normalization of (3.3) follows from (3.2) provided there is a linear combination of the $F_n(\mathbf{r})$ that equals 1. There is no loss of generality in requiring the latter since, if it is not already true, an operator $F_0 \equiv 1$ can always be added to make it true.

When (3.3) is inserted into (3.2), the result is

$$\operatorname{Tr} \{F \exp[-\lambda(t) * F]\} = \langle F \rangle_t. \tag{3.5}$$

This is to be solved for the multiplier $\lambda(t)$ in terms of the expectation $\langle F \rangle_t$. For example, if $F_0 \equiv 1$ and the energy density $F_1(\mathbf{r}) \equiv H(\mathbf{r})$ are the only mechanical variables, the dependence of the expectation of the energy density $\langle H(\mathbf{r}) \rangle_t$ on the multiplier $\lambda_1(\mathbf{r}, t) \equiv \beta(\mathbf{r}, t)$ can in principle be calculated. This dependence can in principle be inverted to obtain $\beta(\mathbf{r}, t)$ as a functional of $\langle H(\mathbf{r}) \rangle_t$. The time dependence of $\beta(\mathbf{r}, t)$ arises only because $\langle H(\mathbf{r}) \rangle_t$ depends on time. In this example the multiplier $\beta(\mathbf{r}, t)$ is the inverse local temperature. In general, (3.5) defines the thermodynamic conjugate $\lambda(t)$ as a functional of $\langle F \rangle_t$.

There always is a physical interpretation for the multipliers $\lambda_n(\mathbf{r}, t)$. They are the thermodynamic conjugates to the mechanical variables $\langle F_n(\mathbf{r}) \rangle_t$. When the system is in equilibrium, the formalism reduces to equilibrium statistical mechanics. For example, with (3.3) inserted into (3.1), the expression $S(t)$ equals the thermodynamic entropy, and the solution $\lambda(t)$ of (3.5) equals the thermodynamic conjugates, whenever the system settles down to equilibrium. We apply the same formalism arbitrarily far from equilibrium. The important difference from equilibrium is that here the $F_n(\mathbf{r})$ need not be constants of the motion so that we have a generalized canonical statistical density instead of the canonical density that applies at equilibrium. We define $S(t)$ to be the entropy of the system and $\lambda(t)$ to be the thermodynamic conjugates (such as reciprocal temperature) whether the system is at equilibrium or not. We use the generalizations of thermodynamic relations[1,2] no matter how far the system is from equilibrium.

In order to obtain a simple initial condition of general form for use in the derivation that follows, we require enough operators to be included in the set $F_1(\mathbf{r})$, $\cdots$, $F_m(\mathbf{r})$ that the Hamiltonian H can be written as a linear combination of these operators. If this is not already true, the operator $F_1 \equiv H$ can always be added to the set in order to make it true. If the system is in equilibrium at $t = 0-$, the linear combination (3.3) then includes βH, and as a result the initial condition (2.4) can be expressed concisely as

$$\rho(0) = \sigma(0). \tag{3.6}$$

The initial condition for $\rho(t)$ can even be a generalized canonical density, which is much more general than an equilibrium statistical density, and (3.6) will still be satisfied. This initial condition is not very restrictive since most systems on which nonequilibrium experiments are performed can be described by a generalized canonical density at one time or another, and we choose the time $t = 0-$ to be one such time.

The generalized canonical density $\sigma(t)$ is not even approximately a solution to the Liouville equation. The statistical density $\rho(t)$ is the operator that satisfies the Liouville equation. In the next section we derive an expression for $\rho(t)$ in terms of $\lambda(t)$. We will use that expression to derive equations of motion that determine the time dependence of $\lambda(t)$ and thus of $\langle F \rangle_t$.

4. Nonequilibrium Statistical Density

The nonequilibrium statistical density $\rho(t)$ can be written as a sum of a reversible part and an irreversible part, where the reversible part is the generalized canonical density $\sigma(t)$, and the irreversible part is whatever is left over. Here no assumption is made that $\rho(t)$ is approximately in local equilibrium or that it is approximately equal to $\sigma(t)$. Indeed, a perturbation expansion of $\rho(t)$ about $\sigma(t)$ may be not at all valid because the irreversible terms in the equations of motion to be derived may not be small compared with the reversible terms, which may be zero. We derive an exact expression for the irreversible part $\rho(t) - \sigma(t)$ and thus an exact expression for $\rho(t)$ itself.

Since (3.3) defines $\sigma(t)$ as a functional of $\lambda(t)$, and (3.5) defines $\lambda(t)$ in turn as a functional of $\langle F \rangle_t$, it follows that $\sigma(t)$ is a functional of $\langle F \rangle_t$. Now $\langle F \rangle_t$ depends on time through (2.1), and so the chain rule for a total derivative gives

$$\frac{\partial \sigma(t)}{\partial t} = P(t)\frac{\partial \rho(t)}{\partial t}, \tag{4.1}$$

where $P(t)$ is a linear operator defined by

$$P(t)A \equiv \frac{\delta\sigma(t)}{\delta\langle F \rangle_t} * \mathrm{Tr}(FA) \tag{4.2}$$

for any quantum-mechanical operator A. Here $\delta/\delta\langle F \rangle_t$ denotes a functional derivative. The dependence of $P(t)$ on time arises because $P(t)$ depends on $\langle F \rangle_t$.

We now can use a generalization of the Zwanzig projection-operator method[13] provided we are careful with the time dependence of $P(t)$. Calculate the time derivative $\partial[\rho(t) -$

$\sigma(t)]/\partial t$ using (4.1) and the Liouville equation (2.2) and write the $\rho(t)$ in the resulting $iL\rho(t)$ as a sum of its reversible part $\sigma(t)$ and its irreversible part $\rho(t) - \sigma(t)$ to get

$$\partial[\rho(t) - \sigma(t)]/\partial t + [1 - P(t)]iL[\rho(t) - \sigma(t)] = -[1 - P(t)]iL\sigma(t). \tag{4.3}$$

This equation is sufficient to determine $\rho(t) - \sigma(t)$ in terms of $\sigma(t)$. It is of first-order and linear in $\rho(t) - \sigma(t)$ and so can be solved by the following well-known method. Introduce an integrating factor $T(t, t')$ defined to be the solution to

$$\partial T(t, t')/\partial t' = T(t, t')[1 - P(t')]iL \tag{4.4}$$

with the initial condition

$$T(t, t) = 1. \tag{4.5}$$

Since both $P(t)$ and L are linear operators, so also is $T(t, t')$, and since $P(t')$ is a functional of $\langle F \rangle_{t'}$, so also is $T(t, t')$. Equations (4.4), (4.1), and (2.2) give

$$\partial \{T(t, t')[\rho(t') - \sigma(t')]\}/\partial t' = -T(t, t')[1 - P(t')]iL\sigma(t'). \tag{4.6}$$

Integrate this over t' from 0 to t and use the initial conditions (4.5) and (3.6) to get

$$\rho(t) = \sigma(t) - \int_0^t dt' \, T(t, t')[1 - P(t')]iL\sigma(t'). \tag{4.7}$$

This is the desired exact expression for the nonequilibrium statistical density operator $\rho(t)$ as a functional of $\langle F \rangle_t$. It comprises two terms: The first term $\sigma(t)$ gives reversible (e.g., convective) effects, and the second term gives irreversible (i.e., dissipative) effects. This structure appears also in the equations of motion to be derived in the next section.

5. Equations of Motion for $\langle F \rangle_t$

Closed equations for $\langle F \rangle_t$ can be derived by substituting (4.7) for $\rho(t)$ on the right of the Liouville equation (2.2), multiplying the result on the left by F, taking the trace, and using (2.1) to get

$$\partial \langle F \rangle_t/\partial t = -\text{Tr}[FiL\sigma(t)] + \int_0^t dt' \, \text{Tr}\{FiLt(t, t')[1 - P(t')]iL\sigma(t')\}. \tag{5.1}$$

This can be written in a more suggestive way as follows. For arbitrary A and B, the identity

$$\text{Tr}(ALB) = -\text{Tr}[(LA)B] \tag{5.2}$$

follows from (2.3) and cyclic invariance of the trace. Thus the first term on the right of (5.1) becomes just $\langle \dot{F} \rangle_t$, where the dot is defined by

$$\dot{A} \equiv iLA, \tag{5.3}$$

and the angular brackets are defined by

$$\langle A \rangle_t \equiv \text{Tr}[A\sigma(t)] \tag{5.4}$$

for any quantum-mechanical operator A.

The last term in (5.1) can be transformed using

$$iL\sigma(t) = -\lambda(t) * \overline{\dot{F}}\sigma(t), \tag{5.5}$$

where the dot is defined by (5.3), and the overbar is defined by

$$\overline{A} \equiv \int_0^1 dx\ \sigma(t)^x A\sigma(t)^{-x} \tag{5.6}$$

for any operator A. Equation (5.5) can be derived by integrating the derivative

$$i\hbar d(\sigma^x H\sigma^{1-x})/dx = -\lambda * \sigma^x \dot{F}\sigma^{1-x} \tag{5.7}$$

over x from 0 to 1.

When (5.2)-(5.6) are applied to the last term of (5.1), we get

$$\frac{\partial \langle F \rangle_t}{\partial t} = \langle \dot{F} \rangle_t + \int_0^t dt'\ K(t,t') * \lambda(t'), \tag{5.8}$$

with the kernel given by

$$K(t,t') = \langle \dot{F}T(t,t')[1 - P(t')]\overline{\dot{F}}' \rangle_{t'}, \tag{5.9}$$

where the prime on the bar over $\dot{F}$ on the right of (5.9) indicates that the t in (5.6) is primed. Note that since P and T in (5.9) operate to the right, $\sigma(t)$ must be on the right in the definition (5.4) of the angular bracket.

Equations (5.8) and (5.9) are the equations of motion for the expectations $\langle F_n(\mathbf{r}) \rangle_t$ and their thermodynamic conjugates $\lambda_n(\mathbf{r}, t)$, which are defined by (2.1) and (3.5), respectively. The first term on the right of (5.8) is a functional of $\lambda(t)$ and hence is a functional of $\langle F \rangle_t$. It depends only on the present values of these functions. If it were the only term on the right, the entropy would not change, and hence this term is called the reversible term. The kernel (5.9) is a generalization of a rate-rate correlation function, the cross correlation of the rate of change of $F_n(\mathbf{r})$. The kernel and thus the last term in (5.8) are functionals of $\lambda(t')$ and hence of $\langle F \rangle_{t'}$ over the interval $0 \le t' \le t$. The last term describes relaxation that changes the entropy and hence is called the irreversible term.

Because (3.3), (4.2), (4.4), (5.4), and (5.9) are defined in terms of $\lambda(t)$, which is defined by (3.5) as a functional of $\langle F \rangle_t$, the latter is the only unknown, and the equations of motion (5.8) are closed. However, explicit calculation of the kernel (5.9) is difficult and requires approximation. References to some approximation methods are given in Ref. 18. For example, a linear expansion about equilibrium and a proof of reciprocity relations are given in Ref. 16. However, the present derivation does not require approximation until the last step—at the end of the next section.

6. Nonequilibrium Statistical Thermodynamics

Often $\dot{F}_n(\mathbf{r})$ can be written as the divergence

$$\nabla \cdot \mathbf{J}_n(\mathbf{r}) = -\dot{F}_n(\mathbf{r}), \qquad \mathbf{r}\ \text{inside}\ \mathcal{R}, \tag{6.1}$$

of a flux $\mathbf{J}_n(\mathbf{r})$, with

$$\mathbf{J}_n(\mathbf{r}) = 0, \qquad \mathbf{r} \text{ outside } \mathcal{R}, \tag{6.2}$$

where $\mathcal{R}$ is the region of space occupied by the system. Integration of (6.1) over space using (6.2) and (5.3) shows that this can be done only when

$$iL \int F_n(\mathbf{r}) \, d^3r = 0, \tag{6.3}$$

i.e. when $F_n(\mathbf{r})$ is the density of a conserved quantity. When this is true, we can obtain a generalization of the equations of nonequilibrium thermodynamics.

Insert (5.5) and (6.1) into (4.7), integrate by parts using (3.4) and (6.2), multiply on the left by $J_n(\mathbf{r})$, and take the trace to get

$$\mathrm{Tr}[\mathbf{J}_n(\mathbf{r})\rho(t)] = \langle \mathbf{J}_n(\mathbf{r})\rangle_t + \sum_{n'=1}^{m} \int_0^t dt' \int d^3r' \; K_{nn'}(\mathbf{r},t,\mathbf{r}',t') \cdot \nabla'\lambda_{n'}(\mathbf{r}',t'),$$

$$n = 1, \cdots, m, \tag{6.4}$$

where the kernels are given by the flux-flux correlation functions

$$K_{nn'}(\mathbf{r},t,\mathbf{r}',t') = \langle \mathbf{J}_n(\mathbf{r})T(t,t')[1 - P(t')]\overline{\mathbf{J}_{n'}(\mathbf{r}')}'\rangle_{t'}. \tag{6.5}$$

Equations (6.4) are the memory-retaining nonlocal generalizations of the equations of nonequilibrium thermodynamics (1.1). The $\mathrm{Tr}[\mathbf{J}_n(\mathbf{r})\rho(t)]$ are the total fluxes, the $\langle \mathbf{J}_n(\mathbf{r})\rangle_t$ are the reversible fluxes, the integrals on the right are the irreversible fluxes, and the $\nabla'\lambda_{n'}(\mathbf{r}',t')$ are the thermodynamic forces $\mathbf{X}_{n'}$. Combining (6.1) and (6.4) with (2.1), (2.2), (5.2), and (5.4) gives the equations of motion

$$\frac{\partial}{\partial t}\langle F_n(\mathbf{r})\rangle_t = -\nabla \cdot \langle \mathbf{J}_n(\mathbf{r})\rangle_t - \nabla \cdot \sum_{n'=1}^{m} \int_0^t dt' \int d^3r' \; K_{nn'}(\mathbf{r},t,\mathbf{r}',t') \cdot \nabla'\lambda_{n'}(\mathbf{r}',t'),$$

$$n = 1, \cdots, m, \tag{6.6}$$

which in principle can be solved simultaneously with (3.5) to obtain $\langle F \rangle_t$.

The derivation up to this point is exact. We now make approximations and show how (6.4) reduces to (1.1). Since the operators $F_n(\mathbf{r})$ are all densities of conserved quantities, their expectations, and hence their thermodynamic conjugates $\lambda_n(\mathbf{r},t)$, may be expected to vary slowly compared with the kernels $K_{nn'}(\mathbf{r},t,\mathbf{r}',t')$, which decay to zero as $t - t'$ increases. Thus the $\nabla'\lambda'$ may be taken out of the t' integral. If the thermodynamic forces $\nabla'\lambda'$ also vary slowly compared with the kernels as $|\mathbf{r} - \mathbf{r}'|$ increases, the $\nabla'\lambda'$ may also be taken out of the d^3r' integral, and (6.4) becomes

$$\mathrm{Tr}[\mathbf{J}_n(\mathbf{r})\rho(t)] = \langle \mathbf{J}_n(\mathbf{r})\rangle_t + \sum_{n'=1}^{m} L_{nn'} \cdot \nabla\lambda_{n'}(\mathbf{r},t), \qquad n = 1, \cdots, m, \tag{6.7}$$

where the transport coefficients are given by

$$L_{nn'} = \int_0^t dt' \int d^3r \; K_{nn'}(\mathbf{r},t,\mathbf{r}',t'). \tag{6.8}$$

Equation (6.7) is the only approximate result in this paper; all other results are exact. In a linear approximation, the kernels (6.5) satisfy reciprocity relations and time-reversal symmetry relations.[15]

For t much larger than the time required for the kernels (6.5) to decay to zero, the transport coefficients (6.8) will be constant. Then (6.7) become the equations of nonequilibrium thermodynamics (1.1), and (6.8) become generalizations of the Green-Kubo,[19,20] expressions for transport coefficients. References to applications are given in Ref. 18.

7. Summary

We have derived the equations of nonequilibrium statistical mechanics (5.8) or (6.6) using a generalization of Zwanzig's projection operator technique.[13] They are valid even in the nonlinear regime arbitrarily far from equilibrium. In these equations of motion the time derivative of the expectations comprises two terms: a reversible (e.g. convective) term plus an irreversible (or dissipative) term. The reversible term is computed using a generalized canonical density, which gives an equilibrium-like formalism (including temperature, entropy, etc.) no matter how far the system is from equilibrium. The irreversible term is a sum of integrals of correlation functions multiplied by thermodynamic forces.

All approximations are delayed to the last step in the derivation. In the short-correlation-time and nonlocal limits, the exact equations (6.4) reduce to the equations of nonequilibrium thermodynamics (1.1). Then the integrals of the flux-flux correlation functions become generalizations of the Green-Kubo expressions for the transport coefficients.

In the derivation, an explicit expression for an operator $P(t)$ is introduced. Use of this operator automatically leads to a subtraction of the fluxes, which is necessary for the correlation functions to approach zero as time approaches infinity. Without the operator $P(t)$ there is a singularity that occurs in the Green-Kubo expression in the thermodynamic and infinite-time limits. References to this and a discussion of the properties and advantages of $P(t)$ are given in Refs 16–18. There the relationship between $P(t)$ and a time-dependent generalization of Mori's projection operator[21] is derived. The same time-dependent projection operator has been used in a similar way by Grabert[22,23] and by Oppenheim and Levine[24] to derive an exact generalized Langevin equation. Their equation is a generalization of Mori's approximate equation, which was obtained by expanding about equilibrium. Others[25,26] have also used these methods. References to applications are given in Ref. 18.

REFERENCES

1. E. T. Jaynes, 'Information Theory and Statistical Mechanics. I. II.,' *Phys. Rev.* **106**, 620; **108**, 171 (1957).

2. E. T. Jaynes, 'Information Theory and Statistical Mechanics,' in *Statistical Physics* (1962 Brandeis Lectures, Vol. 3) edited by K. W. Ford (W. A. Benjamin, 1963).

3. E. T. Jaynes, 'Gibbs vs. Boltzmann Entropies,' *Am. J. Phys.* **33**, 391 (1965).

4. E. T. Jaynes: *Papers on Probability, Statistics, and Statistical Physics*, edited by R. D. Rosenkrantz (D. Reidel, 1983). Reprint of Refs. 1-3 and 6-7.

5. W. T. Grandy, Jr., *Foundations of Statistical Mechanics Vol. I: Equilibrium Theory* (D. Reidel, 1987).

6. E. T. Jaynes, 'Foundations of Probability Theory and Statistical Mechanics,' in *Delaware Seminar on the Foundations of Physics*, M. Bunge, Ed. (Springer-Verlag, 1967).

7. E. T. Jaynes, 'Where Do We Stand on Maximum Entropy,' in *The Maximum Entropy Formalism*, edited by R. D. Levine and M. Tribus (MIT Press, 1979), p 15.

8. D. J. Scalapino, *Irreversible Statistical Mechanics and the Principle of Maximum Entropy*, dissertation (Stanford University, 1961).

9. D. N. Zubarev and V. P. Kalashnikov, 'Extremal Properties of the Nonequilibrium Statistical Operator,' *Teor. Math. Phys.* **1**, 137 (1969) [*Theor. Math. Phys.* **1**, 108 (1969)];

10. D. N. Zubarev and V. P. Kalashnikov, 'Derivation of the Nonequilibrium Statistical Operator from the Extremum of the Information Entropy,' *Physica* **46**, 550 (1970).

11. D. N. Zubarev, *Nonequilibrium Statistical Thermodynamics* (Consultants Bureau, 1974), p. 439.

12. W. T. Grandy, Jr., *Foundations of Statistical Mechanics. Vol. II: Nonequilibrium Phenomena* (D. Reidel, 1987).

13. R. Zwanzig, 'Ensemble Method in the Theory of Irreversibility,' *J. Chem. Phys.* **33**, 1338 (1960).

14. B. Robertson, *Nonequilibrium Statistical Mechanics and Its Application to Nuclear Magnetic Resonance*, dissertation (Stanford University, 1964).

15. B. Robertson, 'Equations of Motion in Nonequilibrium Statistical Mechanics,' *Phys. Rev.* **144**, 151 (1966).

16. B. Robertson, 'Equations of Motion in Nonequilibrium Statistical Mechanics. II. Energy Transport,' *Phys. Rev.* **160**, 175 (1967); Erratum **166**, 206 (1968).

17. B. Robertson, 'Application of Maximum Entropy to Nonequilibrium Statistical Mechanics,' in *The Maximum Entropy Formalism*, edited by R. D. Levine and M. Tribus (MIT Press, 1979), p 289.

18. B. Robertson and W. C. Mitchell, 'Equations of Motion in Nonequilibrium Statistical Mechanics. III. Open Systems,' *J. Math. Phys.* **12**, 563 (1971).

19. R. Kubo, 'Statistical-Mechanical Theory of Irreversible Processes. I. General Theory and Simple Applications to Magnetic and Conduction Problems,' *J. Phys. Soc. Japan* **12**, 570 (1957).

20. M. S. Green, 'Markoff Random Processes and the Statistical Mechanics of Time-Dependent Phenomena. I. II.,' *J. Chem. Phys.* **20**, 1281 (1952); **22**, 398 (1954).

21. H. Mori, 'Transport, Collective Motion, and Brownian Motion,' *Progr. Theor. Phys. (Kyoto)* **33**, 423 (1965); **34**, 399 (1965).

22. H. Grabert, 'Microdynamics and Equations of Motion for Macrovariables,' *Z. Physik B*, **27**, 95 (1977);

23. H Grabert, *Projection Operator Techniques in Nonequilibrium Statistical Mechanics* (Springer-Verlag, 1982).

24. I. Oppenheim and R. D. Levine, 'Nonlinear Transport Processes: Hydrodynamics,' *Physica A 99*, 383 (1979).

25. D. Hölzer and E. Fick, 'Robertson's Formalism with Explicitly Time-Dependent Observables,' *Physica A* **168**, 867 (1990).

26. E. Fick and G. Sauermann, *The Quantum Statistics of Dynamic Processes* (Springer-Verlag, 1990).

A BACKWARD LOOK TO THE FUTURE

E. T. Jaynes
Arthur Holly Compton Laboratory of Physics
Washington University
1 Brookings Drive
St. Louis, Missouri 63130

ABSTRACT. We survey briefly some fifty years of thinking about physics and probability with the aim of explaining: (1) What I did not know then, but know now; (2) What I have been trying to accomplish in science and education, and to what extent these efforts have succeeded; (3) What remains unfinished, but where I think the greatest future opportunities lie; and (4) What personal and professional advice I can now give to young people (and wish someone had given me fifty years ago).

Introduction

A meeting like this is an overwhelming experience! I was overwhelmed not only by the sheer number of people who came here from so far; but even more by the kind sentiments expressed. Of course, I had looked forward to seeing again many former students and colleagues and had expected to have chats with each one, to renew our friendship and bring us both up to date about the other's work. But what one can actually do in two days is so helplessly short of what one wants to do! Some of the participants had to leave without our being able to talk at all; there was simply no time for it. Then perhaps this reminiscence can serve as a substitute channel for conveying my thanks and appreciation to all of you.[†]

But a reminiscence may be boring unless it also conveys something useful to the reader. It must be that fifty years of thinking about science and of observing the successes and failures of many ideas, has given one *some* kind of insight that would be helpful to others. Whilst exuding long–term optimism, the following remarks will appear at some places to be of a negative character, deploring recent trends in both science and education. So please understand that my purpose is not to complain, but rather to seek constructive remedies; one must first know what the problems are. The problems created by the blindness of people of my generation are also the opportunities for the next generation, so my advice for the future has necessarily some warnings about recent trends that need to be resisted and reversed. Those who are at the beginning of their careers today will be obliged to deal with these problems; and it is better to be aware of them now than to learn about them gradually over many years, as I did.

[†] In writing this a peculiar problem arose; the constant use of the personal pronoun "I" which one is not supposed to do, and which looks awkward even to me. But since that happens to be the topic here, I did not see any way to avoid it which does not look even more awkward; so I apologize for this in advance and hope it will not distract you enough to obscure the message.

In this I have only followed what seemed to me a good precedent. When my thesis advisor, Eugene Wigner, retired in 1971, he proceeded to unburden himself, in a series of colloquium talks at many places, of some concerns about science and education that had bothered him for many years. But he felt that he should keep his silence until then. So there is also in the following some of the flavor of *"Now it can be told!"*

Teaching or Research?

This has been one of my major concerns throughout my adult life. We have a serious problem in American science education, for which no realistic solution is yet in sight.

I was always much aware of Einstein's advice to a young theoretical physicist: if you are sure that your ideas are right, and therefore you want to be able to work full time on your research without interruptions, get a job as a light–house keeper. On the other hand, if you are not that confident, then as a matter of insurance get a job as a teacher. Then if the research does not pay off, at least you will have the satisfaction of knowing that you are making a useful contribution to society. The fact that, on entering the job market in 1950 I did not seek out a lighthouse and did get a job as an instructor at Stanford University, perhaps indicates my early level of self–confidence.

But it required a few years before I perceived what a science teacher's job really is. The goal should be, not to implant in the student's mind every fact that the teacher knows now; but rather to implant a *way of thinking* that will enable the student, in the future, to learn in one year what the teacher learned in two years. Only in that way can we continue to advance from one generation to the next.

As I came to realize this, my style in teaching changed from giving a smattering of dozens of isolated details, to analyzing only a few problems, but in some real depth. It doesn't even matter very much what those few problems are; once a student knows what it feels like to analyze something in depth, he can do it for himself on whatever other problems may come his way. Equally important, he can recognize in the work of others the distinction between a superficial study and one that is deep enough to be capable of finding new things.[†]

For Centuries, Universities recognized only the teaching function and "research" was an unknown word. Teachers were not an economic liability to the University; they were paid directly by the students and their income reflected directly the quality of their teaching. So every teacher was under a strong incentive to do better at his job. In the United States, research gained a foothold in the early 20'th Century (in England, about 50 years earlier), and the issue of teaching *vs.* research has been a matter of contention here ever since. My impression is that this is much less of a problem in England.

The following scenario was repeated in essentially the same form at many American Universities; but it was at Stanford that I experienced it first hand. The researchers proceeded to gain influence very aggressively; they created the folklore that a person cannot be a good teacher unless he is first a good researcher. We all know this to be true, necessarily, of the most advanced graduate courses at the boundary of present knowledge; but it is egregiously false for all the other courses where most of the substantive science education

---

[†] As a pertinent example, one sees through expositions of Quantum Electrodynamics which proclaim its wondrous success on the basis of a few accurate numbers – without ever asking what aspect of the formalism led to those numbers, or what changes in the formalism would change those numbers. We return to this later.

takes place. Nevertheless, the idea spread among administrators, and in about 1920 the policy was adopted, very quietly, at several Universities including Stanford, that henceforth faculty appointments and promotions would be based solely on research productivity, not on teaching ability or performance.

Thirty years later this news finally leaked out to the undergraduate students at Stanford and, as one would guess, caused an uproar in the Student Newspaper in the 1950's. I wrote a private memorandum to the Physics faculty expressing my own views, observing in particular that good research can be and is done in many other places (Government and Industrial laboratories, Theoretical Institutes); but that if good teaching is not done in our Universities, it does not get done at all.

The Administration responded to the situation by issuing a public statement denying that any such policy existed on promotions. David L. Webster, who had been Chairman of the Stanford Physics Department at the time this policy was instituted, knew better; and without taking part in the local fracas, he revealed the facts for the record in an article which was ostensibly only a reminiscence of the early days of the American Physical Society (Webster, 1957), in which he quoted a few lines from my memorandum. The physics students were quick to seize upon this and spread it over the campus.

Had the matter been left to internal forces in the University, there would have been a movement back to a condition of something like balance between the factions. But just at this time, what an economist would call an "exogenous variable" appeared and took control of the situation. It was the really big research grant sponsored by various alphabetized Government agencies (ONR, AFOSR, AEC, NSF, DOD, *etc.*). This created the situation that a researcher with the right connections – whether competent or not – paid his own way many times over in the money he brought into the University; while a good teacher became, from the standpoint of the Administration, an economic liability. It was easier to get a million dollars for a big research project than it was to get a thousand dollars for upgrading of the educational facilities. So the researcher was back in firmer control than ever; every faculty member was under constant pressure to push his research and neglect his teaching, under pain of jeopardizing his professional career: *Publish or Perish.*

Of the two biggest spenders on research, one offended our teachers by dismissing teaching as a mere "chore" in a memorandum. At a faculty meeting that I attended, the other actually complained out loud that there were far too many courses and other requirements, and they were interfering with the time that students could spend working on HIS research! As if graduate students were there only as a source of cheap labor to further his ambitions. Seeing that there was some opposition to it, the strategy adopted by the big researchers was to accomplish their goal gradually; there was a steady stream of small changes in the Departmental rules, each one cutting down a little more on the amount of course work, language requirements, and the level of general competence required of graduate students. They received less and less education while spending more and more hours manning the vacuum pumps. I protested against this every step of the way; and was outvoted every time.

This Government intervention (always, of course, held to be essential for National Defense and beneficial to the University) has through this unforeseen side effect done more damage to the cause of advanced science education in the United States than has any other factor in the past 40 years. In my opinion, the University science teaching profession is an honorable and socially necessary one; and no good teacher needs to apologize if he does not also turn out volumes of research. Indeed, the quality of our education today determines the quality of our research tomorrow. It is essential that our Universities carry on both functions, at the highest possible level, simultaneously; and some method of administration and financing must be found that will permit this.

But Why The Move?

When in March 1960 I received an inquiry as to whether I might be interested in a faculty position at Washington University, I visited St. Louis and met Edward U. Condon, the Physics Dept. Chairman. We had a long, serious conversation deep into the night, sitting at his kitchen table, in which I told him – in detail, and naming names – of my dismay at what seemed to me the destructive policies in effect at Stanford. He agreed with my judgment, assured me that he and others at Washington University were well aware of this problem and were determined to resist it; and that no faculty member at W. U. would be placing his professional career in jeopardy by being a good teacher.

Ed Condon was the first Academic administrator I had ever known who was a big enough man to think in terms of the long–run interests of the students and the University, instead of the short–run finances of his own little bailiwick, and was currently in office. I was so impressed that I accepted the position he offered me for the privilege of working with him, even though it meant a loss of income, and moved to St. Louis in September 1960. (But over the next ten years I managed to attract to W. U. about as much research money as anybody else, plus the money that built our present library.)

A year after this move it was quite obvious to me that – in agreement with the general observations of David Webster – Washington University was doing a better job of science education than was Stanford; and I think the record of subsequent accomplishments of our students of that period will bear this out. To my great satisfaction, many of our students became excellent teachers and turned out important research, and a surprising number proceeded quickly to become Department Heads, Deans, and on to higher positions in Academia, Government, and Industry.

Similarly, for many years Reed College in Portland, Oregon was famous for its dedicated undergraduate physics teaching; and it contributed a disproportionate number of the Ph.D. physicists in the United States. The long–run effects of good and indifferent science teaching are not hard to discern.

Today this does not seem to be quite the front–burner issue that it was thirty years ago, but perhaps this signifies only that the dedicated science teachers have been vanquished entirely, and no longer exist at all in major Universities; those who are concerned with the alarming decrease in number of students choosing science majors please take note. In any event, the same factors are in operation and there is nothing obsolete in the maxim: *Publish or Perish*. The pages of our scientific journals continue to be cluttered with bad research whose only function is to gain someone a promotion to Associate Professor.

It seems to me that this was a tragedy that did not have to happen, and is wasteful to everyone concerned. Of course, there are natural differences in talent and inclination of scientists. A few are outstandingly successful at both teaching and research, but most of us tend to do better at one or the other; and we ought to be free to choose. Similar divisions of labor occur naturally in nearly every field. For example, among musicians the talents for composing and for performing seldom occur in the same person, and most tend toward one activity or the other. There is no particular stress between these different "factions" in

Music Departments, because they are all coexisting in the same circumstances and see their functions as complementary rather than in competition. But such stress could be created overnight if some outside agency decided to bestow lavish support only on performers but not on composers; *Perform or Perish* would become, very quickly, the operative principle.

Having recognized the problem, what is the solution? In the future, some readers of these words are going to find themselves in Academic Administrative positions and will face this question. The obvious, simplistic answers are to establish quotas (which nobody really wants), to build up a separate endowment dedicated to support of teaching (which is unrealistic); or to remove the dependence on Government grants (which is even more unrealistic). The latter would seem to require that Universities build up their endowments to the point where they are financially independent, and no longer accept Government grants. But unless the University became so wealthy that money was no consideration at all, the experimental researcher would then become a major economic liability and we would have the same problem in reverse. So here is an opportunity for some really new creative thinking. In the meantime, we can only hope that Universities may find Administrators with the wisdom of David Webster and Ed Condon.

Probability Theory

Turning to technical matters, several people have noted that each of my research papers goes through a twenty–year incubation period, during which it is either ignored or attacked; then somebody starts to take note of what is in it, and it becomes studied intensely. My first work on statistical mechanics appeared in 1957; in 1977 philosophers started their nit–picking analysis and commentary about the meaning of every word in it. The Jaynes–Cummings paper appeared in 1963 and was ignored until 1983, when it suddenly became the most cited work in the quantum optics literature. My exposition of the "fast second law" with quantitative applications in biology appeared only in 1989, so we cannot expect anybody to notice it until the year 2009, while my resolution of the Gibbs paradox is not scheduled to be discovered until the year 2013; and so we need not discuss them here.

Evidently, if we want to find any gratifying successes, we must turn to my earliest works, where more than twenty years have passed. Indeed, today I am far more self–confident about probability theory than was possible forty years ago when I started lecturing on probability and information theory. The reason is that the essential evidence is now in; we are sustained no longer by faith and hope, but by proven theorems and accomplished facts on the level of new useful numerical results. There is no doubt that the formulations of probability theory and statistical mechanics as extensions of logic are here to stay; they will be the universally accepted basis of those fields 100 years hence. Too many things are coming out right to allow any other outcome.

Without giving a lecture course, let me convey a quick impression of the nature of these developments, as they stand today. Theoretically, everything has a single very simple rationale and mathematical form, which makes it easy to teach. The basic rules, from which all else follows, are nothing but the standard product and sum rules of probability theory; but now they are derived uniquely as principles of logic, from very elementary qualitative requirements of consistency and rationality.

Functionally, probability theory as extended logic includes as special cases all the results of the conventional "random variable" theory, and it extends the applications to useful solution of many problems previously considered to be outside the realm of probability theory. The now much strengthened theoretical foundation and continued pragmatic success – and the failure of critics to uncover any defects in it or offer any usable alternatives – justify this confidence in it.*

When applied to problems of parameter estimation or hypothesis testing, probability theory as logic is generally called *Bayesian inference*, on historical grounds explained elsewhere, and it is accomplishing a major house–cleaning in the field of statistics. The "orthodox" methods of inference as taught by statisticians since the 1930's consist of about a dozen intuitive devices (confidence intervals, unbiased estimators, significance tests, *etc.*), without any connected theoretical basis. Each is usable in some small range of problems for which it was invented; but each produces contradictions and absurd conclusions when applied out of its proper range. Now all of these are basically methods for reasoning from incomplete information; that is, for information processing. Yet the orthodox practitioners never thought of probability in terms of information.

Orthodox methods of inference also faced insuperable technical difficulties (nuisance parameters, lack of sufficient or ancillary statistics, *etc.*) which made it impossible to extract all the information from one's data in many problems. Indeed, some textbook authors recommended procedures which amounted to throwing away practically all the relevant information in the data, merely to achieve an unbiased estimate.† When there is cumulative error, an author at the Bureau of Standards advocated estimating the slope of a linear relation from the first and last measurements alone – thus throwing away all the evidence of the intermediate data points – which is easily shown to lead to an order of magnitude less accuracy in the estimate.

When these *ad hoc* devices are replaced by their Bayesian counterparts, all these difficulties disappear. The resulting algorithms have no restricted range of validity; you can apply them to arbitrary extreme conditions and the results continue to make sense. This confirms the theoretical expectation; Bayesian methods are the exact rules for conducting inference, while orthodox methods are only intuitive approximations to them. One expects that an approximation will be valid only in some restricted domain.

In Bayesian inference, nuisance parameters are not only eliminated effortlessly from our final algorithms; now, instead of creating problems, their use becomes a means for improving the precision of our results. In effect, this procedure warns probability theory to be on the lookout for some disturbing but uninteresting effect contaminating the data, and make proper allowances for it. The disturbing effect is not treated as mere "noise"; to do so would be to lose all the highly cogent information we have about its nature.‡

---

* Every objection to this formulation that we have seen arises, very obviously, from the fact that the critic has completely misunderstood what the ideas are; and so is attacking a straw man of his own making. The situation is much like that of Darwin's Theory of Evolution; as the well–known biologist Stephen J. Gould has noted recently, those who still attack Darwin's theory only reveal that they do not understand what the theory is.

† Yet, as R. A. Fisher emphasized already sixty years ago, the criterion of bias has no theoretical justification (it is not even invariant under a change of variables), and a biased estimate can be far more accurate than an unbiased one.

‡ For example, if we want to estimate only the frequency of a sine wave plus noise, then the amplitude and phase are nuisance parameters, which affect the data but whose effects must be eliminated somehow in the final algorithm for processing the data into an estimate of the frequency. Then any prior information about how the amplitude might be varying with time becomes highly cogent for determining the accuracy of our frequency estimates.

New insight into these matters has continued to the present day. A major advance in our understanding and technique occurred five years ago in the thesis of Larry Bretthorst, when it was found that by introducing a nuisance parameter and integrating it out again, one can in some cases achieve orders of magnitude improved resolution in spectrum analysis, over the previous practice of taking the fourier transform of the data. This is having its first impact in analysis of NMR free induction decay data, where the decay rate no longer places the limitation on the accuracy with which we can determine the oscillation frequency; we expect that other applications will profit from it just as much.

Orthodox methods of inference had another distressing property: on the one hand, to get conclusions from data they offer no way to take into account our prior information about the parameters of interest; yet on the other hand they require us to make additional assumptions about the frequency distribution of errors that are arbitrary; that is, not justified by any of our information. In contrast, it is now a proven theorem that, when we apply strict Bayesian principles with initial probabilities assigned by maximum entropy, our subsequent inferences depend only on the data and the circumstances (maximum entropy constraints) about which we had prior information; there is no room for gratuitous assumptions because circumstances about which we have no information automatically cancel out and contribute nothing to our final conclusions. Those simple product and sum rules have concealed, for all these years, a remarkable power and efficiency for information processing.

This is an example of a property, only recently discovered, which is making Bayesian inference exciting today. Because of it we now understand a fact that has been puzzling to workers in probability theory since Augustus de Morgan discovered it in 1838; that Gaussian sampling distributions almost always lead to the most successful inferences *whether or not the frequency distributions of our errors are in fact Gaussian.* To see why this is true it was necessary to think of probability theory as extended logic, because then probability distributions are justified in terms of their demonstrable information content, rather than their imagined – and as it now turns out, irrelevant – frequency connections.*

But in addition to this great power, Bayesian methods are also very simple in principle and easy to apply; therefore they enable us to find useful solutions to problems of reasoning that are so complex that they could not even be formulated in orthodox terms. For many examples, with fully worked–out numerical details, see Bretthorst (1988). I am now trying desperately to complete a large book, *Probability Theory: The Logic of Science*, which is to have all the theoretical background of this, plus many more applications and comparisons with orthodox results. Many of you have seen some preliminary fragments of it, which are issued to interested persons from time to time, as they become available.

New Adhockeries

In recent years the orthodox habit of inventing intuitive devices rather than appealing to any connected theoretical principles has been extended to new problems in a way that makes it appear at first that several new fields of science have been created. Yet all of them are concerned with reasoning from incomplete information; and we believe that we have theorems establishing that probability theory as logic is the *general* means of dealing with all such problems. We note three examples.

---

* The point here – perfectly obvious in retrospect but not noticed for 150 years – is that the actual distribution of the errors is known from the data; whatever distribution we might have expected before seeing the data is made irrelevant by that information.

Fuzzy Sets are – quite obviously, to anyone trained in Bayesian inference – crude approximations to Bayesian prior probabilities. They were created only because their practitioners persisted in thinking of probability in terms of a "randomness" supposed to exist in Nature but never well defined; and so concluded that probability theory is not applicable to such problems. As soon as one recognizes probability as *the general way to specify incomplete information*, the reason for introducing Fuzzy Sets disappears.

Likewise, much of Artificial Intelligence (AI) is a collection of intuitive devices for reasoning from incomplete information which, like the older ones of orthodox statistics, are approximations to Bayesian methods and usable in some restricted class of problems; but which yield absurd conclusions when we try to apply them to problems outside that class. Again, its practitioners are caught in this only because they continue to think of probability as representing a physical "randomness" instead of incomplete information. In Bayesian inference all those results are contained automatically – and rather trivially – without any limitation to a restricted class of problems.

The great new development is Neural Nets, meaning a system of algorithms with the wonderful new property that they are, like the human brain, *adaptive* so that they can learn from past errors and correct themselves automatically (WOW! What a great new idea!). Indeed, we are not surprised to see that Neural Nets are actually highly useful in many applications; more so than Fuzzy Sets or AI. However, present neural nets have two practical shortcomings; (a) They yield an output determined by the present input plus the past training information. This output is really an *estimate* of the proper response, based on all the information at hand, but it gives no indication of its accuracy, and so it does not tell us how close we are to the goal (that is, how much more training is needed); (b) When nonlinear response is called for, one appeals to an internally stored standard "sigmoid" nonlinear function, which with various amplifications and linear mixtures can be made to approximate, to some degree, the true nonlinear function.

But, do we really need to point out that (1) Any procedure which is adaptive is, by definition, a means of taking into account incomplete information; (2) Bayes' theorem is precisely the mother of all adaptive procedures; the *general* rule for updating any state of knowledge to take account of new information; (3) When these problems are formulated in Bayesian terms, a single calculation automatically yields both the best estimate and its accuracy; (4) If nonlinearity is called for, Bayes' theorem automatically generates the exact nonlinear function called for by the problem, instead of trying to construct an approximation to it by another *ad hoc* device.

In other words, we contend that these are not new fields at all; only false starts. If one formulates all such problems by the standard Bayesian prescription, one has automatically all their useful results in improved form. The difficulties people seem to have in comprehending this are all examples of the same failure to conceptualize the relation between the abstract mathematics and the real world. As soon as we recognize that probabilities do not describe reality – only our information about reality – the gates are wide open to the optimal solution of problems of reasoning from that information.

Quantum Theory

Here no such gratifying successes can be reported, because the fate of my ideas about quantum theory and electrodynamics will be determined, necessarily, by Nature and not by any arguments invented by me or by anybody else. There is no way any of us can feel very confident of what the future may bring here. Nevertheless, we do have a little more than faith and hope to sustain us. I am convinced, as were Einstein and Schrödinger, that the major obstacle that has prevented any real progress in our understanding of Nature since 1927, is the Copenhagen Interpretation of Quantum Theory. This theory is now 65 years old, it has long since ceased to be productive, and it is time for its retirement (along with mine).

Just for that reason, this is where the great opportunities for the next generation lie. Let us examine the logical impasse now facing us, and see why I think the aforementioned Bayesian principles may be the key to resolving it, just as they did in cases just cited. But here the situation is far more complex and subtle, so unlike the above relatively trivial examples, it will require much more deep thinking to see exactly how to carry this out.

The more its defenders insist that all is well and there are no contradictions in present quantum theory, the more blatantly those contradictions stare us in the face and tie our hands, making it impossible to proceed. We are familiar with a proposition of elementary logic: that from a false proposition all propositions, true and false, may be deduced. There is a corollary, noted by Källén (1972): from an inconsistent theory any result may be derived. Indeed, we know that to get correct predictions out of Quantum Electrodynamics (QED) required a great deal of art and tact, found only after twenty years of efforts (1927–1946); for the right experimental numbers to emerge one must do the calculation (*i.e.*, subtract off the infinities) in one particular way and not in some other way that appears in principle equally valid.

In space–time, the Feynman propagators have violent singularities on the light–cone, far worse than the delta functions that arise in the Green's functions in other parts of mathematical physics. They guarantee that any integral of the form $\int S_F(x - y) f(y) \, d^4 y$ which might be thought of as a first–order perturbation, diverges if in the integration the separation $(x - y)$ crosses the light–cone. Surely, in a properly formulated theory, there would be no infinities to subtract; yet another 45 years of efforts have not found that formulation – except perhaps in Schwinger's source theory, which seems to be ignored by workers in the field.* Then in what sense can one claim that QED is a great success?

But this is not limited to field theory. As we have noted in some detail elsewhere (Jaynes, 1991), throughout the history of quantum theory, whenever we advanced to a new application it was necessary to repeat this trial–and–error experimentation to find out which method of calculation gives the right answers. Then, of course, our textbooks present only the successful procedure as if it followed from general principles; and do not mention the actual process by which it was found. In relativity theory one deduces the computational algorithm from the general principles. In quantum theory, the logic is just the opposite; one chooses the principle to fit the empirically successful algorithm.

But after all, how can one build rationally from a theory whose basic principles are in this condition: Present quantum theory uses relativistic wave equations, but tries to solve

---

* Schwinger's recent writings have a cryptic character, almost as if he were trying to conceal what is really happening. However, we think that he achieves the correct finite numbers by his use of finite sources with time–like separation. Then with x in the future source, y in the past one, his regions of integration never reach the light–cone. If so, then we see at once that the singularities in the Feynman propagators are actually contributing nothing to the finite experimental numbers of the theory; only the finite terms inside the light cone will be needed in the future correct theory. This speculation makes such good physical sense that we hope it turns out to be true.

them with propagators that – quite aside from the divergences – violate relativity by failing to vanish outside the light–cone, and run backward in time! What can this possibly mean?

On a more elementary level, present quantum theory claims on the one hand that local microevents have no physical causes, only probability laws; but at the same time admits (from the EPR paradox) instantaneous action at a distance! Today we have in full flower the blatant, spooky contradictions that Einstein foresaw and warned us about 60 years ago, and there is no way to reason logically from them. This mysticism *must* be replaced by a physical interpretation that restores the possibility of thinking rationally about the world.

We see the effects of this in the fact that today, a large portion of research in theoretical physics has been reduced to wheel–spinning; random fiddling with the mathematics of the old theory, without giving a thought to its physical foundations. One would think that the folly of this might have been learned from the example of Einstein; yet his repeated warnings go unheeded even as his worst fears are realized before our eyes.

I believe the answer to this must be that our present formalism contains two different things. It represents in part properties of the real world, in part our information about the world; but all scrambled up so that we do not see how to disentangle them. But at least we can see that the spooky things will cease to be spooky as soon as we think of the formalism in terms of inference from incomplete information. Then what is traveling faster than light, and backwards in time, is not a physical influence; but only a logical inference. David Hestenes thinks that his reformulation of the Dirac equation accomplishes this separation into the subjective and objective features of the theory; in our view this is an attractive possibility, but not yet a demonstrated fact. Before we could judge this, one needs to work out the full, explicit solutions to the standard QED problems (Spontaneous Emission, Lamb Shift, Vacuum Polarization, Anomalous Moment, Bethe–Heitler formula, Mott scattering, *etc.*) in the Hestenes formalism and see what this interpretation is saying at every step of the derivations.

What has held up progress in this field for so long? Always our students are indoctrinated about the great pragmatic success of the quantum formalism – with the conclusion that the Copenhagen interpretation of that formalism must be correct. This is the logic of the Quantum Syllogism:

> The present *mathematical formalism* can be made to reproduce many experimental facts very accurately.
> ### Therefore
> The *physical interpretation* which Niels Bohr tried to associate with it must be true; and it is naïve to try to circumvent it.

Compare this with the Pre–Copernican Syllogism:

> The mathematical system of epicycles can be made to reproduce the motions of the planets very accurately.
> ### Therefore
> The theological arguments for the necessity of epicycles as the only perfect motions must be true, and it is heresy to try to circumvent them.

In what way are they different? The difference is only that today everybody knows what is wrong with the Pre–Copernican syllogism; but (from the frequency with which it is still repeated) only a relatively few have yet perceived the error in the Quantum Syllogism. In quantum theory, we are still using pre–scientific standards of logic in a dozen different places, as noted in more detail elsewhere (Jaynes, 1989, 1990).

Both Quantum Electrodynamics (QED) and my "neoclassical" theory (Jaynes, 1973) contain some elements of truth, but also some elements of intolerable nonsense, and thus far nobody – least of all, me – has seen how to unscramble them. But there is great new hope in the fact that their elements of truth and nonsense are so different, and in the spectacular recent advances in experimental technique reported at this meeting by Herbert Walther and Pierre Meystre, which are now making direct contact with these issues.

From a physical rather than formal standpoint, the worst elements of nonsense in QED concern the infinite zero–point energy and vacuum fluctuations. We have given physical arguments (Jaynes, 1990) which seem conclusive to me, showing that (a) these fluctuations are not real and are not necessary to account for the experimental facts; and (b) their removal from QED would also remove at least some of the infinities that have plagued the theory from the start, and a great deal of surplus formalism not actually used.

The underlying idea of neoclassical theory (NCT) is that photons are not real physical objects existing in the electromagnetic field; all the field phenomena of propagation, diffraction, and interference are accounted for correctly by the classical EM theory of Maxwell, Poynting, Lorentz, Larmor, and Poincaré. The alleged particle–like properties of radiation are consequences only of the laws of interaction with matter, and appear only in the reaction of matter to the fields. Those who do not believe this continue to be astonished at the fact that, whenever a new interference effect is observed, it is always puzzling and mysterious from the standpoint of quantum theory; but it is just what classical EM theory would have predicted. Another example of this was presented at this meeting by Leonard Mandel, in which the QED uncertainty is just that due to the unknown phase of the classical idler oscillations.

The worst elements of nonsense in NCT concern the apparently different physical status of longitudinal and transverse fields, although they are partly interconverted by a Lorentz transformation.[†] In addition it faces unsolved computational problems in dealing with free particles [which we view as stable wave packets, held together by the self–interaction forces of the Dirac equation as discussed in Jaynes (1991), that are ignored in the "spreading wave packet" solutions of elementary quantum theory], and really works nicely only for bound states of atoms, where we know the wave functions quite accurately from previous solutions of the Schrödinger or Dirac equation (because then the Coulomb forces are large compared to the self–interaction forces). Some critics have claimed that NCT gives some field correlation effects wrong; I think they have misapplied it and suggest that, as noted in Jaynes (1973), it is QED that gives some field correlation effects wrong (For that work, the mandatory twenty year incubation period is nearly over; we expect that in 1993 somebody will start to take note of what is in it).

But NCT also contains one bit of fundamental truth that is not in QED. This is the simple explanation of the basic reason for the relation $E = h\nu$, which is in present quantum theory only an empirical relation for which – astonishingly – nobody seems ever to have

---

[†] This may be a removable difficulty, because NCT is still far from being cast permanently in concrete; many modifications are still possible while preserving the basic ideas.

sought any theoretical explanation.[‡] We found part of the answer in that this relation follows from an action conservation law of the Hamiltonian; and that, viewed in this way, it need not be a universally valid principle, so there is a possibility of an experimental test.

Presumably, that elusive "future correct theory" of electrodynamics, toward which we have been groping for a Century, will be some kind of mixture of these theories and others, but with some great unifying principle not yet seen (but which will be perfectly obvious to all as soon as it is seen). Surely, it will have to recognize this element of truth, and find some way of incorporating it into quantum theory, while discarding the elements of nonsense of both theories.

In the meantime, those who apply the naïve ideas of present quantum theory confidently to astrophysical problems – conditions extrapolated dozens of orders of magnitude beyond anything that we know to be true – are in my view engaging in speculation that is not necessarily wrong in aim, but is just a Century premature. Attempts at unified force theories also seem premature for the same reason; until we get elementary quantum theory into some kind of rational order, unrestrained speculation from it is highly unlikely to lead to the truth.

One of the principles of scientific inference – which has always been well understood by the greatest scientists – is that it is idle to raise questions prematurely, when they cannot be answered with the resources available. For Isaac Newton it would have been foolish to raise questions that were not foolish for Erwin Schrödinger 250 years later; for Gregor Mendel it would have been foolish to raise questions that were not foolish for Francis Crick 100 years later. By "foolish" we mean "without hope of success". Of course, we all enjoy indulging in a little free speculation about the future of science; but for scientists to expend their serious professional time and effort on idle speculation can only delay any real progress.

Dealing with Critics

Looking back over the past forty years, I can see that the greatest mistake I made was to listen to the advice of people who were opposed to my efforts. Just at the peak of my powers I lost several irreplaceable years because I allowed myself to become discouraged by the constant stream of criticism from the Establishment, that descended upon everything I did. I have never – except in the past few years – had the slightest encouragement from others to pursue my work; the drive to do it had to come entirely from within me. The result was that my contributions to probability theory were delayed by about a decade, and my potential contributions to electrodynamics – whatever they might have been – are probably lost forever.

But I can now see that all of this criticism was based on misunderstanding or ideology. My perceived sin was not in my logic or mathematics; it was that I did not subscribe to the dogmas emanating from Copenhagen and Rothamsted. Yet I submit that breaking those dogmas was the necessary prerequisite to making any further progress in quantum theory and probability theory. If not in my way, then necessarily in some other.

[‡] Contrast this with the reasoning of Einstein, when he noted that the equality of gravitational mass and inertial mass was only an empirical relation without any theoretical explanation; in 1946 he wrote: "It is, however, clear that science is fully justified in assigning such a numerical equality only after this is reduced to an equality of the real nature of the two concepts." To find that explanation he was led to General Relativity. Neoclassical theory was made in quite conscious imitation of Einstein's example; we expect that a full understanding of $E = h\nu$ will be of just as fundamental a nature as is General Relativity, far beyond anything in present quantum theory.

In any field, the Establishment is not seeking the truth, because it is composed of those who, having found part of it yesterday, believe that they are in possession of all of it today. Progress requires the introduction, not just of new mathematics which is always tolerated by the Establishment; but new conceptual ideas which are *necessarily* different from those held by the Establishment (for, if the ideas of the Establishment were sufficient to lead to further progress, that progress would have been made).

Therefore, to anyone who has new ideas of a currently unconventional kind, I want to give this advice, in the strongest possible terms: *Do not allow yourself to be discouraged or deflected from your course by negative criticisms* – particularly those that were invented for the sole purpose of discouraging you – unless they exhibit some clear and specific error of reasoning or conflict with experiment. Unless they can do this, your critics are almost certainly wrong, but to reply by trying to show exactly where and why they are wrong would be wasted effort which would not convince your critics and would only keep you from the far more important, constructive things that you might have accomplished in the same time. Let others deal with them; if you allow your enemies to direct your work, then they have won after all.

Although the arguments of your critics are almost certainly wrong, they will retain just enough plausibility in the minds of some to maintain a place for them in the realm of controversy; that is just a fact of life that you must accept as the price of doing creative work. Take comfort in the historical record, which shows that no creative person has ever been able to escape this; the more fundamental the new idea, the more bitter the controversy it will stir up. Newton, Darwin, Boltzmann, Pasteur, Einstein, Wegener were all embroiled in this. Newton wrote in 1676: "*I see a man must either resolve to put out nothing new, or become a slave to defend it.*" Throughout his lifetime, Alfred Wegener received nothing but attacks on his ideas; yet he was right and today those ideas are the foundation of geophysics. We revere the names of James Clerk Maxwell and J. Willard Gibbs; yet their work was never fully appreciated in their lifetimes, and even today it is still, like that of Darwin, under attack by persons who, after a Century, have not yet comprehended their message (Atkins, 1986).

The recent reminiscences of Francis Crick (1988) make many other important points that we might otherwise have included here, because they apply as well to physics as to biology. For example, he notes that in studying brain function, theoretical work "tended to fall into a number of somewhat separate schools, each of which was rather reluctant to quote the work of the others. This is usually characteristic of a subject that is not producing any definite conclusions." Exactly the same could be said of several areas of physics; we would add only that, even when definite conclusions are proclaimed by one school, they are not often justified or permanent.

Conclusion

From all of the above discussion one generalization stands out: today we are desperately in need of some fundamental reforms in science education. *Scientists need to be trained, not only in the details of their specialty, but also in the principles of scientific inference in general.* As our incomplete information becomes more massive and quantitative, the same disciplined reasoning that Louis Pasteur was able to carry out in his head, now needs to be put into quantitative form.

I submit that this is precisely the Bayesian inference that Arnold Zellner and I have been teaching for several years; and it needs to be consolidated into undergraduate courses that are considered just as essential as is general literacy, for a scientist in any field. A little familiarity with and respect for the principles of rational inference would have muted the old claims of absolute truth and finality for quantum theory, and the sensational recent ones for such things as Chaos, Artificial Intelligence, Neural Nets, and the Big Bang theory.* I have the impression that physicists are in most need of this, because their training has completely neglected probability theory, and so they have become the most naïve of all scientists in the matter of judging what conclusions are and are not justified by some observed facts. Perhaps this is a legacy from the pre–scientific logic that we teach in quantum theory; the Bayesian–trained economists that I know would never commit such errors.

But it must not be supposed that our problems arise only in advanced science education; they commence already in kindergarten. From the educationists one hears constantly such phrases as: "skills for effective living", or "socially desirable responses", "adequate social behavior". Why is it important to be literate? Not because one will need to read with comprehension; but rather because "personal adjustment demands some speech and reading facility." Why is it important for a child to learn arithmetic? Not because he is going to have to know how to add and subtract numbers correctly for all the rest of his life. Of course not! It is important that he learn arithmetic because "a sense of failure here might affect his personality development". This is not material for standup comedians: every one of these quotations was found in course catalogs for Schools of Education. Such phrases as "adequate knowledge" or "rational behavior" do not occur at all. We need look no further to understand what needs to be corrected in American elementary education.

Perhaps it is clear that each of my little scoldings is at the same time an opportunity for the next generation to do something constructive about it. I am optimistic about the future because I know that my former students are capable of this, and suspect that their students are also. But the solutions will not come overnight, and they may need encouragement. Young scientists ought to study these problems most earnestly – just because of the fact that the solutions are not yet obvious (else my generation would have solved them).

Thank you again for all your kindness, and patience with my ramblings. Now I shall return home and do my best to complete those promised books, with all the details that these remarks have only hinted at, rather vaguely.

REFERENCES

Atkins, P. W. (1986), "Entropy in Relation to Complete Knowledge", Contemp. Phys. **27**, pp. 257–259.

Bretthorst, G. L. (1988), *Bayesian Spectrum Analysis and Parameter Estimation*, Lecture Notes in Statistics, Vol. 48, Springer–Verlag, Berlin.

Crick, Francis (1988), *What Mad Pursuit*, Basic Books, Inc., New York.

Einstein, Albert (1946), *The Meaning of Relativity*, Princeton University Press, pp. 56–57.

---

* Recently it has been claimed that directional irregularities in the cosmic microwave background constitute decisive evidence in favor of the Big Bang theory. But the principles of scientific inference tell us that a theory is confirmed by observing things which it predicts *that are otherwise unexpected*; not by observing things that would be expected whatever one's theory of the beginning of the Universe.

Jaynes, E. T. (1973), "Survey of the Present Status of Neoclassical Radiation Theory", in *Coherence and Quantum Optics*, L. Mandel & E. Wolf, editors, Plenum Press, New York.

———— (1989), "Clearing up Mysteries: The Original Goal", in *Maximum Entropy and Bayesian Methods*, J. Skilling, Editor, Kluwer Academic Publishers, Dordrecht, Holland, pp. 1–27.

———— (1990), "Probability in Quantum Theory", in *Complexity, Entropy and the Physics of Information*, W. H. Zurek, editor, Addison–Wesley Pub. Co., Redwood City CA, pp. 381–404.

———— (1991), "Scattering of Light by Free Electrons as a Test of Quantum Theory", in *The Electron*, D. Hestenes & A. Weingartshofer, Editors, Kluwer Academic Publishers, Dordrecht, Holland; pp 1–20.

Källén, Gunnar (1972), *Quantum Electrodynamics*, Springer–Verlag, New York.

Schwinger, Julian (1970), *Particles, Sources, and Fields*, Volume 1, Addison–Wesley Publishing Co., Reading MA. Volume 2 (1973).

Webster, David L. (1957), "Reminiscences of the Early Years of the Association", Am. Jour. Phys. **35**, pp. 131–134.

VITA AND BIBLIOGRAPHY
of
EDWIN T. JAYNES

Ed Jaynes was born in Waterloo, Iowa, on 5 July 1922, the son of a surgeon. He attended Cornell College and the University of Iowa, receiving the B.A. degree in physics from the latter in 1942, and was then engaged in Doppler radar development at the Sperry Gyroscope Company in New York in 1942–43. He was subsequently appointed an Ensign in the U.S. Navy, and worked at the Naval Research Laboratory at Anacostia on microwave IFF equipment.

Lt(jg) Jaynes was discharged from the Navy in 1946 and spent that summer at Stanford working with W.W. Hansen on the design of the first linear electron accelerator. In the 1946–47 school year he was a graduate student at Berkeley, a student of J.R. Oppenheimer. When Oppenheimer left to take over the Institute for Advanced Study in Princeton in the summer of 1947, Jaynes also transferred to Princeton University. He received his Ph.D. in physics there in 1950, with a thesis in solid-state theory supervised by Eugene Wigner.

He then returned to Stanford, where he was Instructor, Assistant Professor, and Associate Professor during 1950–60. At this time he also consulted for Varian Associates on problems of magnetic resonance instrumentation. In 1960 he was appointed Professor of Physics at Washington University in St. Louis, and in 1975 became Wayman Crow professor of Physics.

His research interests have been largely in electromagnetic theory and statistical mechanics, on which he has published numerous articles, while a life-long hobby has been the application of principles of physics to the operation, and proper methods of playing musical instruments. His course on the Physical Basis of Music was taught several times at both Stanford and Washington Universities.

In 1957 he published the first articles on the Maximum Entropy principle as the foundation of statistical mechanics, and since 1956 has been teaching Maximum Entropy and Bayesian methods to two generations of students. He has also given lectures and summer school courses in these subjects at U.C.L.A., Brandeis, Colorado, Purdue, Dartmouth, Minnesota, Maryland, and several industrial and government research laboratories. Jaynes has been Adjunct Professor (1982–83) at the University of Wyoming, and a Fellow of St. John's College, Cambridge (1983–84). At the present time he is in a semi-retired condition due to a heart attack, and is working full time writing books on Probability Theory, Statistical Mechanics, the Foundations of Electrodynamics, and the Physical Basis of Music.

BIBLIOGRAPHY

This bibliography presents the scholarly research publications and book reviews of E.T. Jaynes, but does not reflect numerous reports and unpublished notes he has also authored. A number of articles appear also in a limited collection edited by R.D. Rosenkrantz [*E.T. Jaynes: Papers on Probability, Statistics and Statistical Physics*, D. Reidel, Dordrecht, 1983], identified by a marginal asterisk (*). The editors wish to thank G.L. Bretthorst and C.R. Smith for sharing their compilations with us for this volume.

[1] Jaynes, E.T.: 1950, 'Displacement of Oxygen in BaTiO3', *Phys. Rev.* **79**, 1008.

[2] _____ : 1952, 'The Concept and Measurement of Impedance in Periodically Loaded Waveguides', *J. Appl. Phys.* **23**, 1077.

[3] _____ : 1953, *Ferroelectricity*, Princeton Univ. Press, London.

[4] Hornig, A.W., E.T. Jaynes, and H.E. Weaver: 1954, 'Observation of Paramagnetic Resonances in Single Crystals of Barium Titanate', *Phys. Rev.* **96**, 1703.

[5] Jaynes, E.T.: 1955, 'Matrix Treatment of Nuclear Induction', *Phys. Rev.* **98**, 1099.

[6] _____ : 1955, 'Nonlinear Dielectric Materials', *Proc. IRE* **43**, 1733.

[7] Vartanian, P.H., and E.T. Jaynes: 1956, 'Propagation in Ferrite-Filled Transversely Magnetized Waveguide', *IRE Trans. on Microwave Theory and Techniques* **MTT-4**, 140.

* [8] Jaynes, E.T.: 1957, 'Information Theory and Statistical Mechanics', *Phys. Rev.* **106**, 620.

* [9] _____ : 1957, 'Information Theory and Statistical Mechanics II', *Phys. Rev.* **108**, 171.

[10] _____ : 1958, 'Ghost Modes in Imperfect Waveguides', *Proc. IRE* **46**, 416.

[11] _____ : 1958, 'Relativistic Clock Experiments', *Am. J. Phys.* **26**, 197.

[12] _____ : 1959, 'Note on Unique Decipherability', *IRE Trans. on Information Theory* **IT-5**, 98.

[13] Forrer, M.P., and E.T. Jaynes: 1960, 'Resonant Modes in Waveguide Windows', *IRE Trans. on Microwave Theory and Techniques* **MTT-8**, 147.

[14] Jaynes, E.T.: 1960, 'The Maser as a Parametric Amplifier', in *Quantum Electronics*, C. H. Townes (ed.), Columbia Univ. Press, New York, p.237.

[15] S. Heims, and E.T. Jaynes: 1962, 'Theory of Gyromagnetic Effects and Some Related Magnetic Phenomena', *Rev. Mod. Phys.* **34**, 143.

[16] Jaynes, E.T.: 1963, 'New Engineering Applications of Information Theory', in *Engineering Uses of Random Function Theory and Probability*, J.L. Bogdanoff and F. Kozin (eds.), Wiley, New York, p.163.

[17] Jaynes, E.T., and F.W. Cummings: 1963, 'Comparison of Quantum and Semiclassical Radiation Theory with Application to the Beam Maser', *Proc. IEEE* **51**, 89.

* [18] Jaynes, E.T.: 1963, 'Information Theory and Statistical Mechanics', in *Statistical Physics*, K. Ford (ed.), Benjamin, New York, p.181.

[19] _____ : 1963, 'Comments on an article by Ulric Neisser', *Science* **140**, 216.

* [20] _____ : 1965, 'Gibbs vs Boltzman Entropies', *Am. J. Phys.* **33**, 391.

* [21] _____ : 1967, 'Foundations of Probability Theory and Statistical Mechanics', in *Delaware Seminar in the Foundations of Physics*, M. Bunge (ed.), Springer-Verlag, Berlin, p.77.

[22] _____ : 1968, 'You CAN Parallel Storage Batteries', *Popular Electronics* **33**, 86.

* [23] _____ : 1968, 'Prior Probabilities', *IEEE Trans. on Systems Science and Cybernetics* **SSC-4**, 227.

[24] Crisp, M.D., and E.T. Jaynes: 1969, 'Radiative Effects in Semiclassical Theory', *Phys. Rev.* **179**, 1253.

[25] Stroud, C.R., Jr., and E.T. Jaynes: 1970, 'Long-Term Solutions in Semiclassical Radiation Theory', *Phys. Rev. A* **1**, 106.

[26] Jaynes, E.T.: 1970, 'Reply to Leiter's Comments', *Phys. Rev. A* **2**, 260.

[27] ————— : 1971, 'Violation of Boltzman's *H*-Theorem in Real Gases', *Phys. Rev.* **A4**, 747.

* [28] ————— : 1973, 'The Well-Posed Problem', *Found. Phys.* **3**, 477.

[29] ————— : 1973, 'Survey of the Present Status of Neoclassical Radiation Theory', in *Coherence and Quantum Optics*, L. Mandel and E. Wolf (eds.), Plenum, New York, p.35.

* [30] ————— : 1976, 'Confidence Intervals vs Bayesian Intervals', in *Foundations of Probability Theory, Statistical Inference, and Statistical Theories of Science*, W.L. Harper and C.A. Hooker (eds.), D. Reidel, Dordrecht, p.175.

[31] ————— : 1976, 'Reply to Kempthornes' Comments', in *Foundations of Probability Theory, Statistical Inference and Statistical Theories of Science*, W.L. Harper and C.A. Hooker (eds.), D. Reidel, Dordrecht, p.229.

[32] ————— : 1978, 'Ancient History of Free Electron Devices', in *Novel Sources of Coherent Radiation*, S.F. Jacob, M. Scully, and M. Sargent III (eds.), Addison-Wesley, Reading, MA, p.1.

[33] ————— : 1978, 'Electrodynamics Today', in *Coherence and Quantum Optics IV*, L. Mandel and E. Wolf (eds.), Plenum Press, New York, p.495.

[34] O'Donnell, M., E.T. Jaynes, and J.G. Miller: 1978, 'General Relationships between Ultrasonic Attenuation and Dispersion', *J. Acoust. Soc. Am.* **63**, 1935.

* [35] Jaynes, E.T.: 1979, 'Where do we Stand on Maximum Entropy?', in *The Maximum Entropy Formalism*, R.D. Levine and M. Tribus (eds.), M.I.T. Press, Cambridge, MA, p.15.

* [36] ————— : 1979, 'Concentration of Distributions at Entropy Maxima', in *E.T. Jaynes: Papers on Probability, Statistics and Statistical Physics*, R.D. Rosenkrantz (ed.), D. Reidel, Dordrecht, p.315.

[37] ————— : 1980, 'Quantum Beats', in *Foundations of Radiation Theory and Quantum Electrodynamics*, A.O. Barut (ed.), Plenum Press, New York, p.37.

* [38] ————— : 1980, 'Marginalization and Prior Probabilities', in *Bayesian Analysis in Econometrics and Statistics*, A. Zellner (ed.), North-Holland, Amsterdam, p.43.

[39] ————— : 1980, 'Reply to Dawid, Stone, and Zidek', in *Bayesian Analysis in Econometrics and Statistics*, A. Zellner (ed.), North-Holland, Amsterdam, p.83.

* [40] ————— : 1980, 'What is the Question?', in *Bayesian Statistics*, J.M. Bernardo, M.H. DeGroot, D.V. Lindly, and A.F.M. Smith (eds.), Valencia Univ. Press, Valencia.

* [41] ————— : 1980, 'The Minimum Entropy Production Principle', *Ann. Rev. Phys. Chem.* **31**, 579.

[42] Matthys, D.R., and E.T. Jaynes: 1980, 'Phase-Sensitive Optical Amplifier', *J. Opt. Soc. Am.* **70**, 263.

[43] O'Donnell, M., E.T. Jaynes, and J.G. Miller: 1980, 'Kramers-Kronig Relationship between Ultrasonic Attenuation and Phase Velocity', *J. Acoust. Soc. Am.* **69**, 696.

[44] Jaynes, E.T.: 1982, 'On the Rationale of Maximum-Entropy Methods', *Proc. IEEE* **70**, 939.

[45] ————— : 1984, 'The Intuitive Inadequacy of Classical Statistics', *Epistemologia* **VII**, 43.

[46] ————— : 1984, 'Prior Information and Ambiguity in Inverse Problems', in *Inverse Problems*, D.W. McLaughlin (ed.), Am. Math. Soc., Providence, RI [SIAM-AMS Proceedings **14**, 151].

[47] ————— : 1985, 'Where do we go from Here?', in *Maximum-Entropy and Bayesian Methods in Inverse Problems*, C.R. Smith and W.T. Grandy, Jr. (eds.), D. Reidel, Dordrecht, p.21.

[48] ———— : 1985, 'Generalized Scattering', in *Maximum-Entropy and Bayesian Methods in Inverse Problems*, C.R. Smith and W.T. Grandy, Jr. (eds.), D. Reidel, Dordrecht, p.377.

[49] ———— : 1985, 'Entropy and Search-Theory', in *Maximum-Entropy and Bayesian Methods in Inverse Problems*, C.R. Smith and W.T. Grandy, Jr. (eds.), D. Reidel, Dordrecht, p.443.

[50] ———— : 1985, 'Highly Informative Priors', in *Bayesian Statistics 2*, J.M. Bernardo, M.H. DeGroot, D.V. Lindley, and A.F.M. Smith (eds.), Elsevier, Amsterdam, p.329.

[51] ———— : 1985, 'Macroscopic Prediction', in *Complex Systems – Operational Approaches*, H. Haken (ed.), Springer-Verlag, Berlin, p.254.

[52] ———— : 1985, 'Some Random Observations', *Synthese* **63**, 115.

[53] ———— : 1986, 'Predictive Statistical Mechanics', in *Frontiers of Nonequilibrium Statistical Physics*, G.T. Moore and M.O. Scully (eds.), Plenum Press, New York, p.33.

[54] ———— : 1986, 'Bayesian Methods: General Background', in *Maximum-Entropy and Bayesian Methods in Applied Statistics*, J.H. Justice (ed.), Cambridge Univ. Press, Cambridge, p.1.

[55] ———— : 1986, 'Monkeys, Kangaroos and N', in *Maximum-Entropy and Bayesian Methods in Applied Statistics*, J.H. Justice (ed.), Cambridge Univ. Press, Cambridge, p.26.

[56] ———— : 1986, 'Some Applications and Extensions of the de Finetti Representation Theorem', in *Bayesian Inference and Decision Techniques with Applications: Essays in Honor of Bruno de Finetti*, P.K. Goel and A. Zellner (eds.), North–Holland, Amsterdam, p.31.

[57] Muckenheim, W., G. Ludwig, C. Dewdney, P.R. Holland, A. Kyprianidis, J.P. Vigier, N.C. Petroni, M.S. Bartlett, and E.T. Jaynes: 1986, 'A Review of Extended Probability', *Phys. Repts.* **133**, 337.

[58] ———— : 1987, 'Bayesian Spectrum and Chirp Analysis', in *Maximum Entropy and Bayesian Spectral Analysis and Estimation Problems*, C.R. Smith and G.J. Erickson (eds.), D. Reidel, Dordrecht, p.1.

[59] ———— : 1987, 'Comments on a Review by P.W. Atkins', *Contemp. Phys.* **28**, 501.

[60] ———— : 1988, 'How Does the Brain Do Plausible Reasoning?', in *Maximum-Entropy and Bayesian Methods in Science and Engineering, Vol.1*, G.J. Erickson and C.R. Smith (eds.), Kluwer, Dordrecht, p.1 [This first appeared as a Stanford Microwave Laboratory Report in 1957].

[61] ———— : 1988, 'The Relation of Bayesian and Maximum Entropy Methods', in *Maximum-Entropy and Bayesian Methods in Science and Engineering, Vol.1*, G.J. Erickson and C.R. Smith (eds.), Kluwer, Dordrecht, p.25.

[62] ———— : 1988, 'Detection of Extra-Solar-System Planets', in *Maximum-Entropy and Bayesian Methods in Science and Engineering, Vol.1*, G.J. Erickson and C.R. Smith (eds.), Kluwer, Dordrecht, p.147.

[63] ———— : 1988, 'The Evolution of Carnot's Principle', in *Maximum-Entropy and Bayesian Methods in Science and Engineering, Vol.1*, G.J. Erickson and C.R. Smith (eds.), Kluwer, Dordrecht, p.267.

[64] ———— : 1989, 'Clearing up Mysteries – The Original Goal', in *Maximum-Entropy and Bayesian Methods*, J. Skilling (ed.), Kluwer, Dordrecht, p.1.

[65] ———— : 1990, 'Probability in Quantum Theory', in *Complexity, Entropy, and the Physics of Information*, W.H. Zurek (ed.), Addison-Wesley, Redwood City, CA, p.381.

[66] ———— : 1990, 'Probability Theory as Logic', in *Maximum-Entropy and Bayesian Methods*, P.F. Fougère (ed.), Kluwer, Dordrecht, p.1.

[67] ———— : 1991, 'Notes On Present Status And Future Prospects', in *Maximum Entropy and Bayesian Methods*, W.T. Grandy, Jr. and L.H. Schick (eds.), Kluwer, Dordrecht, p.1.

[68] ──────── : 1991, 'Scattering of Light by Free Electrons', in *The Electron*, D. Hestenes and A. Weingartshofer (eds.), Kluwer, Dordrecht, p.1.

[69] ──────── : 1991, 'Commentary on Two Articles by C.A. Los', *Computers & Math. Appl.* **3**, 267.

[70] ──────── : 1992, 'The Gibbs Paradox', in *Maximum-Entropy and Bayesian Methods*, G. Erickson, P. Neudorfer, and C.R. Smith (eds.), Kluwer, Dordrecht.

[71] ──────── : 1993, 'A Backward Look to the Future', in *Physics and Probablity*, W.T. Grandy, Jr. and P.W. Milonni, Cambridge Univ. Press, Cambridge, England.

BOOK REVIEWS

[1] Jaynes, E.T.: 1952, *Advanced Theory of Waveguides*, by L. Lewin, *Rev. Sci. Inst.* **23**, 373.

[2] ──────── : 1953, *Theory of Electric Polarization*, by C.J.F. Böttcher, *Electronics Magazine*.

[3] ──────── : 1955, *Laplace Transforms for Electrical Engineers*, by B.J. Starkey, *Proc. IRE.* **43**, 898.

[4] ──────── : 1956, *Spheroidal Wave Functions*, by J.A. Stratton, *et al*, *Proc. IRE.* **44**, 951.

[5] ──────── : 1958, *Principles of Electricity (3rd ed.)*, by L. Page and N.I. Adams Jr., *Proc. IRE* **46**, 1664.

[6] ──────── : 1958, *Television Engineering, Vol.IV: General Circuit Techniques*, by S.W. Amos and D.C. Birkinshaw, *Proc. IRE* **46**, 1974.

[7] ──────── : 1959, *Principles of Circuit Synthesis*, by E.S. Kuh and D.O. Pederson, *Proc. IRE* **47**, 1676.

[8] ──────── : 1961, *Statistical Theory of Communication*, by Y.W. Lee, *Am. J. Phys.* **29**, 276.

[9] ──────── : 1962, *The Fermi Surface*, by W.A. Harrison and M.B. Webb (eds.), *Am. J. Phys.* **30**, 231.

[10] ──────── : 1962, *Atomic Theory and the Description of Nature*, by N. Bohr, *Am. J. Phys.* **30**, 658.

[11] Jaynes, E.T. 1962, *Wave Mechanics of Crystalline Solids*, by R.A. Smith, *Am. J. Phys.* **30**, 846.

[12] ──────── : 1963, *The Algebra of Probable Inference*, by R.T. Cox, *Am. J. Phys.* **31**, 66.

[13] ──────── : 1963, *Semiconductor Device Physics*, by A. Nussbaum, *Am. J. Phys.* **31**, 220.

[14] ──────── : 1963, *New Perspectives in Physics*, by L. DeBroglie, *Am. J. Phys.* **31**, 225.

[15] ──────── : 1963, *Lasers*, by B. A. Lengyel, *Am. J. Phys.* **31**, 739.

[16] ──────── : 1963, *Noise and Fluctuations*, by D.K.C. MacDonald, *Am. J. Phys.* **31**, 946.

[17] ──────── : 1964, *An Introduction to Transport Theory*, by G. M. Wing, *Am. J. Phys.* **32**, 235.

[18] ──────── : 1968, *Principles of Statistical Mechanics – The Information Theory Approach*, by A. Katz, *IEEE Trans. on Information Theory*, **IT-14**, 611.

[19] ──────── : 1979, *Inference, Method, and Decision: Towards a Bayesian Philosophy of Science*, by R.D. Rosenkrantz, *J. Am. Stat. Assn.*, **74**, 740.par

[20] ──────── : 1981, *Works on the Foundations of Statistical Physics*, by N.S. Krylov, *J. Am. Stat. Assn.*, **76**, 742.

[21] ──────── : 1982, *Methods of Statistical Physics*, by A.I. Akhiezer and S.V. Peletminskii, *Physics Today*, **35** (8), 57.

INDEX

This is primarily a subject index. Names are included only in the event of a significant quote attributed the person named — or in reference to the principal figure with whom the book is concerned.